T0254824

Fundamental Fluid Mechanics and Magnetohydrodynamics

Roger J. Hosking · Robert L. Dewar

Fundamental Fluid Mechanics and Magnetohydrodynamics

Roger J. Hosking
School of Mathematical Sciences
The University of Adelaide
Adelaide, SA
Australia

Robert L. Dewar
Research School of Physics and Engineering
The Australian National University
Canberra, ACT
Australia

Additional material to this book can be downloaded from http://extras.springer.com/.

ISBN 978-981-10-1291-4 ISBN 978-981-287-600-3 (eBook)
DOI 10.1007/978-981-287-600-3

Springer Singapore Heidelberg New York Dordrecht London

Softcover re-print of the Hardcover 1st edition 2016

Printed on acid-free paper

Springer Science+Business Media Singapore Pte Ltd. is part of Springer Science+Business Media (www.springer.com)

To Kanokwan Hosking and Julie Anne Clegg, without whose patience and understanding this long project would never have been completed

Preface

Mathematicians continue to make major contributions to the classical but continually evolving discipline of Fluid Mechanics, which is often included in advanced undergraduate and postgraduate applied mathematics or engineering programmes. On the other hand, Magnetohydrodynamics (MHD) is more often left to undergraduate and postgraduate physics programmes, and relatively few research mathematicians proceed to work on MHD or plasma theory. However, Fluid Mechanics and MHD are both branches of Continuum Mechanics, often drawing upon much the same mathematics and yielding many closely related results. Moreover, both are fundamental to many fascinating areas in geophysics and astrophysics, and have a wide variety of human and laboratory or engineering applications (from blood flow to power generation).

This book is primarily intended to enable graduate research students in science and engineering to enhance their understanding and expertise in Fluid Mechanics and MHD, which are no longer treated in isolation as in other texts. A previous background or preliminary reading in either could be an advantage, and prior knowledge of multivariate calculus and differential equations is expected. However, we write in a didactic way that reflects our scientific inclination and experience (not least as past research student supervisors), and the mathematical development is largely self-contained. Our presentation could be supplemented by historical or more descriptive material (including relevant basic physics) readily obtained from elsewhere if desired—in particular, research students are now generally well aware that the Internet can be a source of quite accessible ideas and information. Incidentally, the mathematical theory and applications of Fluid Mechanics and MHD are usually presented in the classical Newtonian context as in this book, but there are relativistic formulations in the literature for motion nearer the speed of light.

Common fluids (such as air or water) flow so readily that early theory in Fluid Mechanics, sometimes called "classical" but usually ideal in this book, entirely neglected viscosity. Wave motion is one area where the ideal (inviscid) model is often appropriate, although an observer of water waves on the surface of a lake might notice that some insects exploit the force of surface tension rather well there!

Ideal MHD is the corresponding inviscid model that Hannes Alfvén invented last century, to predict the famous new wave named after him (in a plasma permeated by a magnetic field). Air flow past a streamlined object such as an aerofoil (a cross-section of an aircraft wing) may appear to be well described by the ideal fluid model, but the air does not slip past freely close to its surface as envisaged in that theory—and even more obviously, it does not predict flow past a blunter object, typically associated with trailing swirls and drag. As is now well known, there is a boundary layer near the surface of an obstacle where even quite small fluid viscosity cannot be neglected, and in MHD (including stability theory) non-ideal properties may similarly be expected to modify an otherwise largely acceptable ideal MHD model in particular regions. Mathematically, one may proceed to invoke singular perturbation theory, where an ideal "outer" solution is matched to a non-ideal "inner" solution at the relevant transition layer boundaries. Variational formulations and energy principles also provide a common and often powerful mathematical approach, particularly for ideal theory but sometimes to investigate non-ideal effects.

As indicated above, we have deliberately correlated essential theory in Fluid Mechanics *and* MHD, and we also proceed to some notable applications with which we are most familiar. Three sections indicated by a star (*) are not referenced later, but introduce aspects that some readers may find interesting. Certain fundamental topics are dealt with more briefly than elsewhere, for there are several well written basic textbooks on either fluid mechanics or MHD, such as Acheson ("Elementary Fluid Mechanics", Clarendon Press, Oxford) and Davidson ("An Introduction to Magnetohydrodynamics", Cambridge University Press). We do provide more detail in some areas—e.g. where we have personally contributed or where we believe future research work may prove fruitful, trusting our readers will enjoy that personal touch without in any way losing an appreciation of the essential component. Our orientation in the application of MHD is not to fluids but to plasmas, mainly in the context of magnetic confinement in controlled thermonuclear fusion research, although we also discuss some aspects in astrophysics and solar physics that caught our interest in the past. Incidentally, we recommend Prof. Kulsrud's book on "Plasma Physics for Astrophysics" (Princeton University Press) as an excellent point of departure for research students interested in astrophysical applications.

The mathematical development from Newton's basic description of the motion of a single particle to modern mathematical modelling of complex dynamical systems is a remarkable scientific achievement, and the evolution of Fluid Mechanics and MHD is no small part of that tapestry to which a new generation of young scientists can be expected to contribute. While our mathematical development here is largely self-contained, there are inevitably quite important areas that cannot be included in a book of this size—e.g. nonlinear modelling underlying the description of turbulence or chaos, phenomena that often follow or precede coherent structures in fluids and plasmas. Indeed, given the rapid increase in computing power, numerical simulation has become increasingly important in Fluid Mechanics and MHD, especially in nonlinear calculations. However, the fundamental theory and analytical procedures represented in this book are essentially indispensable for any emerging researcher in Fluid Mechanics and MHD, and we hope that more established members of the research community may also enjoy our correlated presentation. We list various books in the bibliography to cover further material in many areas, ranging from introductory to classical and more specialist treatments. Further, we occasionally include additional comments in footnotes to the text, often with reference to relevant original sources.

Each chapter in this book begins with a brief preamble, but it may be helpful to comment on our overall presentation here. Chapter 1 on Vectors and Tensors incorporates notable material on dyadics and vector and dyadic representations, not only in familiar orthogonal coordinate systems but in general as non-orthogonal coordinates are needed later. The additional topics that follow in this chapter then complete a deliberately chosen near-sufficient "mathematical toolkit" for the rest of the book. After briefly introducing the concept of conservation equations and discussing several basic topics often found elsewhere, in Chap. 2 on the Fundamental Equations we proceed to a unified derivation that in particular clearly distinguishes the viscosity of a neutral fluid from the viscosity of a plasma in a magnetic field (to be subsequently described by MHD). Thus although this detailed derivation of the macroscopic equations from a Boltzmann-like equation may seem quite complicated at first sight, it applies to either a collection of molecules in a fluid or of charged particles in a plasma, and we encourage our readers to follow it (with the help of some associated analysis in related Exercises) as a suitable theoretical foundation at the outset with results for reference later. The adopted closure at thirteen moments also allows for a generalisation of the classical law of heat conduction, but this section is starred as that aspect is not pursued in this book. Chapter 2 concludes with a brief summary of forms of the magnetic induction equation needed to complete any MHD model, either the ideal or some non-ideal model to be discussed.

Only then do we proceed to discuss Basic Fluid Mechanics in Chap. 3 and Waves in Fluids in Chap. 4, before returning again to MHD in Chap. 5 followed by MHD Stability Theory in Chap. 6. Given the diversity of formal background in Fluid Mechanics or MHD previously mentioned, we anticipate that certain topics will already be familiar to our readers but others less so—some in Chaps. 3 and 4 with an entertaining history in fluid mechanics but others more recent, often serving as precursors to our discussion of various topics in Chaps. 5 and 6. The emphasis on waves rather than instabilities in Chap. 4, but on stability theory rather than waves in Chap. 6, is quite deliberate as there are already some excellent books on hydrodynamic stability. Moreover, we consider important aspects of wave propagation in appropriate places in Chap. 5; and in Chap. 6 we emphasise further developments in both ideal and non-ideal MHD stability from the perspective of our personal research experience, given the extensive classical treatise by Chandrasekhar on "Hydrodynamic and Hydromagnetic Stability" (Dover).

The Exercises throughout the book often serve to provide additional quite significant knowledge or develop mathematical skill, and may also fill in certain details or enhance understanding of some essential concept—or even to extend the discussion in the text in some interesting way. We strongly recommend that any student first attempt to answer each Exercise personally and independently, before participating in a group discussion or referring to the Answers we have provided for help (available online). Over many years, we have often found that the independent effort involved significantly assists student learning.

One of us (RJH) acknowledges the hospitality of Prof. Ian Craig and his group at the University of Waikato, and that of Prof. John Howard and Dr. Matthew Hole and his group at the Australian National University, during two important stages in completing this book. Michael Howard and several others provided perceptive and helpful comments that led to a clearer exposition at various places in the discussion, and we dedicate it more generally to the continual stimulation and enthusiasm of so many of our students and colleagues over the years.

March 2015

R.J. Hosking
R.L. Dewar

Contents

Chapter 1
Vectors and Tensors

Vector and dyadic (second-order tensor) fields are basic entities in Fluid Mechanics and MHD. The term "field" in this book means that the quantity is a function of position in three-dimensional space, in addition to any time dependence. This spatial dependence is conveniently represented by the associated position vector $\mathbf{r}$ that may itself be a function of the time t (in a dynamical system), so any field is an invariant function of the form $\mathbf{f}(\mathbf{r}(t), t)$. On the other hand, vector and tensor representations depend upon the reference coordinate system chosen, and we discuss the representation of vectors as a useful prelude to the following sections on dyadics and their representation. The vector differential operator then introduced is used more extensively, from determining basis sets for coordinate systems to its role in so many mathematical expressions throughout the book. Although there is a brief section on the familiar special case of orthogonal curvilinear coordinates, it is notable that the previous and following sections on the integral theorems apply to any three-dimensional coordinate system (non-orthogonal curvilinear coordinate systems are particularly important in MHD). The remaining three sections on Green identities and Heaviside, Dirac and Green functions complete this chapter on relevant mathematical topics. The associated bibliography provides recommended sources on vectors and tensors, on distribution theory and mathematical methods, and on partial differential equations for further reading.

1.1 Introduction

Although certain physical quantities such as temperature and electric charge may be represented by numbers (scalars), over a century ago it was recognised that others with both magnitude and direction—such as position, velocity, force, or electric or magnetic field—are invariant but their *representations* change whenever a new coordinate system is adopted. An early mathematical approach to acknowledge this was the theory of quaternions devised by Hamilton, but the vector algebra and calculus due to Gibbs and Heaviside is mainly used nowadays. A directed line segment is

R.J. Hosking and R.L. Dewar, *Fundamental Fluid Mechanics and Magnetohydrodynamics*, DOI 10.1007/978-981-287-600-3_1

the primitive notion for a vector, defined to be an entity with a representation that transforms under coordinate transformations in the same way as an infinitesimal line element in space. Vector algebra and calculus may be regarded as a branch of tensor analysis developed by Ricci and Levi-Civita, *where any tensor is invariant but its representation changes according to a specific transformation law, when another coordinate system is adopted.*[1]

Modern mathematics and physics textbooks use **bold face** symbols to denote a vector—e.g. $d\mathbf{r}$ for a line element, $\mathbf{B}$ for a magnetic field, etc. Bold face sans serif or calligraphic notation may be used for higher order tensors, including dyadics (second-order tensors) such as the important *pressure tensor* p in this book. The elementary dot and cross product operations involving vectors are well known, including the respective commutative and anti-commutative properties

$$\begin{aligned} \mathbf{a} \cdot \mathbf{b} &= \mathbf{b} \cdot \mathbf{a}, \\ \mathbf{a} \times \mathbf{b} &= -\mathbf{b} \times \mathbf{a}, \end{aligned} \tag{1.1}$$

and various identities such as

$$\begin{aligned} &\mathbf{a} \cdot \mathbf{b} \times \mathbf{c} = \mathbf{b} \cdot \mathbf{c} \times \mathbf{a} = \mathbf{c} \cdot \mathbf{a} \times \mathbf{b}, \\ &\qquad\qquad (invariance\ under\ cyclic\ permutation) \\ &\mathbf{a} \times (\mathbf{b} \times \mathbf{c}) = \mathbf{a} \cdot \mathbf{c}\, \mathbf{b} - \mathbf{a} \cdot \mathbf{b}\, \mathbf{c}, \\ &(\mathbf{a} \times \mathbf{b}) \times \mathbf{c} = \mathbf{a} \cdot \mathbf{c}\, \mathbf{b} - \mathbf{b} \cdot \mathbf{c}\, \mathbf{a}. \end{aligned} \tag{1.2}$$

Dot and cross product operations also apply to dyadics, or to tensors of even higher order, via basis sets appropriate to the tensor order (cf. Sect. 1.4).

In this book, only representations with respect to any three-dimensional coordinate system are needed, although generalisation to higher dimensions is straightforward. After detailing the representation of vectors in the next section, we discuss dyadics (second-order tensors) and their representation in Sects. 1.3 and 1.4. The fundamental vector differential operator ∇ is introduced in Sect. 1.5 and immediately applied in Sects. 1.6–1.9 (on curvilinear coordinates, the integral theorems and Green identities), and then almost everywhere in the equations of subsequent chapters! The preparatory "mathematical toolkit" provided in this chapter is completed with a discussion of the Heaviside step and Dirac delta functions in Sect. 1.10, followed by a short section on Green functions.

Exercises

(Q1) The *unit vector* corresponding to any vector **a**, considered to be a directed line segment in three-dimensional Euclidean space, is denoted and defined by

[1] *Encyclopedia Britannica* provides accurate biographies on many of the mathematicians and physicists mentioned throughout this book, and there are various other (sometimes less reliable) sources of information available nowadays on the internet—e.g. Wikipedia.

$\hat{\mathbf{a}} \equiv \mathbf{a}/|\mathbf{a}|$ where $|\mathbf{a}|$ is the magnitude of $\mathbf{a}$.[2] The unit vector defines the direction of the line segment, and the magnitude is its length. The elementary dot and cross products of a vector $\mathbf{a}$ with a vector $\mathbf{b}$ are then defined as

$$\mathbf{a} \cdot \mathbf{b} \equiv |\mathbf{a}||\mathbf{b}| \cos(\hat{\mathbf{a}}, \hat{\mathbf{b}})$$

$$\text{and} \qquad \mathbf{a} \times \mathbf{b} \equiv |\mathbf{a}||\mathbf{b}| \sin(\hat{\mathbf{a}}, \hat{\mathbf{b}})\hat{\mathbf{e}}$$

respectively, where $(\hat{\mathbf{a}}, \hat{\mathbf{b}})$ denotes the angle between the two vectors, and $\hat{\mathbf{e}}$ is a unit vector perpendicular to both $\mathbf{a}$ and $\mathbf{b}$ such that $\{\mathbf{a}, \mathbf{b}, \hat{\mathbf{e}}\}$ constitute a right-handed set. From these definitions, deduce that

$$\mathbf{a} \cdot \mathbf{b} = \mathbf{b} \cdot \mathbf{a},$$

$$\mathbf{a} \times \mathbf{b} = -\mathbf{b} \times \mathbf{a},$$

$$\text{and} \qquad \mathbf{a} \cdot \mathbf{b} \times \mathbf{c} = \mathbf{b} \cdot \mathbf{c} \times \mathbf{a} = \mathbf{c} \cdot \mathbf{a} \times \mathbf{b}.$$

(Q2) A vector space over K (either real or complex numbers) is a non-empty set V closed under the two operations "addition" (+) and "multiplication by scalars" (elements of K), subject to the following eight axioms:
$\forall\, a, b, c \in \mathrm{V}$ and $\alpha, \beta \in K$,

$$a + b = b + a$$

$$(a + b) + c = a + (b + c)$$

$$\text{there is } 0 \in \mathrm{V} \text{ such that} \quad a + 0 = a$$

$$\text{and } -a \in \mathrm{V} \text{ such that } a + (-a) = 0$$

$$\alpha(\beta a) = (\alpha\beta)a$$

$$\alpha(a + b) = \alpha a + \alpha b$$

$$(\alpha + \beta)a = \alpha a + \beta a$$

$$1 \cdot a = a.$$

Define familiar operations of "addition" and "multiplication by scalars" when V is (i) a set of ordered n-tuples $\{a \equiv (a_1, a_2, \ldots, a_n)\}$ where $a_i \in K$ and (ii) a set of $m \times n$ matrices $\{[a_{ij}]\}$ where $a_{ij} \in K$. Do the eight axioms hold in each case?

1.2 Vector Representations

Although it can be convenient to undertake vector (or tensor) algebra or calculus in terms of the invariant quantities, it is common to resolve them onto some basis set. Thus in the case of vectors, a vector basis set $\{\mathbf{e}_1, \mathbf{e}_2, \mathbf{e}_3\}$ is selected, consisting of

[2]The superposed hat symbol is used to denote unit vectors throughout this book.

three linearly independent (non-coplanar) vectors that span the three-dimensional space. The Cartesian basis set $\{\hat{\mathbf{i}}, \hat{\mathbf{j}}, \hat{\mathbf{k}}\}$, consisting of constant unit vectors directed along orthogonal axes in a chosen system of Cartesian coordinates (x, y, z), is most familiar—but more generally the three basis vectors are functions of position and need not be orthonormal (mutually orthogonal unit) vectors. The *Jacobian* is defined by

$$J \equiv \mathbf{e}_1 \cdot \mathbf{e}_2 \times \mathbf{e}_3, \tag{1.3}$$

such that $J \neq 0$ is necessary and sufficient for linear independence and $J = 1$ if the vectors are orthonormal.

The general *contravariant representation* of any vector $\mathbf{a}$ is thus the ordered triple (a^1, a^2, a^3) such that

$$\mathbf{a} = \sum_{i=1}^{3} a^i \mathbf{e}_i, \tag{1.4}$$

where it is customary to omit the summation symbol under the convention that summation over $i = 1, 2, 3$ is implied by any repeated index. Note that the vector basis set may readily be obtained in any arbitrary curvilinear coordinate system (x^1, x^2, x^3) by taking partial derivatives such that

$$\mathbf{e}_i = \frac{\partial \mathbf{r}}{\partial x^i}, \tag{1.5}$$

where the Cartesian coordinates in the position vector $\mathbf{r} = x\hat{\mathbf{i}} + y\hat{\mathbf{j}} + z\hat{\mathbf{k}}$ are expressed in terms of the curvilinear coordinates under a coordinate transformation—e.g. cylindrical coordinates (r, θ, z) that are polar coordinates in the x, y-plane, $x = r\cos\theta$ and $y = r\sin\theta$, but remain Cartesian in the z-direction (cf. also Sect. 1.6). The complementary set of reciprocal basis vectors

$$\left\{\mathbf{e}^1 \equiv \frac{\mathbf{e}_2 \times \mathbf{e}_3}{\mathbf{e}_1 \cdot \mathbf{e}_2 \times \mathbf{e}_3},\ \mathbf{e}^2 \equiv \frac{\mathbf{e}_3 \times \mathbf{e}_1}{\mathbf{e}_1 \cdot \mathbf{e}_2 \times \mathbf{e}_3},\ \mathbf{e}^3 \equiv \frac{\mathbf{e}_1 \times \mathbf{e}_2}{\mathbf{e}_1 \cdot \mathbf{e}_2 \times \mathbf{e}_3}\right\} \tag{1.6}$$

satisfies the orthogonality property

$$\mathbf{e}_i \cdot \mathbf{e}^j = \delta_i^j \qquad (i, j \in \{1, 2, 3\}), \tag{1.7}$$

where the Kronecker delta symbol is defined by

$$\delta_i^j = \begin{cases} 1, & i = j \\ 0, & 1 \neq j. \end{cases} \tag{1.8}$$

From (1.3) and (1.6) we obtain the reciprocity property

$$\mathbf{e}^1 \cdot \mathbf{e}^2 \times \mathbf{e}^3 = J^{-1}, \tag{1.9}$$

which implies that the reciprocal basis vectors are also linearly independent (non-coplanar). Each component a^i in the contravariant representation (1.4) may be found by resolving vector $\mathbf{a}$ onto the corresponding reciprocal basis vector $\mathbf{e}^i$—i.e.

$$a^i = \mathbf{e}^i \cdot \mathbf{a}. \tag{1.10}$$

The alternative *covariant representation* (a_1, a_2, a_3) is defined such that

$$\mathbf{a} = a_i \mathbf{e}^i \tag{1.11}$$

under the summation convention, where

$$a_i = \mathbf{e}_i \cdot \mathbf{a}. \tag{1.12}$$

The algebraic distributive law produces the dot product representation

$$\mathbf{a} \cdot \mathbf{b} = a^i \mathbf{e}_i \cdot b_j \mathbf{e}^j = a^i b_j \delta_i^j = a^i b_i, \tag{1.13}$$

or alternatively $\mathbf{a} \cdot \mathbf{b} = a_i b^i$, under the summation convention. Similarly, the cross product may be represented by

$$\mathbf{a} \times \mathbf{b} = a^i b^j \mathbf{e}_i \times \mathbf{e}_j = J \epsilon_{ijk} a^i b^j \mathbf{e}^k = J \epsilon_{ijk} a^j b^k \mathbf{e}^i, \tag{1.14}$$

where the Levi-Civita symbol ϵ_{ijk} is $+1$ (-1) for (i, j, k) an even (odd) permutation of $(1, 2, 3)$ and 0 in all other cases (when two or more subscript values are identical).

The *metric* coefficient defined by

$$g^{ij} = \mathbf{e}^i \cdot \mathbf{e}^j \tag{1.15}$$

facilitates the transformation from the covariant to the contravariant representation—thus invoking (1.11) in (1.10) gives $a^i = \mathbf{e}^i \cdot \mathbf{a} = a_j \mathbf{e}^i \cdot \mathbf{e}^j = g^{ij} a_j$, so the metric coefficient is said to raise the index. Likewise, the coefficient

$$g_{ij} = \mathbf{e}_i \cdot \mathbf{e}_j \tag{1.16}$$

facilitates the transformation from the contravariant to the covariant representation—thus invoking (1.4) in (1.12) gives $a_i = \mathbf{e}_i \cdot \mathbf{a} = a^j \mathbf{e}_i \cdot \mathbf{e}_j = g_{ij} a^j$, so the coefficient g_{ij} lowers the index. These raising and lowering operations also extend to indices in dyadic representations (cf. Sect. 1.4) and higher order tensor representations generally.[3]

[3]The adjective "metric" refers to the coefficient g^{ij} defining the length ds of a line element at any point in space, corresponding to $ds^2 = d\mathbf{r} \cdot d\mathbf{r} = dx_i \mathbf{e}^i \cdot dx_j \mathbf{e}^j = g^{ij} dx_i dx_j$ where $\mathbf{r}$ denotes

An orthonormal basis set ($\hat{\mathbf{e}}_i \cdot \hat{\mathbf{e}}_j = \delta_{ij}$) is self-reciprocal ($\hat{\mathbf{e}}^j \equiv \hat{\mathbf{e}}_j,\ \forall\, j \in \{1,2,3\}$), when we need not distinguish between the two representations. (Evidently, $g_{ij} \equiv \delta_{ij}$ for any orthonormal basis set.) The simplest case is of course the Cartesian basis $\{\hat{\mathbf{i}}, \hat{\mathbf{j}}, \hat{\mathbf{k}}\}$, which is convenient to use to prove vector and tensor identities—because there is then no need to distinguish superscripts (only subscripts are needed) nor to carry the basis vectors, and such identities are invariant. For example, from (1.13) and (1.14)

$$\mathbf{a} \cdot \mathbf{b} \times \mathbf{c} = a_i(\epsilon_{ijk} b_j c_k) = (\epsilon_{kij} a_i b_j) c_k = \mathbf{a} \times \mathbf{b} \cdot \mathbf{c},$$

where the brackets are not essential but included to aid interpretation. The other equivalent forms of the scalar triple product ($\mathbf{b} \cdot \mathbf{c} \times \mathbf{a}$ and $\mathbf{c} \cdot \mathbf{a} \times \mathbf{b}$) also follow from the invariance of the symbol ϵ_{ijk} under cyclic permutation of its indices $\{ijk\}$. The vector cross product identities in (1.2) follow with the help of the property

$$\epsilon_{ijk}\epsilon_{klm} = \delta_{il}\delta_{jm} - \delta_{jl}\delta_{im}, \tag{1.17}$$

which can be proven case by case (by choosing all conceivable combinations of the indices). Thus for example, the ith component

$$\begin{aligned}[\mathbf{a} \times (\mathbf{b} \times \mathbf{c})]_i = \epsilon_{ijk} a_j (\mathbf{b} \times \mathbf{c})]_k &= \epsilon_{ijk} a_j \epsilon_{klm} b_l c_m = (\delta_{il}\delta_{jm} - \delta_{jl}\delta_{im}) a_j b_l c_m \\ &= a_j c_j b_i - a_j b_j c_i = [\mathbf{a} \cdot \mathbf{c}\, \mathbf{b} - \mathbf{a} \cdot \mathbf{b}\, \mathbf{c}\,]_i,\end{aligned}$$

on noting that δ_{ij} acts essentially as a "substitution operator" (implies a zero outcome unless $i = j$). This is sufficient to prove the vector identity, since the result is true $\forall\, i \in \{1,2,3\}$ and the vector identity is invariant.

The Cartesian basis set $\{\hat{\mathbf{i}}, \hat{\mathbf{j}}, \hat{\mathbf{k}}\}$ is global, in the sense that its orientation is fixed when the corresponding set of Cartesian coordinates has been chosen, but it is not the only orthonormal basis set. For example, in MHD it is often convenient to choose a local basis set where one basis vector is aligned with the magnetic field and the others are, respectively, tangent and perpendicular to the magnetic surface. Thus in simplified analysis where the magnetic surface is assumed planar with normal $\hat{\mathbf{k}}$, the corresponding basis set $\{\hat{\mathbf{b}}(z), \hat{\mathbf{k}} \times \hat{\mathbf{b}}(z), \hat{\mathbf{k}}\}$ is orthonormal but the direction of the magnetic field $\mathbf{B}(z)$ defined by the unit vector $\hat{\mathbf{b}}(z) \equiv \mathbf{B}(z)/|\mathbf{B}(z)|$ may vary. However, since actual magnetic surfaces are not usually planar, it is more appropriate to adopt non-Cartesian magnetic coordinates (cf. Sect. 5.11). We then return to (1.5) to define the basis set $\{\mathbf{e}_i\}$, and render the reciprocal basis set either as $\{\mathbf{e}^i = \nabla x^i\}$ under (1.41) below or invoke (1.6) with Jacobian $J \neq 1$ given by (1.3).

(Footnote 3 continued)
the position vector of the point. From their definitions, the coefficients are obviously symmetric (i.e. $g^{ij} = g^{ji}$ and $g_{ij} = g_{ji}$); and given the unit dyadic projection $\mathbf{I} \cdot \mathbf{e}^k = \mathbf{e}^k$ noted in the next section, we also have $g_{ij} g^{jk} = \mathbf{e}_i \cdot \mathbf{e}_j \mathbf{e}^j \cdot \mathbf{e}^k = \mathbf{e}_i \cdot \mathbf{I} \cdot \mathbf{e}^k = \mathbf{e}_i \cdot \mathbf{e}^k = \delta_i^k$, a result which is sometimes collected into the matrix form $[g_{ij}][g^{ij}] = I$ where I is the 3×3 unit matrix.

Exercises

(Q1) Use a Cartesian representation to prove the identity

$$(\mathbf{a} \times \mathbf{b}) \times \mathbf{c} = \mathbf{a} \cdot \mathbf{c}\, \mathbf{b} - \mathbf{b} \cdot \mathbf{c}\, \mathbf{a}.$$

(Q2) Given $J \equiv \mathbf{e}_1 \cdot \mathbf{e}_2 \times \mathbf{e}_3$ and $\{\mathbf{e}^1, \mathbf{e}^2, \mathbf{e}^3\}$ satisfies (1.7), show that $\mathbf{e}^1 \cdot \mathbf{e}^2 \times \mathbf{e}^3 = J^{-1}$.

1.3 Dyadics

Although scalar and vector quantities may be more familiar, the pressure tensor dyadic p mentioned in Sect. 1.1 is fundamental in representing so-called surface forces in fluid mechanics and MHD, and there are distinct expressions for p in a fluid or magnetised plasma as discussed in the next chapter. Another important dyadic in MHD is the magnetic stress tensor $\mathcal{T}$, which represents the characteristic additional magnetic force arising in the equation of motion (cf. Sects. 2.7 and 5.7 in particular).

A dyadic is a linear superposition of ordered pairs of vectors, called dyads. Thus if **a** and **b** are different vectors, then **ab** and **ba** are different dyads, and **ab** + **cd** is a dyadic that may be denoted by A. Their fundamental algebraic properties may be summarised by saying that associative and distributive laws hold for both post- and pre-dot and cross multiplication, such as

$$(\mathbf{ab} + \mathbf{cd}) \cdot \mathbf{e} = \mathbf{ab} \cdot \mathbf{e} + \mathbf{cd} \cdot \mathbf{e}, \tag{1.18}$$

$$\mathbf{e} \cdot (\mathbf{ab} + \mathbf{cd}) \times \mathbf{f} = \mathbf{e} \cdot \mathbf{a}\, \mathbf{b} \times \mathbf{f} + \mathbf{e} \cdot \mathbf{c}\, \mathbf{d} \times \mathbf{f}. \tag{1.19}$$

Since the projection of a dyadic A onto a vector such as in (1.18) produces a vector, a further dot product projection of course produces a scalar. As discussed in the next section, this process may be used to identify respective representations of the dyadic, and the dot and cross product operations also extend to two or more dyadics.

The *transpose* A^T of a dyadic A is the sum of its component dyads with their order reversed. For example, if

$$\mathsf{A} = \mathbf{ab} + \mathbf{cd}, \tag{1.20}$$

then

$$\mathsf{A}^T = \mathbf{ba} + \mathbf{dc}. \tag{1.21}$$

Note also that a dyadic may be *symmetric* (such that $\mathsf{A}^T = \mathsf{A}$) or *antisymmetric* (such that $\mathsf{A}^T = -\mathsf{A}$). The post-dot product $\mathsf{A} \cdot \mathbf{v}$ of the dyadic A with any vector **v** is a vector, which differs from the vector produced by the pre-dot product $\mathbf{v} \cdot \mathsf{A}$ unless A is symmetric.

The *inverse* (or reciprocal) A^{-1} of a dyadic A is defined by

$$\mathsf{A}^{-1} \cdot \mathsf{A} = \mathsf{A} \cdot \mathsf{A}^{-1} = \mathsf{I}, \tag{1.22}$$

where I is the *unit dyadic* (or idemfactor) defined by

$$\mathsf{I} \cdot \mathbf{v} = \mathbf{v} \cdot \mathsf{I} = \mathbf{v}, \quad \forall \text{ vector } \mathbf{v}. \tag{1.23}$$

Exercise

(Q1) Show that any dyadic may be written as the sum of a symmetric and antisymmetric part—viz.

$$\mathsf{A} = \frac{1}{2}[\mathsf{A} + \mathsf{A}^T] + \frac{1}{2}[\mathsf{A} - \mathsf{A}^T]. \tag{1.24}$$

1.4 Dyadic Representations

The vectors in any dyadic

$$\mathsf{A} = \mathbf{a}_1\mathbf{b}_1 + \mathbf{a}_2\mathbf{b}_2 + \cdots + \mathbf{a}_n\mathbf{b}_n \tag{1.25}$$

may of course be resolved onto any arbitrary basis set $\{\mathbf{e}_1, \mathbf{e}_2, \mathbf{e}_3\}$. Resolving the vectors $\mathbf{b}_1, \mathbf{b}_2, \ldots, \mathbf{b}_n$ yields

$$\mathsf{A} = \sum_{i=1}^{n} \mathbf{a}_i (b_i^1 \mathbf{e}_1 + b_i^2 \mathbf{e}_2 + b_i^3 \mathbf{e}_3) = \mathbf{f}_1\mathbf{e}_1 + \mathbf{f}_2\mathbf{e}_2 + \mathbf{f}_3\mathbf{e}_3, \tag{1.26}$$

where vectors $\mathbf{f}_j \equiv \sum_{i=1}^{n} b_i^j \mathbf{a}_i$. Thus any dyadic may be expressed as a sum of just three dyads, which respectively incorporate the arbitrary basis vectors. Moreover, resolution of the companion vectors $\mathbf{f}_1, \mathbf{f}_2, \mathbf{f}_3$ onto the same basis set $\{\mathbf{e}_1, \mathbf{e}_2, \mathbf{e}_3\}$ yields

$$\mathsf{A} = \sum_{j=1}^{3} \sum_{k=1}^{3} A^{jk} \mathbf{e}_j \mathbf{e}_k, \tag{1.27}$$

expressing the dyadic in terms of the resulting set of nine basis dyads $\{\mathbf{e}_j\mathbf{e}_k\}$. The 3×3 matrix of coefficients denoted by $[A^{jk}]$ constitutes the corresponding contravariant tensor representation of the dyadic A. It is again customary to omit the summation symbols under the summation convention, in this case for the two repeated indices j and k.

Projecting the dyadic $\mathbf{A}$ onto the corresponding reciprocal basis set $\{\mathbf{e}^1, \mathbf{e}^2, \mathbf{e}^3\}$ yields the vector

$$\mathbf{e}^l \cdot \mathbf{A} = A^{jk}\mathbf{e}^l \cdot \mathbf{e}_j \mathbf{e}_k = A^{lk}\mathbf{e}_k, \tag{1.28}$$

and hence the contravariant representation entries via a second dot product

$$\mathbf{e}^l \cdot \mathbf{A} \cdot \mathbf{e}^m = A^{lk}\mathbf{e}_k \cdot \mathbf{e}^m = A^{lm}. \tag{1.29}$$

The order of the two dot product operations producing the contravariant entries may be reversed, but note that the intermediate vectors $\mathbf{A} \cdot \mathbf{e}^m = A^{jm}\mathbf{e}_j$ (for $m = 1, 2, 3$) are different from the vectors $\mathbf{e}^l \cdot \mathbf{A} = A^{lk}\mathbf{e}_k$ (for $l = 1, 2, 3$) unless the dyadic $\mathbf{A}$ is symmetric (when $A^{lk} = A^{kl}$). With reference to the reciprocal basis set, the above discussion likewise leads to the covariant representation $[A_{jk}]$ corresponding to the expression $\mathbf{A} = A_{jk}\mathbf{e}^j\mathbf{e}^k$, or mixed representations corresponding to either $\mathbf{A} = A_j{}^k\mathbf{e}^j\mathbf{e}_k$ or $\mathbf{A} = A^j{}_k\mathbf{e}_j\mathbf{e}^k$. Corresponding valid projections of $\mathbf{A}$ produce the respective tensor elements $A_{lm} = \mathbf{e}_l \cdot \mathbf{A} \cdot \mathbf{e}_m$, $A_l{}^m = \mathbf{e}_l \cdot \mathbf{A} \cdot \mathbf{e}^m$, and $A^l{}_m = \mathbf{e}^l \cdot \mathbf{A} \cdot \mathbf{e}_m$.

Dyadic products may be interpreted via dyadic representations. Thus the dyadic dot product

$$\begin{aligned}\mathbf{A} \cdot \mathbf{B} &= A^j{}_k\mathbf{e}_j\mathbf{e}^k \cdot B^l{}_m\mathbf{e}_l\mathbf{e}^m \\ &= A^j{}_k B^l{}_m \mathbf{e}^k \cdot \mathbf{e}_l \mathbf{e}_j \mathbf{e}^m \\ &= A^j{}_k B^k{}_m \mathbf{e}_j \mathbf{e}^m\end{aligned} \tag{1.30}$$

is a dyadic; and we define the double dot (or dyadic scalar) product[4]

$$\mathbf{A} : \mathbf{B} = A^j{}_k B^l{}_m \mathbf{e}^k \cdot \mathbf{e}_l \, \mathbf{e}_j \cdot \mathbf{e}^m = A^j{}_k B^k{}_j. \tag{1.31}$$

Further, the algebra is readily extended to pre- or post-cross products of a vector with a dyadic such that

$$\mathbf{a} \times \mathbf{A} = a^i A^{jl}\mathbf{e}_i \times \mathbf{e}_j\mathbf{e}_l = J\epsilon_{ijk}a^i A^{jl}\mathbf{e}^k\mathbf{e}_l \tag{1.32}$$

and

$$\mathbf{A} \times \mathbf{b} = A^{jk}b^l\mathbf{e}_j\mathbf{e}_k \times \mathbf{e}_l = J\epsilon_{klm}A^{jk}b^l\mathbf{e}_j\mathbf{e}^m \tag{1.33}$$

are both dyadics. There are again of course covariant or mixed representations, corresponding to the three other forms of basis dyads.

Further extension of the algebra, for example to render the cross product of two dyadics or some product of more than two dyadics, leads to higher order tensors (triadics in terms of basis triads, etc.)—the process is quite straightforward, but unnecessary for this book. Note also that many textbooks define contravariant and

---

[4]Some authors define this product as $\mathbf{A} : \mathbf{B} = A^j{}_k B_j{}^k$.

covariant "vectors and tensors" by transformation laws, which relate vector and tensor representations in two different coordinate systems. This has the attraction that the transformation laws may refer to n-tuples or $n \times n$ matrices for example, as the respective representations of vectors or dyadics in n-dimensional space—but the essential invariance of any vector or tensor is sometimes overlooked! Examples of the transformation laws for vector and dyadic representations appear in two of the following exercises, with reference to coordinate transformations in both three-dimensional and n-dimensional space. Any dot product reduces the tensor order by two, and any cross product reduces the tensor order by one—cf. (1.30)–(1.33).[5]

Exercises

(Q1) If $\{\mathbf{e}^i\}$ and $\{\mathbf{e}_i\}$ (where $i \in \{1, 2, 3\}$) are mutually reciprocal vector basis sets, show that the unit dyadic

$$\mathsf{I} = \mathbf{e}^i \mathbf{e}_i = \mathbf{e}_i \mathbf{e}^i . \tag{1.34}$$

Hence, deduce the mixed dyadic representation

$$\mathsf{A} = A^i{}_j \mathbf{e}_i \mathbf{e}^j ,$$

where $A^i{}_j \equiv \mathbf{e}^i \cdot \mathsf{A} \cdot \mathbf{e}_j$. Similarly, deduce that

$$\mathsf{A} \cdot \mathsf{B} = A^i{}_j B^j{}_k \mathbf{e}_i \mathbf{e}^k .$$

(Q2) Show that $\mathsf{I} \times \mathbf{a} = \mathbf{a} \times \mathsf{I}$, where $\mathbf{a}$ is any vector and I is the unit dyadic.

(Q3) Show that the *trace* of a dyadic A

$$\mathrm{Tr}\,\mathsf{A} \equiv \sum_{i=1}^{3} A^i{}_i$$

and its *determinant*

$$\det \mathsf{A} \equiv \det(A^i{}_j)$$

are invariant—i.e. their values are independent of the choice of the basis set $\{\mathbf{e}_i\}$.

(Q4) Find $\det(\mathsf{I} + \mathbf{ab})$, where $\mathbf{a}$ and $\mathbf{b}$ are arbitrary vectors and I is the unit dyadic.

(Q5) Show that $(\mathsf{A}^T)_{ij} = A_{ji}$, $(\mathsf{A}^T)^{ij} = A^{ji}$, $(\mathsf{A}^T)_i{}^j = A^j{}_i$, and $(\mathsf{A}^T)^i{}_j = A_j{}^i$.

(Q6) Suppose that the position vector $\mathbf{r}$ of any point in space has the representations $\mathbf{r} = x^j \mathbf{e}_j$ and $\mathbf{r} = x'^k \mathbf{e}'_k$ in two different coordinate systems, related by the coordinate transformation

[5] Tensor order (or rank) refers to the number of juxtaposed vectors in the entity, so that a scalar is a zeroth-order tensor, a vector is a first-order tensor, a dyadic a second-order tensor, etc.—i.e. in each case equivalent to the number of times the coordinate transformation matrix is applied in the transformation law, or the number of free (non-summation) indices in the tensor representation.

$$x'^i = x'^i(x^1, x^2, x^3), \; i \in \{1, 2, 3\}$$

consisting of independent single-valued continuously differentiable functions (a one-to-one invertible mapping). Noting that any infinitesimal line element at the point may be expressed as

$$d\mathbf{r} = \mathbf{e}_j dx^j = \mathbf{e}'_k dx'^k,$$

deduce that

$$dx'^i = \frac{\partial x'^i}{\partial x^j} dx^j, \text{ where } \mathbf{e}'^i \cdot \mathbf{e}_j \equiv \frac{\partial x'^i}{\partial x^j}.$$

Hence show that the representation of any vector field $\mathbf{v}(\mathbf{r})$ transforms under the coordinate transformation in the same way as the infinitesimal line element—i.e.

$$v'^i = \frac{\partial x'^i}{\partial x^j} v^j.$$

(Q7) Suppose the position vector $\mathbf{r}$ of any point in n-dimensional space has the representations $(x^1, x^2, \ldots, x^n)$ and $(x'^1, x'^2, \ldots, x'^n)$ in two different coordinate systems, related by the coordinate transformation

$$x'^k = x'^k(x^1, x^2, \ldots, x^n), \; k \in \{1, 2, \ldots, n\}$$

consisting of independent single-valued continuously differentiable functions (a one-to-one invertible mapping). Given that a mixed representation of any dyadic field $\mathsf{A}(\mathbf{r})$ transforms as

$$A'^j{}_k = \frac{\partial x'^j}{\partial x^p} \frac{\partial x^q}{\partial x'^k} A^p{}_q,$$

deduce the corresponding transformation law for a mixed representation of the dyadic dot product $\mathsf{A} \cdot \mathsf{B}$. Then show that the double dot product $\mathsf{A} : \mathsf{B}$ is a scalar—i.e. a zeroth-order tensor.

1.5 Vector Differential Operator

The familiar *del* (or nabla) vector differential operator

$$\nabla = \hat{\mathbf{i}}\frac{\partial}{\partial x} + \hat{\mathbf{j}}\frac{\partial}{\partial y} + \hat{\mathbf{k}}\frac{\partial}{\partial z} \tag{1.35}$$

in Cartesian coordinates (x, y, z) may be generalised to the form

$$\nabla = \mathbf{e}^i \frac{\partial}{\partial x^i} \tag{1.36}$$

for any curvilinear coordinates $\{x^1(x, y, z), x^2(x, y, z), x^3(x, y, z)\}$, where the vectors $\mathbf{e}^i \equiv \nabla x^i$ $(i \in \{1, 2, 3\})$ belong to the reciprocal basis set complementary to the vector basis set defined by (1.5). The first-order differential operator ∇ acts in turn on every field quantity (function of position vector $\mathbf{r}$) to its right, and obeys the laws of vector algebra.

The expressions $\nabla\phi$, $\nabla\cdot\mathbf{v}$ and $\nabla\times\mathbf{v}$ are called and sometimes, respectively, written grad ϕ, div $\mathbf{v}$ and curl $\mathbf{v}$, where ϕ and $\mathbf{v}$ are scalar and vector fields. The ∇ notation is preferred in this book, since it is then easier to remember or construct identities such as

$$\nabla\times\nabla\phi = 0 \tag{1.37}$$

$$\nabla(\mathbf{u}\cdot\mathbf{v}) = (\nabla\mathbf{u})\cdot\mathbf{v} + (\nabla\mathbf{v})\cdot\mathbf{u} \tag{1.38}$$

$$\nabla\cdot(\mathbf{u}\times\mathbf{v}) = (\nabla\times\mathbf{u})\cdot\mathbf{v} - (\nabla\times\mathbf{v})\cdot\mathbf{u} \tag{1.39}$$

$$\nabla\times(\mathbf{u}\times\mathbf{v}) = (\nabla\cdot\mathbf{v})\,\mathbf{u} + \mathbf{v}\cdot\nabla\mathbf{u} - (\nabla\cdot\mathbf{u})\,\mathbf{v} - \mathbf{u}\cdot\nabla\mathbf{v}. \tag{1.40}$$

Where there is possible ambiguity about the range of action of the differential operator, this is eliminated by enclosing the affected terms in parentheses. The notational advantage in using the symbol ∇ also extends to other expressions and identities involving dyadics or higher order tensor fields. The dyadic $\nabla\mathbf{v}$ involving the velocity field $\mathbf{v}$ is particularly important in fluid mechanics and MHD, not only when contracted to $\nabla\cdot\mathbf{v}$ and $\nabla\times\mathbf{v}$ but also elsewhere—including in the expression for the rate of deformation tensor S and the nonlinear term $\mathbf{v}\cdot\nabla\mathbf{v}$ in the equation of motion, discussed in the next chapter.

Points for which the scalar field $\phi(\mathbf{r}) =$ constant are said to lie on a *level surface*. The increment in the scalar field $d\phi = d\mathbf{r}\cdot\nabla\phi$ at any point depends upon the direction of the infinitesimal displacement vector (directed line element) $d\mathbf{r}$ emanating from that point; and in particular, if $d\mathbf{r}$ is restricted to the tangential plane at any point on a level surface, we have the important result that $\nabla\phi$ is normal to the level surface at that point (corresponding to $d\phi = 0$). Sometimes, we also consider the increment in a vector field $d\mathbf{v} = d\mathbf{r}\cdot\nabla\mathbf{v}(\mathbf{r})$ at any point (e.g. for a velocity field), similarly dependent upon the direction of the infinitesimal displacement vector $d\mathbf{r}$.

Finally, when the position vector $\mathbf{r}(x^1, x^2, x^3)$ is a continuously differentiable function we note the important result (cf. Exercise 1 below)

$$\nabla\mathbf{r} = \frac{\partial\mathbf{r}}{\partial x^i}\nabla x^i = \mathsf{I}, \tag{1.41}$$

where $\{\mathbf{e}_i = \partial\mathbf{r}/\partial x^i\}$ is the general basis set as defined in Sect. 1.2 and now $\{\mathbf{e}^i = \nabla x^i\}$ is consequently identified as the reciprocal basis set.

Exercises

(Q1) (a) If $\mathbf{r} = x\hat{\mathbf{i}} + y\hat{\mathbf{j}} + z\hat{\mathbf{k}}$ is the position vector in a Cartesian coordinate system (x, y, z), show that $\nabla\mathbf{r} = \mathbf{I}$, $\nabla\cdot\mathbf{r} = 3$ and $\nabla\times\mathbf{r} = 0$ where $\mathbf{I}$ is the unit dyadic.
(b) Deduce (1.34) from $\nabla\mathbf{r} = \mathbf{I}$, assuming that the position vector $\mathbf{r}$ may be expressed as a continuously differentiable function of the arbitrary coordinates (x^1, x^2, x^3).
Now consider the level surfaces x_i = constant ($i \in \{1, 2, 3\}$) defined by the arbitrary curvilinear coordinates, to confirm geometrically that the basis set defined by (1.5) and the set $\{\mathbf{e}^i \equiv \nabla x^i\}$ satisfy the orthogonality property (1.7).

(Q2) Show that

$$\nabla\mathbf{v} = \frac{1}{J}\frac{\partial}{\partial x^i}(J\mathbf{e}^i\mathbf{v}) \tag{1.42}$$

for any vector field $\mathbf{v}(\mathbf{r})$, where x^i are arbitrary curvilinear coordinates and $J \equiv \mathbf{e}_1\cdot\mathbf{e}_2\times\mathbf{e}_3 = [\mathbf{e}^1\cdot\mathbf{e}^2\times\mathbf{e}^3]^{-1}$ as before. Deduce forms for $\nabla\cdot\mathbf{v}$ and $\nabla\times\mathbf{v}$.

(Q3) Show that

$$(\nabla\psi)\cdot\nabla\times(\boldsymbol{\xi}\times\mathbf{B}) = \mathbf{B}\cdot\nabla(\boldsymbol{\xi}\cdot\nabla\psi),$$

given that $\nabla\cdot\mathbf{B} = 0$ and $\mathbf{B}\cdot\nabla\psi = 0$.

1.6 Orthogonal Curvilinear Coordinates

Let us recall that cylindrical coordinates (r, θ, z), related to a system of Cartesian coordinates (x, y, z) via $x = r\cos\theta$ and $y = r\sin\theta$, define a familiar three-dimensional curvilinear coordinate system. In this case, we readily obtain the appropriate orthonormal basis set that varies with position—viz. $\{\hat{\mathbf{e}}_r = \nabla r,\ \hat{\mathbf{e}}_\theta = r\nabla\theta,\ \hat{\mathbf{e}}_z = \nabla z\}$, where the position vector is $\mathbf{r} = r\hat{\mathbf{e}}_r + z\hat{\mathbf{e}}_z$ and $|\mathbf{r}| = \sqrt{r^2 + z^2} \neq r$ unless $z = 0$ (cf. Exercise 1 below). Thus Eq. (1.36) becomes

$$\nabla = \hat{\mathbf{e}}_r\frac{\partial}{\partial r} + \frac{\hat{\mathbf{e}}_\theta}{r}\frac{\partial}{\partial\theta} + \hat{\mathbf{e}}_z\frac{\partial}{\partial z}. \tag{1.43}$$

Since the basis vectors $\hat{\mathbf{e}}_r$ and $\hat{\mathbf{e}}_\theta$ are functions of θ, there are non-zero curvature contributions $\hat{\mathbf{e}}_\theta\cdot\nabla\hat{\mathbf{e}}_r = \hat{\mathbf{e}}_\theta/r$ and $\hat{\mathbf{e}}_\theta\cdot\nabla\hat{\mathbf{e}}_\theta = -\hat{\mathbf{e}}_r/r$, which arise in applying the differential operator ∇. This is all summarised by the dyadic set

$$\nabla\hat{\mathbf{e}}_r = \frac{\hat{\mathbf{e}}_\theta\hat{\mathbf{e}}_\theta}{r}, \tag{1.44}$$

$$\nabla\hat{\mathbf{e}}_\theta = -\frac{\hat{\mathbf{e}}_\theta\hat{\mathbf{e}}_r}{r}, \tag{1.45}$$

$$\nabla\hat{\mathbf{e}}_z = 0. \tag{1.46}$$

Thus for an arbitrary vector field $\mathbf{v} = v_r\hat{\mathbf{e}}_r + v_\theta\hat{\mathbf{e}}_\theta + v_z\hat{\mathbf{e}}_z$ in cylindrical coordinates, there are the well-known results

$$\nabla\cdot\mathbf{v} = \frac{1}{r}\frac{\partial}{\partial r}(rv_r) + \frac{1}{r}\frac{\partial v_\theta}{\partial\theta} + \frac{\partial v_z}{\partial z}, \tag{1.47}$$

and

$$\nabla\times\mathbf{v} = \hat{\mathbf{e}}_r\left(\frac{1}{r}\frac{\partial v_z}{\partial\theta} - \frac{\partial v_\theta}{\partial z}\right) + \hat{\mathbf{e}}_\theta\left(\frac{\partial v_r}{\partial z} - \frac{\partial v_z}{\partial r}\right) + \hat{\mathbf{e}}_z\left(\frac{\partial v_\theta}{\partial r} + \frac{v_\theta}{r} - \frac{1}{r}\frac{\partial v_r}{\partial\theta}\right). \tag{1.48}$$

The Laplacian $\nabla^2 \equiv \nabla\cdot\nabla$ operating on a scalar field ϕ is therefore

$$\nabla^2\phi = \frac{1}{r}\frac{\partial}{\partial r}\left(r\frac{\partial\phi}{\partial r}\right) + \frac{1}{r^2}\frac{\partial^2\phi}{\partial\theta^2} + \frac{\partial^2\phi}{\partial z^2}, \tag{1.49}$$

and there are additional curvature contributions when the Laplacian operates on a vector field $\mathbf{v}$. The divergence of a dyadic field $\mathsf{T} = T_{rr}\hat{\mathbf{e}}_r\hat{\mathbf{e}}_r + T_{r\theta}\hat{\mathbf{e}}_r\hat{\mathbf{e}}_\theta + \ldots$ in cylindrical coordinates may also be obtained directly, bearing in mind the curvature contributions that arise.

Cylindrical polar coordinates are one example of an *orthogonal* curvilinear coordinate system—i.e. where the vector basis set defined by (1.5) is orthogonal. Another example is *spherical polar coordinates*, depicted in Fig. 1.1. Cylindrical and spherical coordinates are particularly appropriate for treating systems that are *axisymmetric* in their unperturbed states—i.e. when these states are *rotationally symmetric* about the z-axis (when scalar fields and vector components or tensor elements of dyadics with respect to the cylindrical or spherical coordinate basis set are independent of θ or ϕ, respectively).

The magnitudes $h_i \equiv |\mathbf{e}_i|$ of the basis vectors $\{\mathbf{e}_1, \mathbf{e}_2, \mathbf{e}_3\}$ in any coordinate system (x^1, x^2, x^3) are often called *scale factors*. In a system of orthogonal curvilinear coordinates, we immediately deduce that the Jacobian may be expressed as $J = h_1h_2h_3$ and the gradient as

$$\nabla = \frac{\hat{\mathbf{e}}_1}{h_1}\frac{\partial}{\partial x^1} + \frac{\hat{\mathbf{e}}_2}{h_2}\frac{\partial}{\partial x^2} + \frac{\hat{\mathbf{e}}_3}{h_3}\frac{\partial}{\partial x^3}, \tag{1.50}$$

where $\{\hat{\mathbf{e}}_i \equiv \mathbf{e}_i/h_i\}$ denotes the relevant orthonormal basis set. Thus for example, the scale factors in cylindrical coordinates are evidently $h_1 = 1$, $h_2 = r$ and $h_3 = 1$.

Exercises

(Q1) The coordinate transformation from Cartesian to cylindrical coordinates is of course equivalent to $\mathbf{r}(r, \theta, z) = r\cos\theta\hat{\mathbf{i}} + r\sin\theta\hat{\mathbf{j}} + z\hat{\mathbf{k}}$. Identify the unit orthogonal (orthonormal) basis vectors $\{\hat{\mathbf{e}}_r, \hat{\mathbf{e}}_\theta, \hat{\mathbf{e}}_z\}$ in terms of $\{\hat{\mathbf{i}}, \hat{\mathbf{j}}, \hat{\mathbf{k}}\}$, obtain

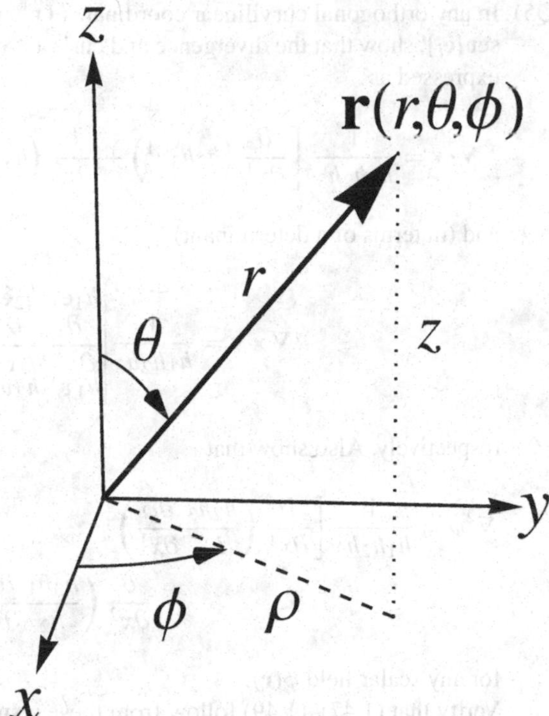

Fig. 1.1 Spherical coordinates (r, θ, ϕ)

the basis set $\{\mathbf{e}_1 = \hat{\mathbf{e}}_r, \mathbf{e}_2 = r\hat{\mathbf{e}}_\theta, \mathbf{e}_3 = \hat{\mathbf{e}}_z\}$ from the definition (1.5), and then obtain the reciprocal basis from (1.6). Hence verify the vector differential operator (1.43) in cylindrical coordinates.

(Q2) Using (1.44)–(1.46), represent $\nabla\mathbf{v}$ in cylindrical coordinates, where the vector field is written $\mathbf{v} = v_r\hat{\mathbf{e}}_r + v_\theta\hat{\mathbf{e}}_\theta + v_z\hat{\mathbf{e}}_z$ as above. Hence obtain (1.47) and (1.48) by contraction, and then deduce (1.49).

(Q3) If a dyadic field is represented as

$$\mathbf{T} = T_{rr}\hat{\mathbf{e}}_r\hat{\mathbf{e}}_r + T_{r\theta}\hat{\mathbf{e}}_r\hat{\mathbf{e}}_\theta + T_{rz}\hat{\mathbf{e}}_r\hat{\mathbf{e}}_z + T_{\theta r}\hat{\mathbf{e}}_\theta\hat{\mathbf{e}}_r + T_{\theta\theta}\hat{\mathbf{e}}_\theta\hat{\mathbf{e}}_\theta + T_{\theta z}\hat{\mathbf{e}}_\theta\hat{\mathbf{e}}_z + T_{zr}\hat{\mathbf{e}}_z\hat{\mathbf{e}}_r + T_{z\theta}\hat{\mathbf{e}}_z\hat{\mathbf{e}}_\theta + T_{zz}\hat{\mathbf{e}}_z\hat{\mathbf{e}}_z,$$

derive $\nabla\cdot\mathbf{T}$ in cylindrical coordinates. Hence or otherwise, represent $\nabla^2\mathbf{v}$ in cylindrical coordinates.

(Q4) Spherical coordinates (r, θ, ϕ) are defined by the coordinate transformation

$$x = \rho\cos\phi,\ \ y = \rho\sin\phi,\ \ z = r\cos\theta$$

where $\rho = r\sin\theta$, as shown in Fig. 1.1. Define a corresponding orthonormal basis set $\{\hat{\mathbf{e}}_r, \hat{\mathbf{e}}_\theta, \hat{\mathbf{e}}_\phi\}$, and directly obtain $\nabla\cdot\mathbf{v}$ in spherical coordinates. Check your result using (1.42).

(Q5) In any orthogonal curvilinear coordinates (x^1, x^2, x^3) with orthonormal basis set $\{\hat{\mathbf{e}}_i\}$, show that the divergence and curl of a vector field $\mathbf{v}(\mathbf{r}) = v^i \hat{\mathbf{e}}_i$ may be expressed as

$$\nabla \cdot \mathbf{v} = \frac{1}{h_1 h_2 h_3} \left[\frac{\partial}{\partial x^1} \left(h_2 h_3 v^1 \right) + \frac{\partial}{\partial x^2} \left(h_3 h_1 v^2 \right) + \frac{\partial}{\partial x^3} \left(h_1 h_2 v^3 \right) \right],$$

and (in terms of a determinant)

$$\nabla \times \mathbf{v} = \frac{1}{h_1 h_2 h_3} \begin{vmatrix} h_1 \hat{\mathbf{e}}_1 & h_2 \hat{\mathbf{e}}_2 & h_3 \hat{\mathbf{e}}_3 \\ \dfrac{\partial}{\partial x^1} & \dfrac{\partial}{\partial x^2} & \dfrac{\partial}{\partial x^3} \\ h_1 v^1 & h_2 v^2 & h_3 v^3 \end{vmatrix},$$

respectively. Also show that

$$\nabla^2 \phi = \frac{1}{h_1 h_2 h_3} \left[\frac{\partial}{\partial x^1} \left(\frac{h_2 h_3}{h_1} \frac{\partial \phi}{\partial x^1} \right) + \frac{\partial}{\partial x^2} \left(\frac{h_3 h_1}{h_2} \frac{\partial \phi}{\partial x^2} \right) + \frac{\partial}{\partial x^3} \left(\frac{h_1 h_2}{h_3} \frac{\partial \phi}{\partial x^3} \right) \right],$$

for any scalar field $\phi(\mathbf{r})$.
Verify that (1.47)–(1.49) follow from these results.

1.7 Stokes Theorem

Stokes Theorem is one of a family of integral theorems that relate surface to line integrals, and a generalisation of the Green Theorem in the Plane

$$\iint_R \left(\frac{\partial Q}{\partial x} - \frac{\partial P}{\partial y} \right) dxdy = \oint_C (Pdx + Qdy). \tag{1.51}$$

Here $P(x, y)$, $Q(x, y)$ are continuous functions with continuous partial derivatives in the region R of the xy-plane bounded by a simple closed curve C, and the line integral on the right-hand side of (1.51) is taken in the anticlockwise direction (curve C is traversed such that region R is always to the left). This result is given in many elementary calculus textbooks, and follows readily by noting the following:
(i) for a regular region R (where any line parallel to either Cartesian axis cuts C in at most two points),

$$\iint_R \frac{\partial Q}{\partial x} dxdy = \int [Q(x_2, y) - Q(x_1, y)] dy$$

on integrating with respect to x from the left component $x_1 = g_1(y)$ to the right component $x_2 = g_2(y)$ of the boundary C, hence the right-hand side of this equation is equivalent to

$$\oint_C Q(x, y)\, dy;$$

(ii) similarly, by integrating with respect to y,

$$\iint_R \frac{\partial P}{\partial y} dxdy = -\oint_C P(x, y)dx; \text{ and}$$

(iii) the extension to any irregular region (where a line parallel to at least one Cartesian axis cuts the boundary curve C in more than two points), and also to a multiply-connected region, is immediate since such regions can always be sub-divided into several regular sub-regions.

The generalised Stokes Theorem can be expressed in the following way. If $\mathbf{f}(\mathbf{r})$ is a continuously differentiable tensor field over any open surface S with boundary curve C, then

$$\int_S \hat{\mathbf{n}} \times \nabla \mathbf{f}\, dS = \oint_C \hat{\mathbf{T}} \mathbf{f}\, ds, \tag{1.52}$$

where $\hat{\mathbf{n}}$ is the unit normal at any point on the surface S and $\hat{\mathbf{T}}$ is the unit tangent at any point on the boundary C, and the integral with respect to s along C is again directed such that the surface S is always to the left. By writing $d\mathbf{S} = \hat{\mathbf{n}}\, dS$ and $d\mathbf{r} = \hat{\mathbf{T}} ds$, (1.52) may be written more succinctly as

$$\int_S d\mathbf{S} \times \nabla \mathbf{f} = \oint_C d\mathbf{r}\, \mathbf{f}. \tag{1.53}$$

In order to prove the generalised Stokes Theorem, first consider the position vector $\mathbf{r}(u, v)$ of any point on the open surface S, where (u, v) is the pair of curvilinear coordinates in the surface for that point [2]. The surface element there is

$$d\mathbf{S} = \mathbf{r}_u du \times \mathbf{r}_v dv = \mathbf{r}_u \times \mathbf{r}_v\, dudv \equiv \hat{\mathbf{n}}\, dS, \tag{1.54}$$

where $\hat{\mathbf{n}} \equiv J^{-1} \mathbf{r}_u \times \mathbf{r}_v$ is the unit normal at the point and it is convenient in this section to follow Ref. [2] in using subscripts to denote partial differentiation with respect to the subscripted variables. The gradient of any differentiable tensor field $\mathbf{f}(\mathbf{r})$ at the surface is $\nabla \mathbf{f} = \nabla u\, \mathbf{f}_u + \nabla v\, \mathbf{f}_v + \hat{\mathbf{n}}\, \mathbf{f}_n$, where

$$\{\nabla u \equiv J^{-1} \mathbf{r}_v \times \hat{\mathbf{n}},\ \ \nabla v \equiv J^{-1} \hat{\mathbf{n}} \times \mathbf{r}_u,\ \ \hat{\mathbf{n}} \equiv J^{-1} \mathbf{r}_u \times \mathbf{r}_v\}$$

is the reciprocal to the basis set $\{\mathbf{r}_u, \mathbf{r}_v, \hat{\mathbf{n}}\}$—cf. (1.5), (1.6) and (1.41). Thus from the vector identity (1.2) and $\hat{\mathbf{n}} \cdot \mathbf{r}_u = \hat{\mathbf{n}} \cdot \mathbf{r}_v = 0$ we have

$$\hat{\mathbf{n}} \times \nabla \mathbf{f} = J^{-1}[\hat{\mathbf{n}} \times (\mathbf{r}_v \times \hat{\mathbf{n}})\mathbf{f}_u + \hat{\mathbf{n}} \times (\hat{\mathbf{n}} \times \mathbf{r}_u)\mathbf{f}_v] = J^{-1}[(\mathbf{r}_v\mathbf{f})_u - (\mathbf{r}_u\mathbf{f})_v],$$

and since $dS = J du dv$ the result

$$\int_S \hat{\mathbf{n}} \times \nabla \mathbf{f}\, dS = \int_{S'} [(\mathbf{r}_v\mathbf{f})_u - (\mathbf{r}_u\mathbf{f})_v]\, du dv, \tag{1.55}$$

where S' is the region in the uv-plane that contains all the points corresponding to the points on the surface S. Then application of the Green Theorem in the Plane to the integral on the right-hand side of (1.55) yields

$$\int_{S'} [(\mathbf{r}_v\mathbf{f})_u - (\mathbf{r}_u\mathbf{f})_v]\, du dv = \oint_{C'} (\mathbf{r}_u\mathbf{f} du + \mathbf{r}_v\mathbf{f} dv), \tag{1.56}$$

where C' is the boundary of the surface S' in the uv-plane. Since $u(s)$ and $v(s)$ are functions of the arc length s along the boundary C of the original surface S, the right-hand side of (1.56) is

$$\oint_{C'} \left(\mathbf{r}_u\mathbf{f}\frac{du}{ds} + \mathbf{r}_v\mathbf{f}\frac{dv}{ds}\right) ds = \oint_C \hat{\mathbf{T}}\mathbf{f}\, ds$$

where

$$\hat{\mathbf{T}} \equiv \frac{d\mathbf{r}}{ds} = \mathbf{r}_u\frac{du}{ds} + \mathbf{r}_v\frac{dv}{ds}$$

is the unit tangent vector along C, so from (1.55) and (1.56) the proof is complete. □

Stokes Theorem is the special case of (1.53) involving a dot product contraction and a vector field $\mathbf{v}(\mathbf{r})$—viz.

$$\int_S (\nabla \times \mathbf{v}) \cdot d\mathbf{S} = \oint_C \mathbf{v} \cdot d\mathbf{r}. \tag{1.57}$$

This integral theorem was applied in fluid mechanics and electromagnetic theory from the late nineteenth century, with $\mathbf{v}$ interpreted as the velocity field or an electromagnetic field, respectively.

Exercises

(Q1) Use (1.53) to show that

$$\oint_C \mathbf{r} \cdot d\mathbf{r} = 0$$

and

$$\oint_C \mathbf{r} \times d\mathbf{r} = 2 \int_S d\mathbf{S},$$

where $\mathbf{r}$ denotes the position vector of any point P and S is any open surface capping the closed curve C.

(Q2) Verify Stokes Theorem (1.57) for the vector

$$\mathbf{v}(\mathbf{r}) = (2x - y)\hat{\mathbf{i}} - yz^2\hat{\mathbf{j}} - y^2z\hat{\mathbf{k}},$$

where S is the upper half surface of the sphere $x^2 + y^2 + z^2 = 1$ with boundary curve C at $z = 0$ (cf. Fig. 1.2 when $a = 1$).

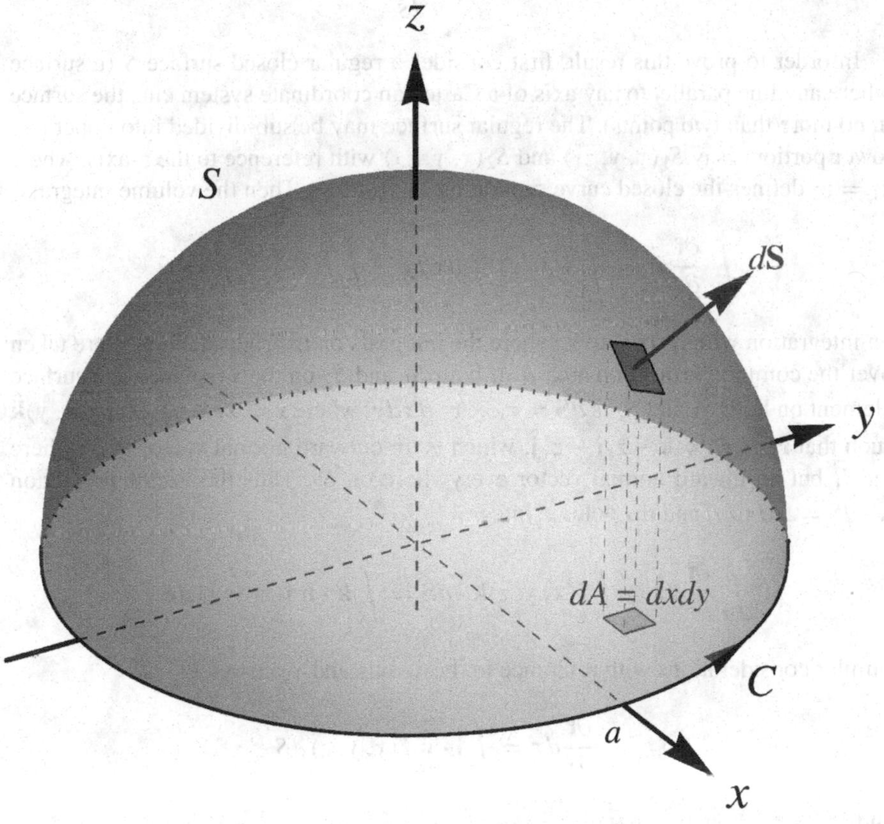

Fig. 1.2 Hemispherical surface of radius a

1.8 Divergence Theorem

The well-known Divergence Theorem is one of a family of integral theorems relating volume to surface integrals, which can be expressed as follows. If $\mathbf{f}(\mathbf{r})$ is a continuously differentiable tensor field over a region V bounded by a closed surface S with a sectionally continuous external (outward) unit normal $\hat{\mathbf{n}}$, then

$$\int_V \nabla\mathbf{f}\, d\tau = \oint_S \hat{\mathbf{n}}\, \mathbf{f}\, dS. \tag{1.58}$$

By again writing $d\mathbf{S} = \hat{\mathbf{n}}\, dS$, (1.58) may be written even more succinctly as

$$\int_V \nabla\mathbf{f}\, d\tau = \oint_S d\mathbf{S}\, \mathbf{f}. \tag{1.59}$$

In order to prove this result, first consider a regular closed surface S (a surface where any line parallel to any axis of a Cartesian coordinate system cuts the surface in no more than two points). The regular surface may be sub-divided into upper and lower portions, say $S_1(x, y, z_1)$ and $S_2(x, y, z_2)$ with reference to the z-axis, where $z_1 = z_2$ defines the closed curve separating S_1 from S_2. Then the volume integral

$$\int_V \frac{\partial\mathbf{f}}{\partial z} d\tau = \int_A \mathbf{f}(x, y, z_1) dxdy - \int_A \mathbf{f}(x, y, z_2) dxdy$$

on integration with respect to z, where the integrals on the right-hand side are taken over the common projection area A of both S_1 and S_2 on the xy-plane. The surface element on both S_1 and S_2 is $d\mathbf{S} = \mathbf{r}_x \times \mathbf{r}_y\, dxdy$, where $\mathbf{r} = x\hat{\mathbf{i}} + y\hat{\mathbf{j}} + z_{1,2}(x, y)\hat{\mathbf{k}}$ such that $\mathbf{r}_x \times \mathbf{r}_y = \hat{\mathbf{k}} - z_x\hat{\mathbf{i}} - z_y\hat{\mathbf{j}}$, which is an outward normal vector everywhere on S_1 but an inward normal vector everywhere on S_2. Thus the vector projection $\hat{\mathbf{k}} \cdot d\mathbf{S} = \pm dxdy$, and the volume integral

$$\int_V \frac{\partial\mathbf{f}}{\partial z} d\tau = \int_S \mathbf{f}(x, y, z)\hat{\mathbf{k}} \cdot d\mathbf{S} = \int_S \hat{\mathbf{k}} \cdot \hat{\mathbf{n}}\, \mathbf{f}(x, y, z)\, dS.$$

Similar considerations with reference to the x-axis and y-axis yield

$$\int_V \frac{\partial\mathbf{f}}{\partial x} d\tau = \int_S \hat{\mathbf{i}} \cdot \hat{\mathbf{n}}\, \mathbf{f}(x, y, z)\, dS$$

and

$$\int_V \frac{\partial\mathbf{f}}{\partial y} d\tau = \int_S \hat{\mathbf{j}} \cdot \hat{\mathbf{n}}\, \mathbf{f}(x, y, z)\, dS,$$

respectively. Combining these last three results appropriately yields (1.59)—i.e.

$$\int_v \nabla \mathbf{f}\, d\tau = \int_V \left(\hat{\mathbf{i}}\frac{\partial \mathbf{f}}{\partial x} + \hat{\mathbf{j}}\frac{\partial \mathbf{f}}{\partial y} + \hat{\mathbf{k}}\frac{\partial \mathbf{f}}{\partial z}\right) d\tau$$

$$= \int_S (\hat{\mathbf{i}}\hat{\mathbf{i}} + \hat{\mathbf{j}}\hat{\mathbf{j}} + \hat{\mathbf{k}}\hat{\mathbf{k}}) \cdot \hat{\mathbf{n}}\, \mathbf{f}\, dS = \int_S \hat{\mathbf{n}}\, \mathbf{f}\, dS,$$

since $\mathsf{I} = \hat{\mathbf{i}}\hat{\mathbf{i}} + \hat{\mathbf{j}}\hat{\mathbf{j}} + \hat{\mathbf{k}}\hat{\mathbf{k}}$ and $\mathsf{I} \cdot \hat{\mathbf{n}} = \hat{\mathbf{n}}$. Any volume V with an irregular surface S or any multiply-connected region can always be sub-divided into volumes with regular surfaces, so extension of the proof to irregular or multiply-connected regions is immediate.

The Divergence Theorem is the special case of (1.59) involving a dot product contraction and a vector field $\mathbf{v}(\mathbf{r})$—viz.

$$\int_V \nabla \cdot \mathbf{v}\, d\tau = \int_S \mathbf{v} \cdot d\mathbf{S}. \tag{1.60}$$

Gauss established many related results. For example, if S is a closed surface and $\mathbf{r}$ denotes the position vector of any point relative to any origin O, then

$$\int_S \frac{d\mathbf{S} \cdot \mathbf{r}}{r^3} = \begin{cases} 0 & \text{if } O \text{ lies outside } S \\ 4\pi & \text{if } O \text{ lies inside } S. \end{cases} \tag{1.61}$$

This corresponds to the Divergence Theorem (1.60) when $\mathbf{v} = \mathbf{r}/r^3$, a vector that may represent the electric field due to a unit charge. Thus when O is outside of V,

$$\int_S \frac{d\mathbf{S} \cdot \mathbf{r}}{r^3} = \int_V \nabla \cdot \frac{\mathbf{r}}{r^3} d\tau = 0,$$

since $\nabla \cdot \mathbf{r}/r^3 = 0$ almost everywhere inside V (i.e. everywhere except at $r = 0$). If O is inside V, then it may be surrounded by a small sphere σ of radius ϵ, such that

$$\int_S \frac{d\mathbf{S} \cdot \mathbf{r}}{r^3} + \int_\sigma \frac{d\mathbf{S} \cdot \mathbf{r}}{r^3} = \int_{V-\sigma} \nabla \cdot \frac{\mathbf{r}}{r^3} d\tau = 0,$$

whence

$$\int_S \frac{d\mathbf{S} \cdot \mathbf{r}}{r^3} = -\int_\sigma \frac{d\mathbf{S} \cdot \mathbf{r}}{r^3}.$$

Now on σ the unit normal pointing outward from $V - \sigma$ is $\hat{\mathbf{n}} = -\mathbf{r}/\epsilon$, hence $d\mathbf{S} \cdot \mathbf{r}/r^3 = -dS\, \mathbf{r} \cdot \mathbf{r}/\epsilon^4 = -dS/\epsilon^2$ and therefore

$$\int_S \frac{d\mathbf{S} \cdot \mathbf{r}}{r^3} = \int_\sigma \frac{dS}{\epsilon^2} = \frac{1}{\epsilon^2}\int_S dS = \frac{1}{\epsilon^2} 4\pi\epsilon^2 = 4\pi.$$

Geometrically, the quantity $d\mathbf{S} \cdot \mathbf{r}/r^3$ is the solid angle subtended at O by the surface element dS, so that (1.61) states that the total solid angle subtended by a closed surface is zero at an exterior point and 4π at an interior point.

The Divergence Theorem is often used to derive field equations in mathematical physics, including the fundamental equations of Fluid Mechanics and MHD introduced in the next chapter.

Exercises

(Q1) If V is a volume bounded by a closed surface S and $\mathbf{a}$ is a constant vector, use (1.59) to show that

$$V\mathbf{a} = \frac{1}{2}\oint_S \hat{\mathbf{n}} \times (\mathbf{a} \times \mathbf{r})dS,$$

where $\hat{\mathbf{n}}$ is the unit normal to S at any point with position vector $\mathbf{r}$ (on the surface S).

(Q2) Verify the Divergence Theorem (1.60) for the following vector fields $\mathbf{F}$ and closed surfaces S:

(a) $\mathbf{F} = x^2\hat{\mathbf{i}} + z\hat{\mathbf{j}} + yz\hat{\mathbf{k}}$, S a unit cube;

(b) $\mathbf{F} = y\hat{\mathbf{i}} + x\hat{\mathbf{j}} + z^2\hat{\mathbf{k}}$, cylinder S bounded by $x^2 + y^2 = a^2$, $z = 0$, $z = h$;

(c) $\mathbf{F} = 2xz\hat{\mathbf{i}} + yz\hat{\mathbf{j}} + z^2\hat{\mathbf{k}}$, hemisphere S where $x^2 + y^2 + z^2 = a^2$ $(z \geq 0)$.

(Q3) When the volume element ΔV with surface ΔS enclosing any point P tends to zero, shows that

$$\frac{\int_{\Delta S} \phi d\mathbf{S}}{\Delta V} \to \nabla\phi,$$

$$\frac{\int_{\Delta S} \mathbf{v} \cdot d\mathbf{S}}{\Delta V} \to \nabla \cdot \mathbf{v}, \text{ and}$$

$$\frac{\int_{\Delta S} d\mathbf{S} \times \mathbf{v}}{\Delta V} \to \nabla \times \mathbf{v}.$$

(Since the ratio on the right-hand side of the second result represents the flux or nett outflow per unit volume if $\mathbf{v}$ is the velocity vector in a fluid, and the limit implies that the volume ΔV shrinks towards a point, the fluid flow in the neighbourhood of a fixed point in space diverges when $\nabla \cdot \mathbf{v} > 0$, and converges when $\nabla \cdot \mathbf{v} < 0$!)

1.9 Green Identities

Application of the Divergence Theorem (1.60) to the vector field $\mathbf{v} = \phi\nabla\psi$, where the scalar fields $\phi(\mathbf{r})$ and $\psi(\mathbf{r})$ respectively have continuous derivatives to first and second order, yields the first Green identity

$$\int_V \nabla\phi \cdot \nabla\psi d\tau + \int_V \phi\nabla^2\psi d\tau = \oint_S \phi\nabla\psi \cdot d\mathbf{S}, \tag{1.62}$$

since $\nabla\cdot(\phi\nabla\psi) = \nabla\phi\cdot\nabla\psi + \phi\nabla^2\psi$. If both ϕ and ψ have continuous derivatives to second order, interchanging ϕ and ψ in this first identity and subtracting yields the second Green identity

$$\int_V (\phi\nabla^2\psi - \psi\nabla^2\phi) d\tau = \int_S (\phi\nabla\psi - \psi\nabla\phi) \cdot d\mathbf{S}. \tag{1.63}$$

Now let $\psi = 1/r$, where r is the distance to any point P from another point O. If O is inside the closed surface S, the singularity in ψ at O may be excluded by surrounding it by a small sphere σ of radius ϵ, so the second identity (1.63) applied to the volume between σ and S gives

$$-\int_V \frac{1}{r}\nabla^2\phi \, d\tau = \oint_\sigma \left(\phi\nabla\frac{1}{r} - \frac{1}{r}\nabla\phi\right) \cdot d\mathbf{S} + \oint_S \left(\phi\nabla\frac{1}{r} - \frac{1}{r}\nabla\phi\right) \cdot d\mathbf{S}. \tag{1.64}$$

The surface integral on σ is

$$\oint_\sigma \left(\frac{\phi}{\epsilon^2} + \frac{1}{\epsilon}\frac{\partial\phi}{\partial r}\right) dS = \oint_\sigma \phi \, d\Omega + \epsilon \oint_\sigma \frac{\partial\phi}{\partial r} d\Omega = 4\pi\bar{\phi} + O(\epsilon), \tag{1.65}$$

where $d\Omega$ and $\bar{\phi}$ denote the solid angle subtended by dS and the value of ϕ at some point of σ, respectively. In the limit $\epsilon \to \infty$, one thus has the third Green identity

$$4\pi\phi(0) = -\int_V \frac{\nabla^2\phi}{r} d\tau + \oint_S \left(\frac{1}{r}\nabla\phi - \phi\nabla\frac{1}{r}\right) \cdot d\mathbf{S}. \tag{1.66}$$

If the point O is outside the closed surface S, this result less the second term on the right-hand side is obtained by setting $\psi = 1/r$ in the second identity (1.63).

The result (1.66) is valid with reference to *any* field point. Thus at any point with position vector $\mathbf{r}_0$,

$$4\pi\phi(\mathbf{r}_0) = -\int_V \frac{\nabla^2\phi}{|\mathbf{r} - \mathbf{r}_0|} d\tau + \oint_S \left(\frac{1}{|\mathbf{r} - \mathbf{r}_0|}\nabla\phi - \phi\nabla\frac{1}{|\mathbf{r} - \mathbf{r}_0|}\right) \cdot d\mathbf{S} \tag{1.67}$$

gives the scalar field $\phi(\mathbf{r}_0)$ in terms of its Laplacian $\nabla^2\phi$ and the values of ϕ and $\nabla\phi$ on the closed surface S. Indeed, since $\mathbf{r}_0$ is any field point, this result may also be used to find derivatives of ϕ. In Sect. 1.11, a formula is derived that defines a potential function $\phi(\mathbf{r}_0)$ in terms of its Laplacian and its values on S, provided a suitable function can be found.

Exercise

(Q1) Consider the Laplacian ∇_0^2 with respect to $\mathbf{r}_0$ operating on (1.67), to show that $\forall \mathbf{r}_0 \in V$ any function $4\pi\rho(\mathbf{r}_0) \equiv \nabla_0^2\phi(\mathbf{r}_0)$ equals the Laplacian of

$$-\int_V \frac{\rho(\mathbf{r})}{|\mathbf{r}-\mathbf{r}_0|}d\tau.$$

Deduce a particular solution of the Poisson equation $\nabla^2\phi = 4\pi\rho$, for the potential ϕ due to a source density ρ.

1.10 Heaviside Step and Dirac Delta Functions

The Heaviside step function can be used to represent instantaneous switching (on or off) in electrical applications—and in an operational calculus later understood to relate to the Laplace transform, Heaviside readily dealt with the step function derivative that is zero everywhere except at a single point. This derivative was already familiar in mechanics, where a finite impulse corresponds to a force acting for an infinitesimally small interval of time (a sudden blow), and is now often called the "unit impulse function" in modern electronics and signal processing. However, this derivative is obviously not a function in the classical sense, and for some time Heaviside's operational approach was not appreciated as a precursor to the modern widespread use of integral transforms in science and engineering. It was not until 1930 that Dirac introduced the notation $\delta(x)$ reflecting the only non-zero value at $z = 0$ that provides the sifting operation below, somewhat reminiscent of the Kronecker delta, and accounted for other of its properties that mathematicians eventually recognised in developing the theory of distributions (or generalised functions). Here we choose to restrict our discussion to a somewhat novel approach to the Heaviside step function $H(x)$ and Dirac delta function $\delta(x)$ together with some relevant properties, in preparing to discuss jump or boundary conditions and also associated "weak solutions" of conservation equations in the theory of shock waves later in this book.

Let us consider a single rectangular pulse of width b and height $1/b$ (and hence of unit area), centred at the value $x = x_0$. When the width of the pulse is reduced by half, the unit area remains if the pulse height is doubled. The delta function $\delta(x - x_0)$ may be regarded as the limit of this process, where the pulse width $b \to 0$ and the pulse height $1/b \to \infty$. It is non-zero only at $x = x_0$, where it is notionally infinite such that the unit area is preserved—i.e. such that

$$\int_{p}^{q} \delta(x - x_0)\, dx = 1 \text{ if } p < x_0 < q,$$

and indeed

$$\int_{x_0-}^{x_0+} \delta(x - x_0)\, dx = 1.$$

This non-zero outcome for an integrand zero almost everywhere (zero everywhere except at $x = x_0$) means that $\delta(x)$ cannot be a classical function. The sifting property

$$\int_{p}^{q} f(x)\delta(x - x_0)\, dx = f(x_0),$$

where $f(x)$ is a function defined on any interval $p < x < q$ and $x_0 \in (p, q)$, is especially useful. Recall for example the Laplace transforms

$$\mathcal{L}\{\delta(t - a)\} = \int_{0}^{\infty} e^{-st}\delta(t - a)\, dt = e^{-as}$$

and

$$\mathcal{L}\{f(t)\delta(t - a)\} = \int_{0}^{\infty} e^{-st} f(t)\delta(t - a)\, dt = f(a)e^{-as},$$

when $a > 0$. As mentioned above, Dirac noted many of the important properties involving $\delta(x)$, but it was another two decades before distribution theory provided a firm mathematical foundation for the delta function.

Although the expression $y = f(x)$ is often regarded as a map of the values of the independent variable x (the domain of f) to the values of the dependent variable y (the range of f), it is possible to characterise a classical function in other ways. A second way is the *inner product (functional)*[6]

$$\langle f, \phi \rangle = \int_{-\infty}^{\infty} f(x)\phi(x)\, dx,$$

regarded as a mapping of $\phi(x)$ into the value $\langle f, \phi \rangle$ by $f(x)$. Consequently, $f(x)$ can be characterised by its moments

$$\int_{-\infty}^{\infty} f(x)\, dx, \quad \int_{-\infty}^{\infty} x f(x)\, dx, \quad \int_{-\infty}^{\infty} x^2 f(x)\, dx, \ldots$$

where $\{\phi(x) = x^n,\ n = 0, 1, 2, \ldots\}$ is a countably infinite set, which is a result that extends to distributions as is well known in statistics (cf. Exercise 5 below). In

[6] A functional is a map between one or more function spaces and the real or complex numbers, typically involving integration. The arguments of functionals are often surrounded by square brackets $[\cdot]$, rather than the parentheses $(\cdot)$ used for those of "ordinary" functions.

another familiar example, if a function $f(x)$ is integrable with bounded total variation on $\Re$ and $\lim_{\varepsilon\to 0}[f(x+\varepsilon)+f(x-\varepsilon)-2f(x)]=0\ \forall x\in\Re$, then $f(x)$ can be characterised by the Fourier transform

$$\left\langle f, e^{ikx}\right\rangle = \int_{-\infty}^{\infty} f(x)\mathrm{e}^{ikx}\,dx$$

where $\{\phi(x)=\mathrm{e}^{ikx}, k\in\Re\}$ is an uncountably infinite set. Laurent Schwartz and others used the inner product in their development of distribution theory,[7] but we prefer to provide a firm foundation for the delta function via the *convolution product*

$$\int_{-\infty}^{\infty} f(x-y)\phi(y)\,dy,$$

which characterises $f(x)$ as a mapping of any function $\phi(x)$ into another function. Nevertheless, if either the inner product or the convolution product is to characterise a distribution unambiguously, the class $\{\phi(x)\}$ must be chosen such that $f_1(x)\neq f_2(x)$ implies there exists $\phi(x)$ such that the inner or convolution product is different.

Let us define the *test function* $\phi(x)$ to be differentiable to any order and to vanish outside some finite interval (the property often referred to as compact support), which may be different for different test functions. For example,

$$\phi(x)=\begin{cases}\exp[1/(x^2-a^2)] & \text{if } |x|<a\\ 0 & \text{if } |x|\geq a\end{cases}$$

belongs to this class. Consequently, if a function $f(x)$ is integrable on $\Re$ and $\lim_{\varepsilon\to 0}[f(x+\varepsilon)+f(x-\varepsilon)-2f(x)]=0\ \forall x\in\Re$, we adopt the convolution product

$$\underline{f(x)}\,\phi \equiv \int_{-\infty}^{\infty} f(x-y)\phi(y)\,dy = \int_{-\infty}^{\infty} f(y)\phi(x-y)\,dy$$

to define the *distribution* $\underline{f(x)}$ *corresponding to the function* $f(x)$ with the following properties:

(1) $\underline{f(x)}$ maps the test function $\phi(x)$ into the function

$$\int_{-\infty}^{\infty} f(x-y)\phi(y)\,dy,$$

such that $f_1(x)=f_2(x)$ almost everywhere $\Longleftrightarrow \underline{f_1(x)}=\underline{f_2(x)}$;

(2) $\underline{f(x)}$ is a linear operator—i.e. $\underline{f_1(x)+f_2(x)}=\underline{f_1(x)}+\underline{f_2(x)}$, $\underline{af_1(x)}=a\,\underline{f_1(x)}$; and

(3) $f(x)=0 \Longleftrightarrow \underline{f(x)}=0$.

[7] The five volume set by Gel'fand and Shilov on *Generalized Functions* (Academic Press, 1964) is often referenced. The book by Vladimirov [10] may also be consulted.

Further, the derivative $\underline{f'(x)}$ of a distribution $\underline{f(x)}$ may be defined by

$$\underline{f'(x)}\,\phi = \underline{f(x)}\,\phi'.$$

If the distribution $\underline{f(x)}$ corresponds to an integrable function $f(x)$ on $\Re$ that satisfies $\lim_{\varepsilon\to 0}[f(x+\varepsilon)+f(x-\varepsilon)-2f(x)] = 0\ \forall x \in \Re$, then

$$\begin{aligned}\underline{f'(x)}\,\phi &= \int_{-\infty}^{\infty} f(x-y)\,\phi'(y)\,dy \\ &= \int_{-\infty}^{\infty} f(y)\,\phi'(x-y)\,dy \\ &= \frac{d}{dx}\int_{-\infty}^{\infty} f(y)\,\phi(x-y)\,dy \\ &= \frac{d}{dx}\,[\underline{f(x)}\,\phi].\end{aligned}$$

Indeed, if the function $f(x)$ is differentiable and its derivative $f'(x)$ also satisfies the condition $\lim_{\varepsilon\to 0}[f'(x+\varepsilon)+f'(x-\varepsilon)-2f'(x)] = 0\ \forall x \in \Re$, then

$$\begin{aligned}\underline{f'(x)}\phi &= \textstyle\int_{-\infty}^{\infty} f(x-y)\,\phi'(y)\,dy \\ &= \textstyle\int_{-\infty}^{\infty} f'(x-y)\,\phi(y)\,dy\end{aligned}$$

—i.e. the distribution $\underline{f'(x)}$ then corresponds to the function $f'(x)$.

However, it is notable that the derivative function $f'(x)$ corresponding to the distribution $\underline{f'(x)}$ as defined above need not exist! This is the case for the Dirac delta function, which is defined as the *distribution* $\underline{\delta(x)} \equiv \underline{H'(x)}$. The Heaviside step function

$$H(x) = \begin{cases} 0 & \text{if } x < 0 \\ \frac{1}{2} & \text{if } x = 0 \\ 1 & \text{if } x > 0 \end{cases} \tag{1.68}$$

is integrable and satisfies $\lim_{\varepsilon\to 0}[H(x+\varepsilon)+H(x-\varepsilon)-2H(x)] = 0$, such that the corresponding distribution $\underline{H(x)}$ is defined by

$$\underline{H(x)}\,\phi = \int_{-\infty}^{\infty} H(x-y)\phi(y)\,dy = \int_{-\infty}^{x} \phi(y)\,dy,$$

and its derivative is

$$\underline{\delta(x)}\,\phi \equiv \underline{H'(x)}\,\phi = \frac{d}{dx}\left[\underline{H(x)}\,\phi\right] = \phi(x).$$

Thus although there is no classical function corresponding to $\underline{\delta(x)}$ everywhere, we may omit the underlining and refer to $\delta(x)$ as the delta function on the understanding that any equation involving $\delta(x)$ refers to distributions.

Differentiation readily extends to the nth order (for any n): thus

$$\underline{f^{(n)}(x)}\,\phi = \underline{f(x)}\,\phi^{(n)},$$

so a distribution is differentiable to the same order as the class of test functions. It similarly follows that

$$\underline{f^{(n)}(x)}\,\phi = \frac{d^n}{dx^n}\left[\underline{f(x)}\,\phi\right],$$

if the distribution $\underline{f(x)}$ corresponds to an integrable function $f(x)$ on $\Re$ satisfying $\lim_{\varepsilon \to 0}[f(x+\varepsilon) + f(x-\varepsilon) - 2f(x)] = 0\ \forall x \in \Re$. Further, if the distribution product $\underline{f(x)g(x)}$ corresponds to an integrable product of two differentiable functions on $\Re$ where this condition is likewise satisfied, then

$$\begin{aligned}\frac{d}{dx}\left[\underline{f(x)g(x)}\,\phi\right] &= \int_{-\infty}^{\infty} f(x-y)g(x-y)\,\phi'(y)\,dy \\ &= \int_{-\infty}^{\infty}\left[f'(x-y)g(x-y) + f(x-y)g'(x-y)\right]\phi(y)\,dy \\ &= \underline{f'(x)g(x)}\,\phi + \underline{f(x)g'(x)}\,\phi.\end{aligned}$$

Moreover, *when* $g(x)$ *is differentiable but* $f(x)$ *is not we may define*

$$\underline{f'(x)g(x)}\,\phi = \frac{d}{dx}\left[\underline{f(x)g(x)}\right]\phi - \underline{f(x)g'(x)}\,\phi$$

provided $f(x)g'(x)$ also satisfies the condition, with the extension under such conditions to

$$\underline{f^{(n)}(x)g(x)}\,\phi = \frac{d}{dx}\left[\underline{f^{(n-1)}(x)g(x)}\right]\phi - \underline{f^{(n-1)}(x)g'(x)}\,\phi$$

for $n = 2, 3, \ldots$. Some consequent properties of the delta function are (cf. also Exercise (Q4)):

(1) $\delta^{(n)}\,\phi = \phi^{(n)}(x)$.
(2) $\delta(x)f(x) = f(0)\delta(x)$.
(3) $\delta'(x)f(x) = f(0)\delta'(x) - f'(0)\delta(x)$.[8]

Finally, if a test function $\phi(x)$ satisfies $\int_{-\infty}^{\infty}\phi(y)dy = 0$ then $\Phi(x) = \int_{-\infty}^{x}\phi(y)dy$ is also a test function, so the indefinite integral

---

[8] In passing, we note that any distribution product $f(x)g(x)$ where the underlining is omitted means $\underline{f(x)g(x)}$, *not* $\underline{f(x)}\ \underline{g(x)}$.

$$\underline{F(x)} = \int^x \underline{f(y)}\,dy$$

of a distribution $\underline{f(x)}$ may be defined by

$$\underline{F(x)}\,\phi = \underline{f(x)}\,\Phi \quad \forall\,\phi \text{ such that } \int_{-\infty}^{\infty} \phi(y)\,dy = 0.$$

Note $\underline{F(x+a)}$ is also an indefinite integral of $\underline{f(x)}$ as $\underline{a}\,\phi = \int_{-\infty}^{\infty} a\phi(y)\,dy = 0$; and $\underline{F'(x)} = \underline{f(x)}$ since $\underline{F'(x)}\,\Phi = \underline{F(x)}\,\Phi' = \underline{F(x)}\,\phi = \underline{f(x)}\,\Phi$. In general, there is no definite integral of a distribution. However, if the distribution $\underline{F(x)}$ corresponds to function $F(x)$, then

$$\int_a^b \underline{f(x)}\,dx = F(b) - F(a)$$

is the definite integral of the distribution $\underline{f(x)}$ over $[a, b]$. Note that $\underline{f(x)}$ need not correspond to a function $f(x)$, but the definite integral is a value (not a distribution). In particular, we readily obtain the following results:

$$\int_a^b \delta(x - x_0)\,dx = H(b - x_0) - H(a - x_0) = \begin{cases} 0 & \text{if } x_0 < a \text{ or } x_0 > b \\ \frac{1}{2} & \text{if } x_0 = a \text{ or } x_0 = b\,, \\ 1 & \text{if } a < x_0 < b \end{cases}$$

since $\underline{H(x - x_0)}$ is the indefinite integral of $\delta(x - x_0)$; and the sifting property

$$\begin{aligned} \int_{-\infty}^{\infty} \delta(x) f(x)\,dx &= \int_{-\infty}^{\infty} f(0)\delta(x)\,dx \\ &= f(0) \int_{-\infty}^{\infty} \delta(x)\,dx = f(0)\,[H(x)]_{-\infty}^{\infty} = f(0), \end{aligned}$$

which is extended in Exercise Q4 below.

Exercises

(Q1) For

$$f(x) = \operatorname{sgn}(x) = \begin{cases} d|x|/dx & \text{if } x \neq 0 \\ 0 & \text{if } x = 0 \end{cases}\text{'}$$

show that $\underline{f(x)} = 2\underline{H(x)} - 1$ and $\underline{f'(x)} = 2\,\delta(x)$.

(Q2) Show that the general solution of the algebraic equation $\underline{x^k y} = \underline{f(x)}$ is

$$y = x^{-k} f(x) + \sum_{j=1}^{k} c_j \delta^{(j-1)}(x), \quad \text{where } \delta^{(0)}(x) = \delta(x).$$

(Q3) Provided $f'(x)$ exists, show that $\delta(x-x_0)f(x) = f(x_0)\delta(x-x_0)$ where $x_0 \in \Re$ is a constant, and deduce the sifting property

$$\int_{-\infty}^{\infty} \delta(x-x_0)f(x)dx = f(x_0).$$

Similarly, show that $\delta'(x-x_0)f(x) = f(x_0)\,\delta'(x-x_0) - f'(x_0)\,\delta(x-x_0)$ if $f''(x)$ exists, and deduce that

$$\int_{-\infty}^{\infty} \delta'(x-x_0)f(x)dx = -f'(x_0).$$

(Q4) Solve the following boundary value problems:

(a)

$$\frac{dy}{dx} = \delta(x-1), \quad y(0) = 0;$$

(b)

$$\frac{d^2y}{dx^2} = \delta(x-1) + \delta'(x-1), \quad y = 0 \text{ and } \frac{dy}{dx} = 0 \text{ when } x = 0;$$

(c)

$$x\frac{d^2(xy)}{dx^2} = \delta(x), \quad y(\pm\infty) = 0.$$

(Q5) Show that

$$\int_{-\infty}^{\infty} e^{ixy}\,dy = 2\pi\delta(x),$$

and hence

$$\int_{-\infty}^{\infty} e^{ixy}y^n\,dy = 2\pi(-i)^n\delta^{(n)}(x).$$

Deduce that any distribution $f(x)$ can be constructed from its moments—i.e.

$$f(x) = \sum_{n=0}^{\infty}(-1)^n f_n\,\delta^{(n)}(x)/n!, \quad \text{where } f_n = \int_{-\infty}^{\infty} x^n f(x)\,dx.$$

1.11 Green Functions

Potential theory is an important branch of mathematics.[9] Solutions of the Laplace equation $\nabla^2\phi = 0$, commonly called harmonic functions, frequently arise in applied mathematics—e.g. gravitational, velocity, electrostatic or magnetohydrostatic potentials discussed in this book. The real and imaginary parts of a complex analytic function also satisfy the Laplace equation. The Poisson equation $\nabla^2\phi = 4\pi\rho$ defines the potential due to sources—e.g. the gravitational potential due to a mass distribution, or the electric or magnetic potential due to charge or current distributions.

Closed solutions of boundary value problems involving the Laplace and Poisson equations, and indeed many other ordinary or partial differential equations, may be written in terms of a Green function. Thus let $G(\mathbf{r}, \mathbf{r}_0)$ be a function such that

$$\nabla^2 G(\mathbf{r}, \mathbf{r}_0) = \delta(\mathbf{r} - \mathbf{r}_0) \tag{1.69}$$

where $\delta(\mathbf{r})$ is a three-dimensional Dirac delta function.[10] Then taking $\psi = G(\mathbf{r}, \mathbf{r}_0)$ in (1.63) yields

$$\phi(\mathbf{r}_0) = \int_S [\phi\nabla G(\mathbf{r}, \mathbf{r}_0) - G(\mathbf{r}, \mathbf{r}_0)\nabla\phi] \cdot d\mathbf{S} + \int_V G(\mathbf{r}, \mathbf{r}_0)\nabla^2\phi d\tau, \tag{1.70}$$

where $\mathbf{r}_0 \in V$. Now if ϕ is an harmonic function in V ($\nabla^2\phi = 0$ in V) and Dirichlet boundary conditions apply on S (ϕ is specified on S), and G is chosen such that

$$G(\mathbf{r}, \mathbf{r}_0) = 0 \text{ for any } \mathbf{r} \in S, \tag{1.71}$$

then (1.70) yields the solution to the Laplace equation $\nabla^2\phi = 0$ in the form

$$\phi(\mathbf{r}_0) = \int_S [\phi\nabla G(\mathbf{r}, \mathbf{r}_0)] \cdot d\mathbf{S}, \quad \mathbf{r}_0 \in V. \tag{1.72}$$

In the case of the Poisson equation $\nabla^2\phi = 4\pi\rho$, the comparable solution is of course

$$\phi(\mathbf{r}_0) = \int_S [\phi\nabla G(\mathbf{r}, \mathbf{r}_0)] \cdot d\mathbf{S} + 4\pi \int_V G(\mathbf{r}, \mathbf{r}_0)\rho(\mathbf{r})d\tau, \quad \mathbf{r}_0 \in V. \tag{1.73}$$

The Green function $G(\mathbf{r}, \mathbf{r}_0)$ specified here, for the Dirichlet problem involving either the Laplace or Poisson equation, is the solution of the boundary value problem for a point source located at the position $\mathbf{r} = \mathbf{r}_0$. The second term on the right-hand side

---

[9] Potential theory was developed by many famous mathematicians—including not only Green and Gauss but also Riemann, Poincaré and Hilbert.

[10] In Euclidean n-space $\mathcal{R}^n$ we have $\delta(\mathbf{x}) = \delta(x_1)\delta(x_2)\cdots\delta(x_n)$, where the distribution product must be suitably interpreted as discussed in Sect. 1.10. Alternatively, we may choose to define the delta function as the measure such that $\int_{\mathcal{R}^n} \phi(\mathbf{x})\delta(d\mathbf{x}) = \phi(\mathbf{0})$ $\forall$ compactly supported continuous function $\phi(\mathbf{x})$, when $\delta(\mathbf{x}) = \delta(x_1)\delta(x_2)\cdots\delta(x_n)$ corresponds to the product measure of the $\{\delta(x_i)\}$.

of (1.73) is the contribution to the scalar field $\phi(\mathbf{r}_0)$ from the total distributed source (mass or charge sources or whatever) represented by $\rho(\mathbf{r})$. In passing, we note that there is no standard notation for the Green function—and that a factor of $\pm 4\pi$ may be introduced on the right-hand side of (1.69), or the respective roles of $\mathbf{r}$ and $\mathbf{r}_0$ may be interchanged.

Since $\nabla^2\phi = \nabla\cdot\nabla\phi$, integration of (1.69) over any volume V containing the field point $\mathbf{r}_0$ yields

$$\int_V \nabla^2 G(\mathbf{r}, \mathbf{r}_0)\, d\tau = \oint_S d\mathbf{S}\cdot\nabla G(\mathbf{r}, \mathbf{r}_0) = 1,$$

from the Divergence Theorem and the integral property of the delta function. Thus integration over a spherical surface of radius r, for convenience centred at $\mathbf{r}_0 = 0$, yields

$$4\pi r^2 \frac{dG}{dr} = 1 \quad \text{and hence} \quad G(\mathbf{r}, 0) = -\frac{1}{4\pi|\,\mathbf{r}|} + \text{ constant},$$

where the constant may be chosen to satisfy the Dirichlet condition (1.71).

The familiar Coulomb potential function $-1/(4\pi|\mathbf{r}-\mathbf{r}_0|)$ due to a unit charge at $\mathbf{r}_0$ (on relocating the origin) satisfies (1.71) when the volume V is infinite (i.e. the surface S is at infinity). The potential difference

$$G(\mathbf{r}, 0) = -\frac{1}{4\pi}\left(\frac{1}{|\mathbf{r}|} - \frac{1}{a}\right) \tag{1.74}$$

between a unit charge at the origin ($\mathbf{r}_0 = 0$) and a finite spherical surface S of radius a (centred on the unit charge) satisfies (1.71) for a finite volume V. In this case, Eq. (1.70) becomes

$$\phi(0) = -\frac{1}{4\pi}\int_S \left[\phi\,\nabla\left(\frac{1}{|\mathbf{r}|} - \frac{1}{a}\right)\right]\cdot d\mathbf{S} - \frac{1}{4\pi}\int_V \left(\frac{1}{|\mathbf{r}|} - \frac{1}{a}\right)\nabla^2\phi\, d\tau$$

or

$$\phi(0) = \frac{1}{4\pi a^2}\int_S \phi\, dS - \frac{1}{4\pi}\int_V \left(\frac{1}{|\mathbf{r}|} - \frac{1}{a}\right)\nabla^2\phi\, d\tau, \tag{1.75}$$

since the unit outward normal $\hat{\mathbf{n}}$ equals $\mathbf{r}/|\mathbf{r}| = \mathbf{r}/a$ on S, and hence

$$\left[\nabla\left(\frac{1}{|\mathbf{r}|}\right)\right]\cdot d\mathbf{S} = -\frac{\mathbf{r}}{|\mathbf{r}|^3}\cdot\frac{\mathbf{r}}{|\mathbf{r}|}dS = -\frac{1}{a^2}dS \quad \text{on S.}$$

The first term on the right-hand side of (1.75) is the average value of ϕ on the spherical surface S, hence the difference between the value of any scalar field ϕ at any point (since the choice of origin is arbitrary) and its average over surrounding points is determined by its Laplacian $\nabla^2\phi$. In particular, if ϕ is an harmonic function ($\nabla^2\phi = 0$), its value at a point is the average of its values over a sphere centred at that point. Thus for example, an harmonic temperature field corresponds to zero heat flux, since the temperature at any point is the average temperature of its neighbourhood.

The volume V may be multiply-connected, with one or more internal surfaces. For example, the Green function to determine the electrostatic field in the presence of a conducting spherical surface at $r = a$ (with the second surface at infinity) is

$$G(\mathbf{r}, \mathbf{r}_0) = \begin{cases} \dfrac{-1}{4\pi|\mathbf{r}-\mathbf{r}_0|} + \dfrac{a}{4\pi|\mathbf{r}_0||\mathbf{r}-\mathbf{r}_0'|} & \text{if } |\mathbf{r}| > a \\ \dfrac{-1}{4\pi|\mathbf{r}-\mathbf{r}_0'|} + \dfrac{|\mathbf{r}_0|}{4\pi a|\mathbf{r}-\mathbf{r}_0|} & \text{if } |\mathbf{r}| < a \end{cases} \tag{1.76}$$

corresponding to two "image" source points (electrical charges of opposite sign) situated at $|\mathbf{r}_0| > a$ and $|\mathbf{r}_0'| < a$ respectively, such that (1.71) is not only satisfied at infinity but also on the spherical surface $r = a$.

The reciprocity relation

$$G(\mathbf{r}_0, \mathbf{r}) = G(\mathbf{r}, \mathbf{r}_0), \tag{1.77}$$

evident in all of the above real Green functions, reflects the symmetry (in $\mathbf{r}$ and $\mathbf{r}_0$) of the delta function $\delta(\mathbf{r} - \mathbf{r}_0)$ in the definitive equation (1.69). Physically, the response at $\mathbf{r}_0$ due to a point source at $\mathbf{r}$ is the same as the response at $\mathbf{r}$ due to a point source at $\mathbf{r}_0$.

As previously mentioned, Green functions may also be used for boundary value problems involving ordinary differential equations or other partial differential equations. Moreover, Green functions may be found when the derivative of the function rather than the function is specified on the boundary (Neumann conditions), or for mixed homogeneous boundary conditions. Thus the solution of a more general boundary value problem involving the non-homogeneous equation $L\phi = f$, where L is a known differential operator and f is a known input function, may be found by defining a Green function such that

$$LG(\mathbf{r}, \mathbf{r}_0) = \delta(\mathbf{r} - \mathbf{r}_0), \quad \mathbf{r}_0 \in V \tag{1.78}$$

where $\delta(\mathbf{r} - \mathbf{r}_0)$ is again a three-dimensional delta function and demanding $G(\mathbf{r}, \mathbf{r}_0)$ satisfy the same boundary conditions as ϕ. Then subtracting (1.78) multiplied by ϕ from the equation $L\phi = f$ multiplied by G, and integrating over volume V, yields ($\forall \mathbf{r}_0 \in V$)

$$\begin{aligned} \int_V [\phi(\mathbf{r})LG(\mathbf{r}, \mathbf{r}_0) - G(\mathbf{r}, \mathbf{r}_0)L\phi(\mathbf{r})]d\tau &= \int_V [\phi(\mathbf{r})\delta(\mathbf{r} - \mathbf{r}_0) - G(\mathbf{r}, \mathbf{r}_0)f(\mathbf{r})]d\tau \\ &= \phi(\mathbf{r}_0) - \int_V G(\mathbf{r}, \mathbf{r}_0)f(\mathbf{r})d\tau. \end{aligned} \tag{1.79}$$

The left-hand side may simplify as in the discussion above; or may even be zero under appropriate boundary conditions such that

$$\phi(\mathbf{r}_0) = \int_V G(\mathbf{r}, \mathbf{r}_0) f(\mathbf{r}) d\tau, \tag{1.80}$$

when the operator L is said to be self-adjoint. This result has a form similar to the sifting property

$$f(\mathbf{r}_0) = \int_V \delta(\mathbf{r} - \mathbf{r}_0) f(\mathbf{r}) d\tau,$$

where $\mathbf{r}_0$ is any point in V. However, in contrast to the delta function, the Green function is typically differentiable and hence continuous in the classical sense, so it "spreads out" the effect of the input function $f(\mathbf{r})$ in closed integral solutions such as (1.73) and (1.80) above. It is also notable that

$$L\phi(\mathbf{r}_0) = \int_V LG(\mathbf{r}, \mathbf{r}_0) f(\mathbf{r}) d\tau = \int_V \delta(\mathbf{r} - \mathbf{r}_0) f(\mathbf{r}) d\tau = f(\mathbf{r}_0),$$

if the operator L (interpreted as an operator with reference to $\mathbf{r}_0$) is formally applied to (1.80).

Green functions, and indeed desired closed solutions analogous to (1.73) and (1.80), are often conveniently obtained using integral transform (Laplace, Fourier, Mellin, etc.) techniques, where the kernel function and hence the choice of integral transform depends upon the particular differential operator L and the domain of integration (as is typical of the Green function). Note also that any problem where an homogeneous equation is subject to non-homogeneous boundary conditions may be re-expressed as $L\phi = f$ subject to homogeneous conditions, or an adjoint boundary value problem may be defined to deal with inhomogeneity.

Exercises

(Q1) Assuming (1.76) and that positions $\mathbf{r}_0$ and $\mathbf{r}'_0$ of the "image" point sources satisfy

$$\frac{|\mathbf{r} - \mathbf{r}'_0|}{|\mathbf{r} - \mathbf{r}_0|} = \frac{a}{|\mathbf{r}_0|},$$

derive the vector form of the Poisson integral formulae for the potential:

$$\phi(\mathbf{r}_0) = \begin{cases} \dfrac{|\mathbf{r}_0|^2 - a^2}{4\pi a} \displaystyle\int_{|\mathbf{r}|=a} \frac{\phi(\mathbf{r})}{|\mathbf{r} - \mathbf{r}_0|^3} dS & \text{if } |\mathbf{r}_0| > a \\ \dfrac{a^2 - |\mathbf{r}_0|^2}{4\pi a} \displaystyle\int_{|\mathbf{r}|=a} \frac{\phi(\mathbf{r})}{|\mathbf{r} - \mathbf{r}'_0|^3} dS & \text{if } |\mathbf{r}_0| < a \end{cases}.$$

(Q2) A Green function $G(\mathbf{r}, t; \mathbf{r}_0, t_0)$ satisfies

$$\nabla^2 G - \frac{1}{c^2}\frac{\partial^2 G}{\partial t^2} = \delta(\mathbf{r} - \mathbf{r}_0)\delta(t - t_0), \quad \forall\, \mathbf{r} \text{ and } \forall\, t.$$

Noting that G is consequently a function of $\mathbf{r} - \mathbf{r}_0$ and $t - t_0$, such that $\mathbf{r}_0$ and t_0 can be set zero for convenience, use the Fourier transform

$$\hat{G}(\mathbf{k}, \omega) = \int G(\mathbf{r}, t)e^{-i(\mathbf{k}\cdot\mathbf{r}-\omega t)}d\mathbf{r}dt$$

to obtain on inversion

$$G(\mathbf{r}, t) = \frac{1}{(2\pi)^4}\int \frac{c^2}{\omega^2 - k^2c^2}e^{i(\mathbf{k}\cdot\mathbf{r}-\omega t)}d\mathbf{k}d\omega.$$

Given that after contour integration in the complex ω-plane the essential part of the Green function is

$$G(\mathbf{r}, t) = \begin{cases} 0 & \text{if } t < t_0 \\ -\frac{c}{4\pi|\mathbf{r}|}\delta(|\mathbf{r}| - ct) & \text{if } t > t_0 \end{cases},$$

obtain the "retarded potential" solution

$$\phi(\mathbf{r}_0, t_0) = -\frac{c}{4\pi}\int \frac{f(\mathbf{r}, t_0 - |\mathbf{r} - \mathbf{r}_0|/c)}{|\mathbf{r} - \mathbf{r}_0|}d\mathbf{r}$$

to the forced wave equation

$$\nabla^2\phi - \frac{1}{c^2}\frac{\partial^2\phi}{\partial t^2} = f(\mathbf{r}, t).$$

Bibliography

1. G.B. Arfken, H.J. Weber, F.E. Harris, *Mathematical Methods for Physicists*, 7th edn. (Academic Press, Waltham, 2012). (Popular textbook on essential mathematical methods for graduate students and beginning researchers)
2. L. Brand, *Vector and Tensor Analysis* (Wiley, New York, 1947). (Recommended extensive and thorough presentation, which inter alia covers the key topics of dyadics and general differential geometry as discussed in this chapter—available online)
3. R. Courant, D. Hilbert, *Methods of Mathematical Physics* (Interscience Publishers, New York, 1953, 1962). (Widely referenced survey of mathematical methods in two volumes, developed by Courant and colleagues in New York from the German original of 1924 derived from Hilbert's lectures—first volume available online)
4. L.C. Evans, *Partial Differential Equations* (American Mathematical Society, Providence, 1998). (Graduate textbook, surveying both classical and weak solutions via Sobolev spaces)

5. L.P. Lebedev, M.J. Cloud, *Tensor Analysis* (World Scientific, River Edge, 2003). (Another presentation that includes dyadics and general differential geometry)
6. M.J. Lighthill, *Introduction to Fourier Analysis and Generalised Functions* (Cambridge University Press, Cambridge, 1959). (Another approach to distribution theory, in this case via limits of sequences of functions)
7. J. Mathews, R.L. Walker, *Mathematical Methods of Physics* (Addison-Wesley, Reading, 1971). (Physical intuition encourages students to further develop their mathematical ability—available online)
8. P.M. Morse, H. Feshbach, *Methods of Theoretical Physics, Part I* (McGraw-Hill, New York, 1953). (Another useful source on mathematical methods)
9. J. Rauch, *Partial Differential Equations* (Springer, New York, 2001). (Graduate textbook, with notable reference to the theory of distributions)
10. V.S. Vladimirov, *Methods of the Theory of Generalized Functions* (CRC Press, Boca Raton, 2002). (On the Sobolev-Schwarz concept of distributions with extensions to Fourier, Laplace, Mellin, Hilbert, Cauchy-Bochner and Poisson integral transforms and operational calculus)
11. C.E. Weatherburn, *An Introduction to Riemannian Geometry and the Tensor Calculus* (Cambridge University Press, Cambridge, 2008). (Emphasises tensor representations and coordinate transformations)
12. E.B. Wilson, *Vector Analysis* (Dover, New York, 1960). (Based on the lectures of J. Willard Gibbs and first published in 1901, the seven editions of this classic book helped standardise modern vector notation—available online)

Chapter 2
Fundamental Equations

Conservation equations are the foundation for Fluid Mechanics and MHD, but others are needed to close the mathematical models. Although fluid pressure was at first assumed to be isotropic, when viscous stress was considered early in the eighteenth century it was evident that the assumption of incompressibility or the inclusion of a simple equation of state was no longer sufficient. The classical macroscopic equations (for mass, momentum and energy) follow from underlying microscopic theory, which also provides the relevant pressure tensor to incorporate viscosity. Except near magnetic null points, the pressure tensor for a plasma in a magnetic field is found to differ significantly from the classical shear viscosity form for a neutral fluid. There is also a brief introduction to the additional equation of magnetic induction required in MHD. The bibliography includes some references that provide further background to our presentation in this chapter, a worthy source on thermodynamics, and two books by Lamb and Prager particularly recommended for supplementary reading (more books on Fluid Mechanics are listed for Chap. 3).

2.1 Conservation Equations

The conservation of physical quantities such as mass, momentum and energy is an essential feature of any fluid mechanics or MHD model. A conservation equation is a differential equation of the form

$$\frac{\partial \sigma}{\partial t} + \nabla \cdot \Gamma = s, \tag{2.1}$$

where σ, Γ and s are fields of appropriate tensor rank.

To appreciate the important notion of "conservation" introduced here, consider the integral of (2.1) over any fixed volume V in space bounded by a closed surface S, and use a dot contraction of (1.59) to get

R.J. Hosking and R.L. Dewar, *Fundamental Fluid Mechanics and Magnetohydrodynamics*, DOI 10.1007/978-981-287-600-3_2

$$\frac{d}{dt}\int_V \sigma\, d\tau = -\int_S d\mathbf{S}\cdot\Gamma + \int_V s\, d\tau. \tag{2.2}$$

Equation (2.2) states that the quantity $\int_V \sigma\, d\tau$ would be conserved, if it were not for the flux $\int_S d\mathbf{S}\cdot\Gamma$ of the quantity Γ across the boundary S and the contribution from any term s (a source if $s > 0$, or a sink if $s < 0$). Thus if some closed surface S exists such that Γ vanishes on S, and there is no source or sink term in V, then $\int_V \sigma\, d\tau$ is *conserved* in the corresponding volume V and σ may be called the *density* of this conserved quantity. In passing, we note that the fixed volume in space is often called a "control volume" in the engineering literature.

We first consider the identification of σ with the field $\rho(\mathbf{r}, t)$ denoting the mass density of a fluid, with the mass flux vector $\rho\mathbf{v}$ where $\mathbf{v}(\mathbf{r}, t)$ is the fluid velocity, to produce the mass conservation equation

$$\frac{\partial\rho}{\partial t} + \nabla\cdot(\rho\mathbf{v}) = 0 \tag{2.3}$$

if there is no source or sink ($s = 0$).[1] This scalar conservation equation is often called the continuity equation, due to an alternative derivation where the evolution of any finite volume of the fluid is considered (cf. Sect. 2.5). Equations of energy conservation constitute another important class, where σ is seen to be identified with some scalar field later.

A vector conservation equation has the form (2.1) with σ and s vectors and Γ a dyadic. Of specific interest is the momentum conservation equation

$$\frac{\partial(\rho\mathbf{v})}{\partial t} + \nabla\cdot\mathbf{T} = \mathbf{F}, \tag{2.4}$$

where the dyadic $\mathbf{T}$ defines the total momentum flux, with both advective and pressure components as discussed in the following section, and $\mathbf{F}$ denotes the external force density.[2] The corresponding form of (2.2) in this case is of course

$$\frac{d}{dt}\int_V \rho\mathbf{v} d\tau = -\int_S d\mathbf{S}\cdot\mathbf{T} + \int_V \mathbf{F} d\tau. \tag{2.5}$$

The volume integral on the left-hand side of (2.5) is the total momentum of the fluid in V, so the mass flux vector $\rho\mathbf{v}$ is alternatively called the fluid momentum density. The surface integral on the right-hand side represents the internal surface forces, and the volume integral represents the total external force acting on the fluid in the volume V. Note that the corresponding vector quantity $-\hat{\mathbf{n}}\cdot\mathbf{T}$ in the surface integral,

---

[1] This outcome may be recognised by considering the motion of an infinitesimal cylinder of fluid, slanted in the direction of $\mathbf{v}$, crossing the surface S. The cylindrical volume $\Delta\tau = dS\,\hat{\mathbf{n}}\cdot\mathbf{v}\,\Delta t$ carries the mass $\Delta m = \rho\Delta\tau = \rho\mathbf{v}\cdot d\mathbf{S}\Delta t$, with (2.2) obtained from the Divergence Theorem (1.60) in this case.

[2] In solid mechanics the dyadic $\mathbf{T}$ in the equation for the displacement field corresponding to (2.5) is usually called the stress tensor—but in Fluid Mechanics and MHD we prefer to emphasise the pressure tensor $\mathbf{p}$, which is the pressure component of $\mathbf{T}$.

where the unit normal $\hat{\mathbf{n}}$ such that $d\mathbf{S} = \hat{\mathbf{n}} dS$ points outward from the volume V, usually has both tangential and normal components at any point on the surface S.

The fluid pressure tensor in the total momentum flux dyadic T is first discussed in the next section. Some basic concepts (gravity, the two classical alternative viewpoints, and material rates of change) are then considered in Sects. 2.3–2.5, before we proceed to derive the macroscopic equations of Fluid Mechanics and MHD in Sects. 2.6 and 2.7. We explore extended mathematical models in Sect. 2.8, and distinguish the pressure tensor for magnetised plasma from the fluid pressure tensor in Sects. 2.9 and 2.10. There is then a brief discussion on the heat flux in the starred Sect. 2.11, before the final Sect. 2.12 on the additional equation of magnetic induction introduced in MHD.

2.2 Fluid Pressure Tensor

Fluid is often transported across the boundary S of a fixed volume V in space (i.e. $\mathbf{n} \cdot \mathbf{v}$ is not zero), when one part of the surface term $-\hat{\mathbf{n}} \cdot \mathsf{T}$ above represents advection of momentum across the boundary. (We prefer to use the term "advection" rather than "convection", which more strictly applies to transport due to buoyancy.) The microscopic fluid constituents (atoms or molecules) carry momentum, as they random walk back and forth between collisions with each other. As discussed later in this chapter, the microscopic particle motion is often described by kinetic theory, and the conservation equations of fluid mechanics or MHD may be derived in detail from a resultant mathematically exact basic equation of change. However, let us now first identify the advective and pressure components of the tensor T for a typical fluid in an heuristic fashion.

If the atoms or molecules of a fluid with velocities in a narrow range are grouped to constitute a beam with density ρ_i and average velocity $\mathbf{u}_i$, then summing over many such beams yields the aggregate quantities

$$\rho = \sum_i \rho_i \text{ (total density)} \tag{2.6}$$

$$\rho\mathbf{v} = \sum_i \rho_i \mathbf{u}_i \text{ (total mass flux)} \tag{2.7}$$

$$\mathsf{T} = \sum_i \rho_i \mathbf{u}_i \mathbf{u}_i \text{ (total momentum flux).} \tag{2.8}$$

Note that the velocity $\mathbf{v}$ is actually the mass-weighted mean of the various average velocities of the constituents. It is often convenient to think of a composite fluid element moving with this velocity $\mathbf{v}$ and the related average *peculiar* velocity of the ith beam

$$\mathbf{w}_i \equiv \mathbf{u}_i - \mathbf{v}, \tag{2.9}$$

which immediately splits out the advective and pressure components. Thus substituting (2.9) in (2.8) we have

$$\mathsf{T} = \underset{\substack{\text{advective}\\\text{component}\\\downarrow}}{\rho\mathbf{v}\mathbf{v}} + \underset{\substack{\text{pressure}\\\text{component}\\\downarrow}}{\mathsf{p}} \tag{2.10}$$

where

$$\mathsf{p} \equiv \sum_i \rho_i \mathbf{w}_i \mathbf{w}_i, \tag{2.11}$$

since from (2.6) and (2.7)

$$\sum_i \rho_i \mathbf{w}_i = 0. \tag{2.12}$$

The total momentum conservation equation (2.5) may therefore be rewritten as

$$\frac{\partial(\rho\mathbf{v})}{\partial t} + \nabla\cdot(\rho\mathbf{v}\mathbf{v} + \mathsf{p}) = \mathbf{F}, \tag{2.13}$$

where the term $\nabla\cdot\mathsf{p}$ represents the *surface force* density within the fluid. As discussed below, the diagonal components of the pressure tensor combine to produce a normal total *hydrostatic pressure*, and tangential viscous forces arise from the off-diagonal components in its representation—cf. the integrand of the surface integral in the corresponding integral form (2.5).

Since the *pressure tensor* p is symmetric, there is an orthonormal set $\{\hat{\mathbf{e}}_1, \hat{\mathbf{e}}_2, \hat{\mathbf{e}}_3\}$ of principal axes such that p is diagonal—i.e.

$$\mathsf{p} = \lambda_1\hat{\mathbf{e}}_1\hat{\mathbf{e}}_1 + \lambda_2\hat{\mathbf{e}}_2\hat{\mathbf{e}}_2 + \lambda_3\hat{\mathbf{e}}_3\hat{\mathbf{e}}_3,$$

where λ_i are the eigenvalues of p. Now let us suppose that $\lambda_{\text{mfp}} \ll L$, where λ_{mfp} is the mean free path length between the microscopic particle collisions and L is the characteristic length scale for gradients in the macroscopic fields, such as the density ρ and the fluid velocity $\mathbf{v}$ say. Thus for a typical fluid, to the zeroth order in λ_{mfp}/L the $\mathbf{w}_i$ distribution is isotropic—i.e. there is no preferred direction, which implies that *any* orthonormal set $\{\mathbf{e}_i\}$ defines principal axes such that $\lambda_1 = \lambda_2 = \lambda_3$ ($= p$ say). Let us therefore anticipate a pressure tensor of form $\mathsf{p} = p\mathsf{I} + \mathsf{t}$, in which the second term t of order λ_{mfp}/L is linear in the velocity gradient $\nabla\mathbf{v}$ ($\sim v/L$ by dimensional analysis) that defines the rate of deformation, because $d\mathbf{v} = d\mathbf{r}\cdot\nabla\mathbf{v}$ is the relative velocity of neighbouring fluid elements distance $d\mathbf{r}$ apart.

Now the only two symmetric dyadics that can be formed from $\nabla\mathbf{v}$ are $\nabla\cdot\mathbf{v}\,\mathsf{I}$ and $\nabla\mathbf{v} + (\nabla\mathbf{v})^T$, so it follows that t must be a linear combination of them. In solid mechanics, the two "constants" (scalar fields) in an analogous linear combination are the Lamé parameters, although two other related coefficients (viz. Young's modulus and Poisson's ratio) are often preferred. We may write the corresponding relation for fluids as

$$\mathbf{t} = -2\mu\mathbf{s} - \mu_v \nabla\cdot\mathbf{v}\,\mathbf{I} \tag{2.14}$$

on defining the *rate of deformation tensor*

$$\mathbf{s} \equiv \{\nabla\mathbf{v}\} = \frac{1}{2}[\nabla\mathbf{v} + (\nabla\mathbf{v})^T] - \frac{1}{3}\nabla\cdot\mathbf{v}\,\mathbf{I}, \tag{2.15}$$

where μ and μ_v denote the *shear viscosity coefficient* and the *volume viscosity coefficient* (or "coefficient of expansive friction"), respectively.[3] Thus the fluid pressure tensor is

$$\mathbf{p} = p\,\mathbf{I} - \mu\left[\nabla\mathbf{v} + (\nabla\mathbf{v})^T - \frac{2}{3}\nabla\cdot\mathbf{v}\,\mathbf{I}\right] - \mu_v\nabla\cdot\mathbf{v}\,\mathbf{I}, \tag{2.16}$$

where the combination $p - \mu_v\nabla\cdot\mathbf{v}$ is the total hydrostatic pressure.

The traceless nature of the rate of deformation tensor $\mathbf{s}$ ensures it makes no contribution to the pressure tensor if the fluid dilates (expands or contracts) isotropically, and the associated shear viscosity coefficient μ is often small enough for that viscosity contribution to be negligible except in regions of strong velocity shear (boundary layers—cf. Sect. 3.8). Although the volume viscosity coefficient μ_v in a fluid other than a monatomic gas is non-zero, many authors do not explicitly include the volume viscosity term $\mu_v\nabla\cdot\mathbf{v}\,\mathbf{I}$ that corresponds to dilatation (cf. Sect. 2.5), as we now do too—i.e. henceforth, we also adopt the Newtonian fluid relation $\mathbf{t} = -2\mu\mathbf{s}$, where the volume viscosity may be regarded as implicit in the total hydrostatic pressure field p. Finally, we note that any variation in the coefficient μ (and μ_v if the volume viscosity is retained) is usually assumed to be negligible over the characteristic timescale for the flow. There are fluids where this is not an appropriate approximation, such as a good house paint!

Exercise

Determine the shear and volume viscosity contributions to the pressure tensor if:

(a) the fluid velocity $\mathbf{v} = ky\hat{\mathbf{i}}$, where y is a Cartesian coordinate in the direction perpendicular to the basis vector $\hat{\mathbf{i}}$, and k is a constant; and
(b) the fluid velocity $\mathbf{v} = k\mathbf{r}/3$, where $\mathbf{r}$ is the position vector and k is a constant.

---

[3]The operator { } introduced here and often invoked later is defined by

$$\{\mathbf{F}\} \equiv \frac{1}{2}(\mathbf{F} + \mathbf{F}^T) - \frac{1}{3}\mathbf{F} : \mathbf{I}\,\mathbf{I}$$

for any dyadic $\mathbf{F}$, where $\mathbf{F}^T$ denotes its transpose. Since Tr $\mathbf{F} \equiv \mathbf{F} : \mathbf{I}$ and Tr $\mathbf{I} = 3$, the resulting symmetric dyadic is also traceless.

2.3 Gravity

A gravitational field can exert a force on a fluid. An external force $(\Delta m)\mathbf{g}$ acting on each fluid element of mass $\Delta m = \rho\Delta\tau$ corresponds to the force density $\mathbf{F} = \rho\mathbf{g}$, when the momentum conservation equation (2.13) becomes

$$\frac{\partial(\rho\mathbf{v})}{\partial t} + \nabla\cdot(\rho\mathbf{v}\mathbf{v} + \mathsf{p}) = \rho\mathbf{g}. \tag{2.17}$$

In a self-gravitating system, such as an interstellar gas cloud, the force density $\rho\mathbf{g}$ may be written as $-\nabla\cdot\mathsf{G}$, where

$$\mathsf{G} = \frac{\mathbf{g}\mathbf{g}}{4\pi G} - \frac{|\mathbf{g}|^2}{8\pi G}\mathsf{I} \tag{2.18}$$

denotes the gravitational field stress dyadic and G is the gravitational constant. Thus we have $\nabla\times\mathbf{g} = 0$ since $\mathbf{g}$ is a conservative field, and Newton's law of gravitation in differential form is $\nabla\cdot\mathbf{g} = -4\pi G\rho$, so that

$$\begin{aligned}\nabla\cdot\mathsf{G} &= \frac{(\nabla\cdot\mathbf{g})\mathbf{g}}{4\pi G} + \frac{\mathbf{g}\cdot\nabla\,\mathbf{g}}{4\pi G} - \frac{(\nabla\mathbf{g})\cdot\mathbf{g}}{4\pi G}\\ &= -\rho\mathbf{g} - \frac{\mathbf{g}\times(\nabla\times\mathbf{g})}{4\pi G} = -\rho\mathbf{g}.\end{aligned}$$

Consequently, the momentum equation can be expressed in conservation form

$$\frac{\partial}{\partial t}(\rho\mathbf{v}) + \nabla\cdot(\rho\mathbf{v}\mathbf{v} + \mathsf{p} + \mathsf{G}) = 0. \tag{2.19}$$

Note that this is a conservation equation for the *total* system, consisting of both the fluid subsystem and the gravitational subsystem, which just happens to have a vanishing momentum density! The previous form (2.17), where the interaction with the gravitational subsystem is represented as an external force density, describes the fluid subsystem alone.

Exercise

(Archimedes' Principle.) Show that when a body is partly or completely immersed in a static incompressible fluid (a "liquid"), the resultant force due to the fluid ("upthrust") is equal and opposite to the weight of fluid displaced. Assume constant gravity, and ignore the weight of the air above the fluid.

2.4 Eulerian and Lagrangian Descriptions

The density ρ in the frequently occurring combination $\partial_t(\rho\mathbf{F}) + \nabla\cdot(\rho\mathbf{v}\mathbf{F})$, where ∂_t denotes $\partial/\partial t$ and $\mathbf{F}$ is an arbitrary tensor field, can be commuted outside the derivatives upon invoking the continuity equation (2.3). Thus although the momentum equation (2.17) is in conservation form, it is often replaced by the equation of motion

$$\rho\left(\frac{\partial}{\partial t} + \mathbf{v}\cdot\nabla\right)\mathbf{v} + \nabla\cdot\mathbf{p} = \rho\mathbf{g}, \tag{2.20}$$

since from (2.3)

$$\frac{\partial}{\partial t}(\rho\mathbf{v}) + \nabla\cdot(\rho\mathbf{v}\mathbf{v}) = \rho\left(\frac{\partial}{\partial t} + \mathbf{v}\cdot\nabla\right)\mathbf{v} \tag{2.21}$$

for the velocity field $\mathbf{v}(\mathbf{r}, t)$. The differential operator $(\partial_t + \mathbf{v}\cdot\nabla)$ is called the *material derivative* (or convective derivative, although strictly this term should be "advective derivative"), or referred to as "differentiation following the motion". It is often denoted by D/Dt, but the notation

$$\frac{d}{dt} \equiv \frac{\partial}{\partial t} + \mathbf{v}\cdot\nabla \tag{2.22}$$

is adopted in this book. The various terminology and notation originates from two distinct views of fluid behaviour, as described below and illustrated in the case of two-dimensional steady flow in Fig. 2.1.

In the *Eulerian* view, an observer at any fixed point in space $\mathbf{r}$ watches the fluid flow by, so that $\mathbf{r}$ is an independent coordinate and the partial time derivative

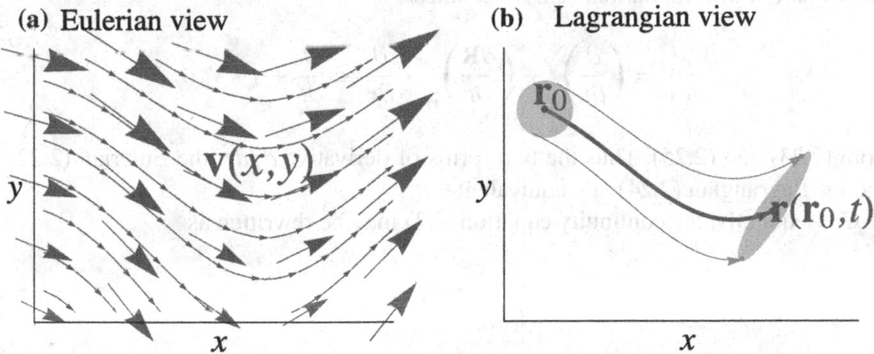

Fig. 2.1 **a** Eulerian picture of a fluid flow—a velocity field $\mathbf{v}$ indicated by the larger arrows. In this simple example $\mathbf{v}$ is tangential to streamlines, the continuous curves with small arrows. **b** The more dynamical Lagrangian picture is based on the map taking an ensemble of fluid elements from their positions $\mathbf{r}_0$ at the initial time, $t = 0$, along their trajectories to their positions $\mathbf{r}$ at time t

$$\frac{\partial}{\partial t} \equiv \left(\frac{\partial}{\partial t}\right)_{\mathbf{r}} \tag{2.23}$$

is used, where the subscript $\mathbf{r}$ means that $\mathbf{r}$ is held fixed. On the other hand, in the *Lagrangian* view the observer moves with a given fluid element and watches changes "as seen by the fluid". Thus $\mathbf{r}$ is no longer an independent variable, since it changes with time as the fluid element moves [7]. The Lagrangian time derivative is defined by

$$\frac{d}{dt} = \left(\frac{\partial}{\partial t}\right)_{\mathbf{r}_0}, \tag{2.24}$$

where $\mathbf{r}_0$ is the initial position of the fluid element at some fixed initial time, say t_0.

However, suppose $\mathbf{r}_0$ lies in a volume of fluid occupying the region V_0 at time t_0, and advected to the region V at time t. Consider the trajectory of the fluid element as it moves from $\mathbf{r}_0$ at time t_0 to another position $\mathbf{r} = \mathbf{R}(t; \mathbf{r}_0, t_0)$ at time t. The trajectory function $\mathbf{R}$ is defined by

$$\dot{\mathbf{R}} \equiv \left(\frac{\partial \mathbf{R}}{\partial t}\right)_{\mathbf{r}_0} = \mathbf{v}(\mathbf{R}, t) \tag{2.25}$$

subject to the initial condition

$$\mathbf{R}(t_0; \mathbf{r}_0, t_0) = \mathbf{r}_0. \tag{2.26}$$

Provided the velocity field is smooth and finite, and fluid elements are neither created nor destroyed (with no two fluid elements occupying the same region in space at the same time), the mapping $\mathbf{r} = \mathbf{R}(t; \mathbf{r}_0, t_0)$ is a diffeomorphism between V_0 and V that is parameterised by t (and also by t_0, regarded as a constant). Applying the chain rule of partial differentiation, (2.24) becomes

$$\frac{d}{dt} = \left(\frac{\partial}{\partial t}\right)_{\mathbf{r}} + \left(\frac{\partial \mathbf{R}}{\partial t}\right)_{\mathbf{r}_0} \cdot \frac{\partial}{\partial \mathbf{r}} \equiv \frac{\partial}{\partial t} + \mathbf{v} \cdot \nabla$$

from (2.23) and (2.25). Thus the two forms of derivative d/dt, the Eulerian (2.22) and the Lagrangian (2.24), are equivalent.

Consequently, the continuity equation (2.3) may be rewritten as

$$\frac{d\rho}{dt} + \rho \nabla \cdot \mathbf{v} = 0; \tag{2.27}$$

and since the material derivative $d\mathbf{v}/dt$ is the acceleration of the fluid element moving with velocity $\mathbf{v}$ at the position $\mathbf{r} = \mathbf{R}(t; \mathbf{r}_0, t)$ at time t, the equation of motion (2.20) may be rendered as

$$\rho \frac{d\mathbf{v}}{dt} = \rho \mathbf{g} - \nabla \cdot \mathbf{p}, \tag{2.28}$$

where the terms deliberately written on the right-hand side may be interpreted as the total force density acting on the fluid element, in analogy with the well-known Newton equation of motion for a particle. It is also often useful to envisage discrete fluid particles, each represented by the mass in a localised fluid element of infinitesimally small volume from the macroscopic viewpoint, such that the density field corresponds to their mass to volume ratios.

Sometimes, it is appropriate to consider motion relative to a moving system of coordinates—e.g. a Cartesian reference frame $Ox'y'z'$ rotating with (instantaneous) angular velocity ω_0 relative to another $Oxyz$ with common origin O. The velocity of a particle is

$$\mathbf{v} = \left.\frac{d\mathbf{r}}{dt}\right|_F = \left.\frac{d\mathbf{r}}{dt}\right|_M + \omega_0 \times \mathbf{r} = \left(\left.\frac{d}{dt}\right|_M + \omega_0 \times\right)\mathbf{r}$$

where $\mathbf{r}(t)$ is the position vector of a fluid particle at time t relative to O as seen in the moving frame, the subscript M referring to the moving reference frame $Ox'y'z'$ and F to the "fixed" (inertial) reference frame $Oxyz$. Moreover, the acceleration in the rotating reference frame is likewise

$$\begin{aligned} \left.\frac{d\mathbf{v}}{dt}\right|_F &= \left.\frac{d}{dt}\right|_F \left.\frac{d}{dt}\right|_F \mathbf{r} \\ &= \left(\left.\frac{d}{dt}\right|_M + \omega_0 \times\right)^2 \mathbf{r}, \end{aligned} \tag{2.29}$$

which may be expanded if desired. A moving reference frame with another origin O', not coincident with the origin O of the inertial reference frame, is likewise readily considered (cf. Sect. 3.14).

Exercises

(Q1) A cylindrical container rotates about its vertical axis with constant angular velocity ω_0. If there is liquid in the container, and assuming the atmospheric pressure is uniform over the surface of the liquid, by integrating the equation of motion show that the surface of the liquid is a paraboloid of revolution about the vertical axis.
(Assume constant gravity and a centrifugal acceleration.)

(Q2) A region of homogeneous self-gravitating fluid rotates about a fixed axis with constant angular velocity ω_0. Assuming there is no external pressure, show that there is an upper limit to $|\omega_0|$.
(Introduce the gravitational potential V such that $\mathbf{g} = -\nabla V$, and consider $\nabla^2 p$ over the fluid volume.)

(Q3) Since p is a symmetric dyadic, from the momentum conservation equation (2.17) derive the angular momentum conservation equation

$$\frac{\partial}{\partial t}(\mathbf{r} \times \rho \mathbf{v}) + \nabla \cdot (\mathbf{v}\mathbf{r} \times \rho \mathbf{v} - \mathsf{p} \times \mathbf{r}) = \mathbf{r} \times \rho \mathbf{g}, \tag{2.30}$$

where $\mathbf{r}$ is the position vector.

2.5 Rates of Change of Material Integrals

The material derivative d/dt may also be applied to any integral over a collection of fluid elements that remain identifiable and usually well connected during their motion [8]. Indeed, let us assume that the fluid is a continuum such that any closed fluid surface always consists of the same fluid elements, and any fluid element in the enclosed fluid volume always remains inside that fluid volume. (The explicit reference is to a material volume, in contrast to a fixed volume in space that is often called a control volume in the engineering literature.) Thus for any tensor field $\mathsf{F}(\mathbf{r}, t)$, the integral

$$I(\mathsf{F}) = \int_V \mathsf{F}(\mathbf{r}, t)\, d\tau \tag{2.31}$$

over a simply connected fluid volume V has the rate of change

$$\frac{dI}{dt} = \lim_{\Delta t \to 0} \frac{\int_{V'} \mathsf{F}(\mathbf{r}', t + \Delta t) d\tau' - \int_V \mathsf{F}(\mathbf{r}, t) d\tau}{\Delta t}. \tag{2.32}$$

Here any element in the fluid volume V, with position vector $\mathbf{r}(t)$ and velocity $\mathbf{v}(\mathbf{r}, t)$ at time t, has the position vector $\mathbf{r}' = \mathbf{r} + \mathbf{v}\Delta t$ to linear order in Δt at time $t + \Delta t$; and the fluid volume V occupies the volume V in space (with closed surface S say) at time t, but then volume V' in space (with closed surface S' say) at time $t + \Delta t$.

Now if the spatial volume element $d\tau$ is not only an element of V but also of V', there is a contribution $\partial_t \mathsf{F}\, d\tau$ in the integrand appearing on the right-hand side of (2.32), where the partial derivative is taken at the position of the element $d\tau$ at time t. However, if a spatial volume element $d\tau'$ in V' has no counterpart in V, then $d\tau' = d\mathbf{S} \cdot \mathbf{v} dt$ and this element contributes $\mathsf{F}\mathbf{v} \cdot d\mathbf{S}$ to the integrand arising from the volume integral over V'; and a spatial element $d\tau$ in V with no counterpart in V' contributes $\mathsf{F}\mathbf{v} \cdot d\mathbf{S}$ to the integrand arising from the volume integral over V, where an opposite sign in $\hat{\mathbf{n}} \cdot \mathbf{v}$ (recall that $\hat{\mathbf{n}}$ is the unit outward normal to any closed surface) compensates for the negative sign on the volume integral over V. In summary, on taking the limit $\Delta t \to 0$ in (2.32) we obtain the Reynolds transport theorem as applied to material elements—i.e.

$$\frac{d}{dt} \int_V \mathsf{F}(\mathbf{r}, t)\, d\tau = \int_V \frac{\partial \mathsf{F}}{\partial t}\, d\tau + \int_S \mathsf{F}\, \mathbf{v} \cdot d\mathbf{S}, \tag{2.33}$$

or from the relevant dot contraction of (1.59)

$$\frac{d}{dt}\int_V \mathbf{F}(\mathbf{r},t)\,d\tau = \int_V \left[\frac{\partial \mathbf{F}}{\partial t} + \nabla\cdot(\mathbf{v}\,\mathbf{F})\right] d\tau, \tag{2.34}$$

or on invoking (2.22)

$$\frac{d}{dt}\int_V \mathbf{F}(\mathbf{r},t)\,d\tau = \int_V \left(\frac{d\mathbf{F}}{dt} + \mathbf{F}\nabla\cdot\mathbf{v}\right) d\tau. \tag{2.35}$$

Identifying $\mathbf{F} \equiv 1$ in either (2.34) or (2.35) yields

$$\frac{d}{dt}\int_V d\tau = \int_V \nabla\cdot\mathbf{v}\,d\tau, \tag{2.36}$$

defining the rate of dilatation (expansion or contraction) of the fluid volume V, measured by the divergence $\nabla\cdot\mathbf{v}$ as mentioned in Sect. 2.2. In particular,

$$\frac{d}{dt}\int_V d\tau = 0$$

for any arbitrary fluid volume V implies the incompressibility condition $\nabla\cdot\mathbf{v} = 0$ everywhere in the fluid, which from (2.35) is evidently necessary and sufficient for the material derivative and volume integral operations to commute. Identifying $\mathbf{F}$ in (2.34) with the fluid density $\rho(\mathbf{r},t)$ yields

$$\frac{d}{dt}\int_V \rho(\mathbf{r},t)\,d\tau = \int_V \left[\frac{\partial\rho}{\partial t} + \nabla\cdot(\rho\mathbf{v})\right] d\tau = 0, \tag{2.37}$$

if the fluid body is conserved (there is no internal source or sink). Thus (2.37) implies the continuity equation (2.3) everywhere in the fluid, if (2.37) applies to any arbitrary fluid volume V. Identifying $\mathbf{F}$ in (2.35) with the fluid density $\rho(\mathbf{r},t)$ likewise yields the alternative form (2.27).

The time rate of change of momentum of an arbitrary fluid volume corresponds to setting $\mathbf{F} = \rho\mathbf{v}$ in (2.34). Then by postulating that this rate of change of momentum equals the sum (over the volume elements) of the total force acting on each as in (2.28), on applying the relevant dot contraction of (1.59) we have

$$\frac{d}{dt}\int_V \rho\mathbf{v}\,d\tau = \int_V \left[\frac{\partial}{\partial t}(\rho\mathbf{v}) + \nabla\cdot(\rho\mathbf{v}\mathbf{v})\right] d\tau = \int_V \rho\mathbf{g}\,d\tau - \int_S \mathbf{p}\cdot d\mathbf{S}, \tag{2.38}$$

which identifies the gravity $\rho\mathbf{g}$ as a "body force" density and the pressure tensor $\mathbf{p}$ with a "surface force" acting on the surface S of the fluid volume V. Conversely, since the fluid volume V is arbitrary, (2.38) may be regarded as the origin of (2.20) on invoking (2.21)—or of course (2.28), after also invoking (2.22).

The field F may be identified with other quantities such as the energy density, but there is often a simple relation between the fluid pressure p and density ρ that leads to a relatively simple complete dynamical description. This is discussed further in the following sections, which are devoted to the derivation of the macroscopic equations of Fluid Mechanics and MHD from a kinetic description of the motion of the microscopic constituents of fluids and plasmas.

For later reference, let us also note here that the rate of change of the material line integral of a vector field $\mathbf{f}$, taken over a collection of fluid elements forming a closed curve C that remains simple and connected, is

$$\begin{aligned} \frac{d}{dt}\oint_C \mathbf{f}\cdot d\mathbf{r} &= \oint_C \left(\frac{d\mathbf{f}}{dt} + (\nabla\mathbf{v})\cdot\mathbf{f}\right)\cdot d\mathbf{r} \\ &= \oint_C \left(\frac{d\mathbf{f}}{dt} - (\nabla\mathbf{f})\cdot\mathbf{v}\right)\cdot d\mathbf{r} \end{aligned} \tag{2.39}$$

since $\nabla(\mathbf{v}\cdot\mathbf{f}) = (\nabla\mathbf{v})\cdot\mathbf{f} + (\nabla\mathbf{f})\cdot\mathbf{v}$. In particular, identifying $\mathbf{f}\equiv\mathbf{v}$ in (2.39) yields the rate of change of the fluid circulation

$$\frac{d}{dt}\oint_C \mathbf{v}\cdot d\mathbf{r} = \oint_C \frac{d\mathbf{v}}{dt}\cdot d\mathbf{r} \tag{2.40}$$

since $(\nabla\mathbf{v})\cdot\mathbf{v} = \nabla(\mathbf{v}\cdot\mathbf{v}/2) = \nabla(v^2/2)$—i.e. the material derivative and line integral operations commute in this case. There is also the rate of change result for a material surface, discussed and invoked in Sect. 5.8.

Exercise

(Q1) Sketch a proof for the result (2.39), noting that values of the velocity field at the end-points of any infinitesimal directed material line segment $d\mathbf{r}'$ are respectively $\mathbf{v}$ and $\mathbf{v}+d\mathbf{v}$, where $d\mathbf{v} = \mathbf{dr}\cdot\nabla\mathbf{v}$.

2.6 Equations of Change

As mentioned earlier, on a microscopic scale the motion of gases and liquids (fluids) is described in terms of their constituent atoms or molecules—and in the case of ionised gases (plasmas), their constituent ions and electrons [3]. A kinetic equation that allows for encounters between the microscopic particles was first considered by Boltzmann in 1872, to describe the classical dynamics of gases. Thus for fluids it is normally assumed that the atoms or molecules are correlated due to binary collisions over distances much less than the macroscopic length scale and at frequencies much greater than any characteristic macroscopic frequency, as previously indicated in Sect. 2.2. The rapid mobility of electrons in plasma, due to their significantly smaller mass relative to any other species, ensures that they are almost instantaneously positioned such that the electric field of any charge is screened out on a scale characterised by the Debye length—and consequently, any collision between the charged particles in a plasma is also predominantly binary over distances bounded below by the Coulomb length and above by the Debye length [10]. This notion of charged particle screening was introduced in the theory of electrolytes developed by Debye and Hückel.[4] The term plasma normally implies that the Debye length scale is much less than the macroscopic length scale, and that electric charge oscillations due to small perturbations from the typical quasi-neutral configuration in plasmas occur at frequencies much greater than any characteristic macroscopic frequency.

For each constituent species s, a Boltzmann-type kinetic equation may be introduced to define the evolution in time t of a velocity distribution function $f_s(\mathbf{r}, \mathbf{c}, t)$ in phase space with coordinates the particle position $\mathbf{r}$ and velocity $\mathbf{c}$:

$$\frac{\partial f_s}{\partial t} + \mathbf{c} \cdot \nabla f_s + \mathbf{a}_s(\mathbf{r}, \mathbf{c}, t) \cdot \nabla_{\mathbf{c}} f_s = \frac{\delta f_s}{\delta t}, \tag{2.41}$$

where the acceleration of the particle of mass m_s and charge e_s is

$$\mathbf{a}_s(\mathbf{r}, \mathbf{c}, t) = \mathbf{g} + \frac{e_s}{m_s}[\mathbf{E}(\mathbf{r}, t) + \mathbf{c} \times \mathbf{B}(\mathbf{r}, t)],$$

if both gravitational and electromagnetic components are included ($\mathbf{g}$, $\mathbf{E}$ and $\mathbf{B}$ denote the respective gravitational, electric and magnetic fields) [3, 9]. The symbol $\nabla_{\mathbf{c}}$ denotes the gradient operator relative to the independent velocity vector $\mathbf{c}$, analogous to ∇ relative to the position vector $\mathbf{r}$ in phase space, and the right-hand side of (2.41) represents the time rate of change of the velocity distribution function due to the microscopic particle collisions.

Any macroscopic field equation of interest corresponds to taking some moment of Eq. (2.41). If $\mathbf{w} = \mathbf{c} - \mathbf{c_0}$ denotes the peculiar velocity for the species relative to some reference velocity $\mathbf{c_0}$, the moment corresponding to any related function $\Psi(\mathbf{w})$ is defined by

$$n_s \langle \Psi \rangle = \int f_s(\mathbf{r}, \mathbf{c}, t)\, \Psi(\mathbf{w})\, d\mathbf{c}, \tag{2.42}$$

including the particle number density (number of particles in a unit volume)

$$n_s(\mathbf{r}, t) = \int f_s(\mathbf{r}, \mathbf{c}, t)\, d\mathbf{c}.$$

We prefer to identify $\mathbf{c_0}$ with the mass-weighted mean velocity $\mathbf{v}$, defined via the total density and total mass flux

$$\rho = \Sigma_s \rho_s\,, \quad \rho \mathbf{v} = \Sigma_s \rho_s \mathbf{v}_s,$$

[4] P. Debye and E. Hückel, Physikalische Zeitschrift **24**, 183 and 305 (1923).

analogous to (2.6) and (2.7) in Sect. 2.2 except that the summation is now over all species present (rather than beams)—i.e. where the summations involve the density ρ_s and the mean flow velocity $\mathbf{v}_s = \langle \mathbf{c} \rangle$ of the particular species (cf. below). The consequent general equation of change that follows from (2.41) may be written in the form

$$\frac{\partial(n_s\langle\boldsymbol{\Psi}\rangle)}{\partial t} + \boldsymbol{\nabla}\cdot(n_s\langle(\mathbf{v}+\mathbf{w})\,\boldsymbol{\Psi}\rangle) - n_s\left\langle\left(\mathbf{f}_s + \frac{e_s}{m_s}\mathbf{w}\times\mathbf{B} - \mathbf{w}\cdot\boldsymbol{\nabla}\mathbf{v}\right)\cdot\boldsymbol{\nabla}_{\mathbf{w}}\boldsymbol{\Psi}\right\rangle = \mathcal{C}_s(\boldsymbol{\Psi}), \tag{2.43}$$

which involves the acceleration

$$\mathbf{f}_s = \mathbf{g} + \frac{e_s}{m_s}[\mathbf{E} + \mathbf{v}\times\mathbf{B}] - \frac{d\mathbf{v}}{dt}$$

and the collision integral

$$\mathcal{C}_s(\boldsymbol{\Psi}) = \int \frac{\delta f_s}{\delta t}\boldsymbol{\Psi}\, d\mathbf{c}. \tag{2.44}$$

This form is a simplification of the Enskog equation, analogous to an equation of change due to Maxwell, because $\boldsymbol{\Psi}$ is only explicitly dependent on $\mathbf{w}$ (cf. Chapman and Cowling [3]). Equation (2.43) replaces (2.41) as our starting point. In passing, we note that the subscript s has been omitted from both $\mathbf{c}$ and $\mathbf{w} = \mathbf{c} - \mathbf{v}$, since $\mathbf{c}$ becomes a dummy integration variable.

Fundamental field quantities in Fluid Mechanics and MHD are related to the lower moments defined by (2.42) for each species s of particle mass m_s—viz.

$$\begin{aligned}
\rho_s &\equiv n_s m_s && \text{(density)}\\
\mathbf{u}_s &\equiv \langle \mathbf{w} \rangle && \text{(mean relative velocity)}\\
p_s &\equiv n_s\left\langle \frac{1}{3} m_s w^2 \right\rangle && \text{(pressure)}\\
\mathbf{p}_s &\equiv n_s \langle m_s \mathbf{w}\mathbf{w} \rangle && \text{(pressure tensor)}\\
\mathbf{q}_s &\equiv n_s\left\langle \frac{1}{2} m_s w^2 \mathbf{w} \right\rangle && \text{(thermal flux).}
\end{aligned}$$

In passing, we note that $\mathbf{u}_s = \mathbf{v}_s - \mathbf{v}$ is the mean species velocity relative to the reference velocity $\mathbf{v}$, the species pressure tensor $\mathbf{p}_s = p_s\mathbf{I} + \mathbf{t}_s$ has the symmetric traceless part $\mathbf{t}_s \equiv \{\mathbf{p}_s\} = n_s\langle m_s\{\mathbf{w}\mathbf{w}\}\rangle$, and the pressure p_s and thermal flux vector $\mathbf{q}_s$ respectively determine the species contribution to the pressure and heat transfer. (The fundamental macroscopic equations discussed in Sect. 2.7 involve the reference velocity $\mathbf{v}$ and the respective total fields ρ, p, $\mathbf{p}$ and $\mathbf{q}$ obtained by summation over the species present, together with the current density and electromagnetic fields in the case of MHD as considered there.)

Other unnamed higher moments arising in the basic equations of change below are the third-degree $\mathsf{H}_s = n_s \langle m_s \mathbf{www} \rangle = (2/3)\,\mathbf{q}_s\,\mathsf{I} + \mathsf{h}_s$ where $\mathsf{h}_s = n_s \langle m_s \mathbf{w}\{\mathbf{ww}\}\rangle$, and the fourth-degree $\boldsymbol{\ell}_s = n_s \langle \frac{1}{2} m_s w^2 \mathbf{ww} \rangle$ or

$$\mathsf{L}_s = \boldsymbol{\ell}_s - \frac{5}{2}\frac{p_s}{\rho_s}\mathbf{p}_s . \tag{2.45}$$

Thus if $\boldsymbol{\Psi}$ is identified with $m_s, m_s\mathbf{w}, \frac{1}{2} m_s w^2, m_s\{\mathbf{ww}\}, \frac{1}{2} m_s w^2 \mathbf{w}$, etc., successively, then for each species the general equation of change (2.43) produces the basic equations for the mass (continuity), momentum, thermal energy, $\mathbf{t}_s$ and $\mathbf{q}_s$, etc.:

$$\frac{\partial \rho_s}{\partial t} + \nabla \cdot (\rho_s \mathbf{v} + \rho_s \mathbf{u}_s) = \mathcal{C}_s(m_s) \tag{2.46}$$

$$\frac{\partial (\rho_s \mathbf{u}_s)}{\partial t} + \nabla \cdot (\rho_s \mathbf{v}\mathbf{u}_s + \mathbf{p}_s) \tag{2.47}$$

$$- \rho_s\, \mathbf{f}_s - \rho_s \frac{e_s}{m_s} \mathbf{u}_s \times \mathbf{B} + \rho_s \mathbf{u}_s \cdot \nabla \mathbf{v} = \mathcal{C}_s(m_s \mathbf{w})$$

$$\frac{\partial (\frac{3}{2} p_s)}{\partial t} + \nabla \cdot \left(\frac{3}{2} p_s\, \mathbf{v} + \mathbf{q}_s\right) - \rho_s\, \mathbf{f}_s \cdot \mathbf{u}_s + \mathbf{p}_s : \nabla \mathbf{v} = \mathcal{C}_s\left(\frac{1}{2} m_s w^2\right) \tag{2.48}$$

$$\frac{\partial \mathbf{t}_s}{\partial t} + \nabla \cdot (\mathbf{v}\mathbf{t}_s + \mathsf{h}_s) - 2\rho_s\{\mathbf{f}_s\, \mathbf{u}_s\} + 2\{\mathbf{p}_s \cdot \nabla \mathbf{v}\} \tag{2.49}$$

$$-2\frac{e_s}{m_s}\{\mathbf{t}_s \times \mathbf{B}\} = \mathcal{C}_s(m_s\{\mathbf{ww}\})$$

$$\frac{\partial \mathbf{q}_s}{\partial t} + \nabla \cdot (\mathbf{v}\mathbf{q}_s + \boldsymbol{\ell}_s) - \mathbf{f}_s \cdot \mathbf{p}_s - \frac{3}{2} p_s \mathbf{f}_s - \frac{e_s}{m_s}\mathbf{q}_s \times \mathbf{B} \tag{2.50}$$

$$+ \mathbf{q}_s \cdot \nabla \mathbf{v} + \mathsf{H}_s : \nabla \mathbf{v} = \mathcal{C}_s\left(\frac{1}{2} m_s w^2 \mathbf{w}\right)$$

etc. On writing $p_s = n_s k T_s$, where k denotes the Boltzmann constant and T_s the kinetic temperature of the species, Eq. (2.50) for $\mathbf{q}_s$ may be replaced by subtracting $\frac{5}{2} p_s/\rho_s$ times equation (2.48) from (2.50)—i.e.

$$\frac{\partial \mathbf{R}_s}{\partial t} + \nabla \cdot (\mathbf{v}\mathbf{R}_s + \mathsf{L}_s) - \mathbf{f}_s \cdot \mathbf{t}_s - \frac{e_s}{m_s}\mathbf{R}_s \times \mathbf{B} + \frac{5}{2}\frac{k}{m_s}\mathbf{p}_s \cdot \nabla T_s \tag{2.51}$$

$$+ \frac{5}{2} n_s k \mathbf{u}_s \frac{dT_s}{dt} + \mathbf{R}_s \cdot \nabla \mathbf{v} + \mathsf{H}_s : \nabla \mathbf{v} = \mathcal{C}_s\left(\frac{1}{2} m_s \left(w^2 - \frac{5kT_s}{m_s}\right)\mathbf{w}\right),$$

which we invoke to analyse thermal effects (cf. Sect. 2.11).

Each equation in the hierarchy of basic equations of change introduces a higher moment. The necessary closure for the subsequent derivation of any related fluid mechanics or MHD model corresponds to either simply ignoring a higher moment at some level or approximating the velocity distribution function in some acceptable way, as discussed further in the following sections. Subsets of these basic equations of change, or an extension to even higher levels in the hierarchy, could be used as the macroscopic description for each species in terms of the associated variables one chooses to retain. In this book we continue to adopt the more traditional classical macroscopic equations obtained from (2.46)–(2.48) as discussed in the next section—but then we use (2.49) and (2.51) to derive appropriate explicit relations for $\mathbf{t}_s$ in Sect. 2.9 and $\mathbf{R}_s$ (and hence $\mathbf{q}_s$) in Sect. 2.11, in our subsequent discussion of the extended macroscopic models we do consider.

Finally, we remark that following Braginskii many plasma theorists have adopted the mean flow velocity $\mathbf{v}_s$ of the particular species as the reference velocity $\mathbf{c}_0$ in defining the moments [5], rather than the universal mass-weighted mean velocity $\mathbf{v}$ as in Chapman and Cowling [3] that we prefer. Thus recalling $\mathbf{w} = \mathbf{c} - \mathbf{v}$ and $\mathbf{v}_s = \langle \mathbf{c} \rangle$, from (2.42) we have

$$n_s\langle \mathbf{w}\rangle = \int f_s(\mathbf{r}, \mathbf{c}, t)(\mathbf{c} - \mathbf{v})\, d\mathbf{c} = \int f_s(\mathbf{r}, \mathbf{c}, t)\mathbf{c}\, d\mathbf{c} - \mathbf{v}\int f_s(\mathbf{r}, \mathbf{c}, t)\, d\mathbf{c} = n_s\langle \mathbf{c}\rangle - n_s\mathbf{v}$$

such that (as mentioned above) $\mathbf{u}_s = \mathbf{v}_s - \mathbf{v}$ or $\mathbf{v}_s = \mathbf{v} + \mathbf{u}_s$, so Eq. (2.46) may be rewritten as

$$\frac{\partial \rho_s}{\partial t} + \nabla\cdot(\rho_s \mathbf{v}_s) = \mathcal{C}(m_s). \tag{2.52}$$

The corresponding momentum equation for the species is

$$m_s n_s\left(\frac{\partial \mathbf{v}_s}{\partial t} + \mathbf{v}_s \cdot \nabla \mathbf{v}_s\right) + \nabla\cdot \mathsf{P}_s - n_s e_s(\mathbf{E} + \mathbf{v}_s \times \mathbf{B}) - m_s n_s \mathbf{g} = \mathbf{F}_s^{\text{coll}}, \tag{2.53}$$

involving the pressure tensor

$$\mathsf{P}_s(\mathbf{r}, t) = \int m_s f_s(\mathbf{r}, \mathbf{c}, t)(\mathbf{c} - \mathbf{v}_s)(\mathbf{c} - \mathbf{v}_s)\, d\mathbf{c}$$

defined using the mean flow velocity $\mathbf{v}_s$ of the particular species. Noting that $\mathsf{P}_s = \mathsf{p}_s - m_s n_s \mathbf{u}_s \mathbf{u}_s$ where p_s is our form for the pressure tensor, Eq. (2.53) is equivalent to (2.47) on writing $\mathbf{F}_s^{\text{coll}}$ for $\mathcal{C}_s(m_s\mathbf{w})$. Moreover, the heat balance equation for each species is written as

$$\frac{3}{2} n_s k\left(\frac{\partial T_s}{\partial t} + \mathbf{v}_s \cdot \nabla T_s\right) + \nabla\cdot \mathbf{Q}_s + \mathsf{P}_s : \nabla \mathbf{v}_s = G_s^{\text{coll}} \tag{2.54}$$

where T_s denotes the kinetic temperature defined such that $P_s = n_s k\, T_s$ is the associated hydrostatic pressure, the thermal flux

$$\mathbf{Q}_s = \int \frac{1}{2} m_s\, f_s(\mathbf{r}, \mathbf{c}, t)\, (\mathbf{c} - \mathbf{v}_s) \cdot (\mathbf{c} - \mathbf{v}_s)\, (\mathbf{c} - \mathbf{v}_s)\, d\mathbf{c}$$

is likewise defined using the mean flow velocity $\mathbf{v}_s$ of the species, and G_s^{coll} is the relevant collision integral.[5]

Exercises

(Q1) Verify $\mathsf{P}_s = \mathsf{p}_s - m_s n_s \mathbf{u}_s \mathbf{u}_s$, and show that (2.53) is equivalent to (2.47).

(Q2) Show that

$$\frac{3}{2} n_s k \left(\frac{\partial T_s}{\partial t} + \mathbf{v}_s \cdot \nabla T_s \right) = \frac{\partial \left(\frac{3}{2} P_s \right)}{\partial t} + \nabla \cdot \left(\frac{3}{2} P_s \mathbf{v}_s \right),$$

and hence that the heat balance equation (2.54) may be rewritten as

$$\frac{\partial \left(\frac{3}{2} P_s \right)}{\partial t} + \nabla \cdot \left(\frac{3}{2} P_s \mathbf{v}_s + \mathbf{Q}_s \right) + \mathsf{P}_s : \nabla \mathbf{v}_s = G_s^{\text{coll}}.$$

Then show that this equation is consistent with (2.48), when the collision term G_s^{coll} is suitably identified.

Hint: Use (2.46), and an energy transport equation given by the dot product of (2.47) with $\mathbf{u}_s$.

(Q3) When the mean flow velocity of the species $\mathbf{v}_s$ is again used as the reference velocity, the next equation in the hierarchy involves the corresponding even higher moment $\mathsf{Q}_s = \int m_s f_s(\mathbf{r}, \mathbf{c}, t)\, (\mathbf{c} - \mathbf{v}_s)\, (\mathbf{c} - \mathbf{v}_s)\, d\mathbf{c}$ (note the heat flux $\mathbf{Q}_s$ defined in the text is an immediate contraction of Q_s):

$$\frac{\partial \mathsf{P}_s}{\partial t} + \nabla \cdot (\mathbf{v}_s \mathsf{P}_s + \mathsf{Q}_s) + \mathbf{v}_s \overleftarrow{\nabla} \cdot \mathsf{P}_s + \mathsf{P}_s \cdot \nabla \mathbf{v}_s - 2 \frac{e_s}{m_s} \{ \mathsf{P}_s \times \mathbf{B} \} = \mathsf{G}_s^{\text{coll}},$$

where the leftward pointing arrow on the ∇ operator introduced here means it *applies to the left*, $\mathsf{G}_s^{\text{coll}}$ is the appropriate collisional term, and the other notation is the same as in the previous two Exercises. Derive this equation directly by expanding $\nabla \cdot \mathsf{Q}_s$.

Hint: Invoke the Boltzmann-type Eq. (2.41) in the tensor calculus. (We note the traceless part of this higher moment equation is analogous to (2.49), and the analogy to (2.50) is derived in the answer to this Exercise.)[6]

---

[5] In the case of a plasma consisting of electrons and only one ion species, the mean velocity is effectively the ion velocity ($\mathbf{v} \simeq \mathbf{v}_i$ corresponding to $m_e \ll m_i$ and $\mathbf{u}_i \simeq 0$) so the defined fields such as P_i and $\mathbf{Q}_i$ for example are essentially the same as p_i and $\mathbf{q}_i$, respectively. However, this is not so for the electrons, since the alternative reference velocities for the electron component are quite distinct corresponding to inter-species (electron) diffusion—i.e. $\mathbf{u}_e \neq 0$.

[6] The corresponding extension of the hierarchy of equations (where the moments are defined using the species mean flow velocity $\mathbf{v}_s$ as the reference velocity) was originally obtained by J.J. Ramos (*Physics of Plasmas* **14**, 052506, 2007).

2.7 Classical Macroscopic Equations

When the sum over all of the species present in the fluid or plasma is taken and it is assumed the combined contributions from the respective collision integrals $\mathcal{C}_s(m_s), \mathcal{C}_s(m_s\mathbf{w})$ and $\mathcal{C}_s(\frac{1}{2}m_s w^2)$ vanish (i.e. the binary collisions summed over all species preserve mass, momentum and energy), noting that $\Sigma_s \rho_s \mathbf{u}_s = 0$ we obtain the classical macroscopic equations of mass conservation (continuity), motion and energy (cf. also [3]):

$$\frac{d\rho}{dt} + \rho\nabla\cdot\mathbf{v} = 0 \tag{2.55}$$

$$\rho\frac{d\mathbf{v}}{dt} + \nabla\cdot\mathbf{p} = \rho\mathbf{g} + \mathbf{j}\times\mathbf{B} \tag{2.56}$$

$$\frac{d}{dt}\left(\frac{3}{2}p\right) + \frac{3}{2}p\nabla\cdot\mathbf{v} + \mathbf{p}:\nabla\mathbf{v} + \nabla\cdot\mathbf{q} = \mathbf{j}\cdot(\mathbf{E}+\mathbf{v}\times\mathbf{B}). \tag{2.57}$$

Here the material derivative $d/dt = \partial_t + \mathbf{v}\cdot\nabla$ appropriately refers to the mass-weighted mean flow velocity $\mathbf{v}$, the variable subset $\{\rho,\ \mathbf{p} = p\,\mathbf{I} + \mathbf{t},\ \mathbf{q}\}$ and implicitly p are total quantities (the respective sums over all of the species present), and

$$\mathbf{j} \equiv \Sigma_s n_s e_s \langle\mathbf{c}_s\rangle = \Sigma_s \frac{e_s}{m_s}\rho_s\langle\mathbf{c}_s\rangle$$

denotes and defines the *total current density*. Equation (2.55) is identical to (2.27), and the equation of motion (2.56) is comparable with (2.28). An electric force term does not appear in (2.56) because the medium is assumed to be quasi-neutral on the macroscopic scale, due to the charged particle screening outlined in the previous section in the case of a plasma, but the gravitational force $\rho\mathbf{g}$ in the equation of motion is supplemented with the electromagnetic body force $\mathbf{j}\times\mathbf{B}$ that is of major interest in MHD—i.e. whenever the system is electrically conducting.

Noting (2.55), the first two terms on the left-hand side of (2.57) are sometimes re-expressed as $(3/2)n\,k\,dT/dt$, where T denotes a systemic temperature. Alternatively, from (2.55) the energy equation (2.57) may be rewritten as

$$\frac{3}{2}\rho^{\frac{5}{3}}\frac{d}{dt}(p\rho^{-\frac{5}{3}}) = -\mathbf{t}:\nabla\mathbf{v} - \nabla\cdot\mathbf{q} + \mathbf{j}\cdot(\mathbf{E}+\mathbf{v}\times\mathbf{B}), \tag{2.58}$$

where the first term on the right-hand side represents viscous dissipation, the second term the heat transfer (thermal conduction), and the third term the electromagnetic heating. The well-known *adiabatic equation of state*

$$\frac{d}{dt}(p\rho^{-\gamma}) = 0 \quad \text{or} \quad \frac{dp}{dt} + \gamma p\nabla\cdot\mathbf{v} = 0 \quad \left(\text{identifying } \gamma = \frac{5}{3}\right) \tag{2.59}$$

corresponds to neglecting all the terms on the right-hand side of (2.58)—i.e. we assume all three dissipation contributions are negligible. The term "adiabatic" (no heat transfer) used above is more common, but a more precise term is isentropic—i.e. (2.59) corresponds to no change in entropy (any *reversible* adiabatic process is isentropic) [2]. Another useful term is *barotropic*, which means that there is some relation between the fluid pressure p and fluid density ρ, but no third thermodynamic variable—e.g. $p\rho^{-\gamma}=$ constant (following the motion), corresponding to (2.59).

In ideal fluid mechanics and various inviscid MHD models, the traceless component $\mathbf{t}$ of the pressure tensor is simply ignored such that $\nabla\cdot\mathbf{p}=\nabla p$, when the system is described in terms of the field variables $\{\rho, \mathbf{v}, p\}$ by (2.55) and (2.56) together with a barotropic equation of state. There can be similar acceptable dynamical descriptions in viscous Fluid Mechanics and MHD too, if an appropriate expression for $\mathbf{p}$ is adopted (cf. Sects. 2.2 and 2.9). An expression for the thermal flux vector $\mathbf{q}$ may also be included if there is significant heat conduction (cf. Sect. 2.11). They are often called constitutive relations in the literature, and sometimes viewed as phenomenological inputs, but we emphasise that the explicit expressions for $\mathbf{t}_s$ and $\mathbf{R}_s$ (and hence $\mathbf{q}_s$) presented below are obtained from their respective higher basic equations of change. In the case of MHD, where the additional body force $\mathbf{j}\times\mathbf{B}$ is significant, supplementary electromagnetic equations must be included (cf. Sect. 2.12).

2.8 Extended Macroscopic Models

As indicated, ideal fluid mechanics and various inviscid MHD models correspond to truncation of the hierarchy of basic equations of change where only (2.46) and (2.47) are retained, to consider just five scalar field variables $\{\rho_s, \mathbf{v}_s, p_s\}$. This is sometimes referred to as the "five-moment" description, albeit with additional electromagnetic variables in MHD. However, the extension to non-ideal Fluid Mechanics and MHD corresponds to including $\mathbf{t}_s$ in a "ten-moment" description, or both $\mathbf{t}_s$ and $\mathbf{q}_s$ in a "thirteen-moment" description, and so on. Thus if the term $\nabla\cdot\mathbf{t}$ is to be retained in expressing $\nabla\cdot\mathbf{p}$ in (2.56), one may proceed to derive an extended macroscopic model by augmenting the lowest level Eqs. (2.46) and (2.47) with the additional basic equation of change (2.49) for $\mathbf{t}_s$ in the underlying hierarchy; and if thermal effects are to be considered, then the extension also involves including (2.50) for $\mathbf{q}_s$, or equivalently (2.51) for $\mathbf{R}_s$.

In collisionless theory, the terms on the right-hand sides of the fundamental equations of change are omitted entirely, but various representations have also been considered. Here we retain the collisional terms for each species s in the simplest way—viz. linear expressions for the integrals in (2.49) and (2.51). On tensor rank alone, these expressions must involve their respective moments and are thus

$$\mathcal{C}_s(m_s\{\mathbf{ww}\}) = -\sum_j \frac{\vartheta_{sj}}{\tau_{sj}}\mathbf{t}_j, \tag{2.60}$$

$$\mathcal{C}_s\left(\frac{1}{2}m_s\left(w^2-\frac{5kT_s}{m_s}\right)\mathbf{w}\right)=-\sum_j\left(\frac{1}{\tau_{sj}}\mathbf{R}_j+\zeta_{sj}\rho_j\mathbf{u}_j\right), \quad (2.61)$$

with the coefficients $\{\tau_{sj}, \vartheta_{sj}, \zeta_{sj}\}$ functions of number density and temperature. As previously indicated, the "ten moment" description is closed if we may ignore terms in the two higher moments $\mathbf{q}_s$ and $\mathbf{h}_s$ from (2.49), in truncating at one level beyond the "five moment" description to include the term $\nabla\cdot\mathbf{t}_s$ from $\nabla\cdot\mathbf{p}_s$ in (2.47). The "thirteen moment" description, corresponding to also including Eq. (2.50) for $\mathbf{q}_s$ or equivalently Eq. (2.51) for $\mathbf{R}_s$ together with Eqs. (2.46)–(2.49) as previously mentioned, could likewise be achieved at that level in the hierarchy of basic equations by omitting terms in $\mathbf{h}_s$ and $\boldsymbol{\ell}_s$—and yet higher "n moment" ($n > 13$) closure could be considered by including even more of the hierarchy and neglecting terms in moments beyond the corresponding set.

Although the ultimate test of the accuracy of the closure at any level is of course experimental or observational agreement with the consequent mathematical model, there have been two major theoretical approaches to this issue. In deriving macroscopic field equations from the kinetic equation (2.41), various asymptotic schemes based upon some small parameter in a particular physical context have been proposed in order to estimate the approximation involved [5]. There is also an elegant procedure where the velocity distribution function is represented as a series expansion, appropriately truncated for an assumed level of closure of the hierarchy of exact equations of change. A well-known example consistent with the expressions (2.60) and (2.61) is Grad's expansion in multidimensional Hermite polynomials, where the truncated expression

$$f_s \simeq f_s^0\left[a_s^0+\mathbf{a}_s^1\cdot\mathbf{w}+\mathbf{a}_s^2:\mathbf{ww}+\mathbf{a}_s^3\cdot\mathbf{w}\left(\alpha_s w^2-\frac{5}{2}\right)\right] \quad (2.62)$$

with the Maxwellian zeroth-order form

$$f_s^0=n_s\left(\frac{\alpha_s}{\pi}\right)^{\frac{3}{2}}\exp(-\alpha_s w^2) \quad (2.63)$$

provides his "thirteen moment approximation".[7] Particle collisions tend to drive the distribution function towards this Maxwellian form, which is associated with "local thermodynamic equilibrium". Consequently, on adopting (2.62) and (2.63) and substituting into (2.42), from the respective lower moments $\langle 1\rangle$, $\langle\mathbf{w}\rangle$, $\langle\frac{1}{2}m_s w^2\rangle$, $\langle\{m_s\mathbf{ww}\}\rangle$ and $\langle\{\frac{1}{2}m_s\mathbf{w}w^2\}\rangle$ we obtain

$$\begin{aligned} &a_s^0=1, \qquad \mathbf{a}_s^1=2\alpha_s\mathbf{u}_s,\\ \alpha_s=m_s/2kT,\quad &\\ &\mathbf{a}_s^2=(\alpha_s/p_s)\,\mathbf{t}_s, \qquad \mathbf{a}_s^3=\tfrac{4}{5}(\alpha_s/p_s)\,\mathbf{R}_s, \end{aligned} \quad (2.64)$$

[7] H. Grad (*Communications in Pure and Applied Mathematics* **2**, 231– 407, 1949—cf. also *Handbuch der Physik*, S. Flügge (Editor), Volume 12, Chapter X, Springer, Berlin, 1958).

together with the relevant results for the higher moments

$$\nabla\cdot\mathbf{h}_s=\frac{4}{5}\{\nabla\mathbf{q}_s\}\,,\ \mathbf{H}_s:\nabla\mathbf{v}=\frac{2}{5}[\mathbf{q}_s\cdot\nabla\mathbf{v}+\mathbf{q}_s\nabla\cdot\mathbf{v}+(\nabla\mathbf{v})\cdot\mathbf{q}_s]\,,\ \mathbf{L}_s=\frac{p_s}{\rho_s}\mathbf{t}_s \tag{2.65}$$

and the coefficients in the expressions (2.60) and (2.61) for the collision integrals. Closure therefore occurs at a level involving no more than the thirteen scalar fields corresponding to the variables $\rho_s=m_s n_s, \mathbf{u}_s, p_s=n_s kT_s, \mathbf{t}_s$, and $\mathbf{q}_s$. Retaining only the first three expansion coefficients listed in (2.64) produces the related "ten moment" description in $\{\rho_s, \mathbf{u}_s, p_s, \mathbf{t}_s\}$, and the heat flux vector $\mathbf{R}_s$ (and hence $\mathbf{q}_s$) included in the "thirteen moment" description is completely determined by the fourth coefficient. Balescu [1] followed Grad by expanding f_s in terms of Hermite polynomials, and other authors have considered truncated expansions involving alternative basis functions (e.g. Sonine-Laguerre polynomials in a "fifteen moment" approximation). The expansion could also be taken about some other zeroth-order distribution function. Indeed, an appropriate number of leading terms in an integrable series expansion of orthogonal functions about the most suitable zeroth-order form (not necessarily the Maxwellian) probably represents almost everywhere the best possible expression for the relevant velocity distribution function, given the corresponding number of basic equations of change for the species one chooses to retain. Thus the fluid or MHD models eventually obtained, upon truncation of the underlying hierarchies of exact equations of change, may be more widely applicable than is traditionally thought.

2.9 Plasma Pressure Tensor

2.9.1 *Invariant Form for the Traceless Component*

Let us now proceed to consider the traceless pressure tensor component $\mathbf{t}_s$, implicit in (2.49) for any species, with the ultimate objective to render an appropriate form for the pressure tensor in plasma. In a magnetic field, it emerges that the consequent explicit relation for $\mathbf{t}_s$ is usually quite different for charged species than for neutrals. Thus the total pressure tensor $\mathbf{p}$ in (2.56) for a magnetised plasma differs from the form for a fluid, which was previously obtained in Sect. 2.2 using an heuristic argument. We adopt the linear representation (2.60), *but for notational simplicity now suppress the subscript s denoting the chosen species where there is no chance of confusion* (including writing $\vartheta/\tau=\vartheta_{ss}/\tau_{ss}$), so (2.49) may be re-expressed as

$$\frac{d\mathbf{t}}{dt} + \frac{\vartheta}{\tau}\mathbf{t} + \mathbf{t}\nabla\cdot\mathbf{v} - 2\rho\{\mathbf{f}\,\mathbf{u}\} + 2\{\mathbf{p}\cdot\nabla\mathbf{v}\}$$

$$- 2\frac{e}{m}\{\mathbf{t}\times\mathbf{B}\} + \nabla\cdot\mathbf{h} = -\sum_{j\neq s}\frac{\vartheta_{sj}}{\tau_{sj}}\mathbf{t}_j. \quad (2.66)$$

It is convenient to introduce a characteristic frequency ω to represent the time derivative term $d\mathbf{t}/dt$ as $\omega\mathbf{t}$ in Eq. (2.66), which may then be rewritten in the notationally convenient form as

$$\mathbf{t} - 2\{\mathbf{t}\times\mathbf{a}\} = -2\mu\mathbf{s} \quad (2.67)$$

where the symmetric and traceless generalised rate of deformation (or strain) tensor $\mathbf{s}$ is given by

$$2p\,\mathbf{s} = -2\rho\{\mathbf{f}\,\mathbf{u}\} + 2p\{\nabla\mathbf{v}\} + \mathbf{t}\nabla\cdot\mathbf{v} + 2\{\mathbf{t}\cdot\nabla\mathbf{v}\} + \nabla\cdot\mathbf{h} + \sum_{j\neq s}\frac{\vartheta_{sj}}{\tau_{sj}}\mathbf{t}_j. \quad (2.68)$$

The coefficient $\mu = \dfrac{p\,\tau}{\omega\tau + \vartheta}$ and the vector $\mathbf{a} = \dfrac{e\mathbf{B}}{m}\dfrac{\tau}{\omega\tau + \vartheta}$ (proportional to the gyrofrequency $\omega_c = eB/m$) introduced in (2.67) reflect the first time derivative and the second collisional terms in (2.66), with one or the other predominant if the magnitude of the parameter $\omega\tau$ is, respectively, sufficiently large or small. Note that we refer here to collisions between particles of the particular species under consideration (with its appropriate factor ϑ typically of order one), and any collisional coupling between the different species is represented by the summation for $j \neq s$ incorporated in $\mathbf{s}$. The form (2.67) isolates the term $\{\mathbf{t}\times\mathbf{a}\}$ proportional to the magnetic field **B**, which enables us to obtain $\mathbf{t}$ as an explicit function of $\mathbf{s}$ below—an important outcome even though there are two residual terms involving $\mathbf{t}$ in the generalised deformation tensor defined by (2.68), a point we examine further in the supplementary discussion for a simple ion-electron plasma in Sect. 2.10.

Using tensor identities, we may re-express (2.67) as (cf. Exercises)

$$\mathbf{t} = -\frac{2\mu}{(1+|\mathbf{a}|^2)(1+4|\mathbf{a}|^2)}\,[\,(\mathbf{s} + 2\{\mathbf{s}\times\mathbf{a}\})(1+|\mathbf{a}|^2)$$

$$+\ 6\,(\{\mathbf{s}\cdot\mathbf{aa}\} + 2\{\{\mathbf{s}\cdot\mathbf{aa}\}\times\mathbf{a}\}) + 6\,\mathbf{s} : \mathbf{aa}\{\mathbf{aa}\}\,]. \quad (2.69)$$

The Cartesian representation of this exact form (2.69) appears in Chapman and Cowling [3], who assumed a uniform magnetic field **B**—but we now recognise that (2.69) is an invariant result (i.e. valid in any system of coordinates) without restriction on the magnetic field. Equation (2.67) and likewise the result (2.69) obviously always reduce to $\mathbf{t} = -2\mu\mathbf{s}$ for a neutral species and for every species if there is no magnetic field (i.e. for $\mathbf{a} = 0$), when we recover the shear viscosity form familiar from Sect. 2.2

on identifying $\mathbf{s} = \{\nabla \mathbf{v}\}$. However, the terms involving the vector $\mathbf{a}$ usually dominate for charged particles in the presence of a magnetic field, producing characteristically anisotropic plasma viscosity contributions as follows.

2.9.2 Parallel, Cross and Perpendicular Components

For any species of charged particles in magnetised plasma (i.e. with a permeating magnetic field), typically $|\mathbf{a}| \gg 1$ except in the near neighbourhood of magnetic null points. Thus on noting

$$-\frac{2\mu}{(1+|\mathbf{a}|^2)(1+4|\mathbf{a}|^2)} = \frac{\mu}{2}\left[-|\mathbf{a}|^{-4} + \frac{5}{4}|\mathbf{a}|^{-6} + \cdots\right],$$

the general explicit form (2.69) may consequently be expanded to obtain

$$\mathbf{t} = \mathbf{t}_{\parallel} + \mathbf{t}_{g} + \mathbf{t}_{\perp} + \cdots, \tag{2.70}$$

where the *parallel*, *cross* (or *gyroviscous*) and *perpendicular* components

$$\mathbf{t}_{\parallel} = -3\mu\,\mathbf{s} : \hat{\mathbf{b}}\hat{\mathbf{b}}\{\hat{\mathbf{b}}\hat{\mathbf{b}}\}, \tag{2.71}$$

$$\mathbf{t}_{g} = -\frac{\mu}{|\mathbf{a}|}\left\{\mathbf{s} \times \hat{\mathbf{b}} + 6\{\mathbf{s} \cdot \hat{\mathbf{b}}\hat{\mathbf{b}}\} \times \hat{\mathbf{b}}\right\} \quad \text{and} \tag{2.72}$$

$$\mathbf{t}_{\perp} = -\frac{\mu}{2|\mathbf{a}|^2}\left[\mathbf{s} + 6\{\mathbf{s} \cdot \hat{\mathbf{b}}\hat{\mathbf{b}}\} - \frac{15}{2}\mathbf{s} : \hat{\mathbf{b}}\hat{\mathbf{b}}\{\hat{\mathbf{b}}\hat{\mathbf{b}}\}\right] \tag{2.73}$$

are named with reference to the magnetic field direction $\hat{\mathbf{b}} \equiv \mathbf{B}/|\mathbf{B}|$. The coefficient $\mu/|\mathbf{a}|$ in (2.72) is entirely independent of the collision time τ, and this second-order component is often called the gyroviscous or alternatively the "finite Larmor radius" (FLR) contribution due to an association with the charged particle gyration in the magnetic field. It has been derived elsewhere from the Vlasov ("collisionless Boltzmann") equation, on assuming that the microscopic particles are correlated through their interaction with the magnetic field. However, the leading parallel component (2.71) remains in the low collisional limit (cf. μ as defined above and also Sect. 2.10), and the magnetic field need not be strong to satisfy $|\mathbf{a}| \gg 1$ nor otherwise restricted in any way.

The parallel component (2.71) produces an anisotropic plasma pressure on its own, for the resulting pressure tensor may be written as $\mathbf{p} = p_{\parallel}\hat{\mathbf{b}}\hat{\mathbf{b}} + p_{\perp}\mathbf{I}_{\perp}$ with $p_{\parallel} = p + t_{\parallel,\parallel}$ and $p_{\perp} = p - (1/2)t_{\parallel,\parallel}$ where the projector into the plane locally perpendicular to $\mathbf{B}$ is defined by

$$\mathbf{I}_{\perp} = \mathbf{I} - \hat{\mathbf{b}}\hat{\mathbf{b}}, \tag{2.74}$$

since we have $t_{\parallel,\parallel} \equiv \mathbf{t} : \hat{\mathbf{b}}\hat{\mathbf{b}} = \mathbf{t}_\parallel : \hat{\mathbf{b}}\hat{\mathbf{b}} = -2\mu \mathbf{s} : \hat{\mathbf{b}}\hat{\mathbf{b}}$ and

$$\{\hat{\mathbf{b}}\hat{\mathbf{b}}\} = \hat{\mathbf{b}}\hat{\mathbf{b}} - \frac{1}{3}\mathbf{I} = -\frac{1}{3}(\mathbf{I}_\perp - 2\hat{\mathbf{b}}\hat{\mathbf{b}}). \tag{2.75}$$

Further, on noting $\mathbf{s} : \hat{\mathbf{b}}\hat{\mathbf{b}} = \hat{\mathbf{b}} \cdot \mathbf{s} \cdot \hat{\mathbf{b}}$ we may re-express (2.71) as

$$\mathbf{t}_\parallel = \mu\, \hat{\mathbf{b}} \cdot \mathbf{s} \cdot \hat{\mathbf{b}}\, (\mathbf{I}_\perp - 2\hat{\mathbf{b}}\hat{\mathbf{b}}). \tag{2.76}$$

The right-hand sides of (2.72) and (2.73) may also be expanded to yield alternative expressions for the gyroviscous (cross) and perpendicular components (Exercise Q3):

$$\mathbf{t}_\mathrm{g} = \frac{\mu}{2|\mathbf{a}|}\left[\hat{\mathbf{b}} \times \mathbf{s} \cdot \mathbf{I}_\perp - \mathbf{I}_\perp \cdot \mathbf{s} \times \hat{\mathbf{b}} + 4\left(\hat{\mathbf{b}} \times \mathbf{s} \cdot \hat{\mathbf{b}}\hat{\mathbf{b}} - \hat{\mathbf{b}}\hat{\mathbf{b}} \cdot \mathbf{s} \times \hat{\mathbf{b}}\right)\right], \tag{2.77}$$

$$\mathbf{t}_\perp = -\frac{\mu}{2|\mathbf{a}|^2}\left[\mathbf{I}_\perp \cdot \mathbf{s} \cdot \mathbf{I}_\perp + \frac{1}{2}\hat{\mathbf{b}} \cdot \mathbf{s} \cdot \hat{\mathbf{b}}\ \mathbf{I}_\perp + 4\left(\mathbf{I}_\perp \cdot \mathbf{s} \cdot \hat{\mathbf{b}}\hat{\mathbf{b}} + \hat{\mathbf{b}}\hat{\mathbf{b}} \cdot \mathbf{s} \cdot \mathbf{I}_\perp\right)\right]. \tag{2.78}$$

In summary, except in the near neighbourhood of a magnetic null, in a magnetic field the components (2.71)–(2.73)—or their equivalents (2.76)–(2.78)—provide the appropriate representation for the traceless pressure tensor component $\mathbf{t}_s$ for either ions or electrons, in any coordinate system.[8]

Exercises

(Q1) If $\mathbf{t}$ is a symmetric traceless dyadic and $\mathbf{a}$ is a vector, show that

$$2\{\{\mathbf{t} \times \mathbf{a}\} \times \mathbf{a}\} = 3\{\mathbf{t} \cdot \mathbf{aa}\} - 2|\mathbf{a}|^2\mathbf{t},$$

$$6\{\mathbf{t} \cdot \mathbf{aa}\} \cdot \mathbf{a} = 3\,\mathbf{t} \cdot \mathbf{a}\,|\mathbf{a}|^2 + \mathbf{a}\,\mathbf{t} : \mathbf{aa}$$

and

$$\{\{\mathbf{t} \times \mathbf{a}\} \cdot \mathbf{aa}\} = \{\mathbf{t} \cdot \mathbf{aa}\} \times \mathbf{a}.$$

(Q2) Take a post-cross product of (2.67) with $\mathbf{a}$, followed by the bracket operator defined in the footnote in Sect. 2.2, to obtain

[8] The result (2.69) was originally obtained by B.S. Liley, and the successive terms (2.71)–(2.73) for a sufficiently large magnetic field (when $|\mathbf{a}| \gg 1$) were identified by R.J. Hosking and G.M. Marinoff (*Plasma Physics* **15**, 327–341, 1971). The alternative forms (2.76)–(2.78) were presented by J.D. Callen, W.X. Qu, K.D. Siebert, B.A. Carreras, K.C. Shaing and D.A. Spong (*Plasma Physics and Controlled Nuclear Fusion Research*, Volume II, IAEA, Vienna, 1987), and essentially earlier by S.I. Braginskii (*Reviews of Plasma Physics* **1**, 205–311, 1965) in the collisional limit—cf. the traceless component of his pressure tensor. In the absence of collisions, an anisotropic plasma pressure $\mathbf{p} = p_\parallel \hat{\mathbf{b}}\hat{\mathbf{b}} + p_\perp \mathbf{I}_\perp$ was first discussed by G.E. Chew, M.L. Goldberger and F.E. Low (*Proceedings of the Royal Society of London* **A 236**, 112–118, 1956).

$$-4\mu\{\mathbf{s} \times \mathbf{a}\} + 4\{\{\mathbf{t} \times \mathbf{a}\} \times \mathbf{a}\} = 2\{\mathbf{t} \times \mathbf{a}\} = 2\mu\mathbf{s} + \mathbf{t} \quad (*).$$

Then use the first identity above, to obtain

$$-4\mu\{\mathbf{s} \times \mathbf{a}\} + 6\{\mathbf{t} \cdot \mathbf{aa}\} - 2\mu\mathbf{s} = (1 + 4|\mathbf{a}|^2)\mathbf{t};$$

and remove the dot product $\mathbf{t} \cdot \mathbf{a}$ using a post-dot product of (*) with $\mathbf{a}$ and the second identity, to obtain

$$(1 + |\mathbf{a}|^2)\,\mathbf{t} \cdot \mathbf{a} = -4\mu\{\mathbf{s} \times \mathbf{a}\} \cdot \mathbf{a} - 2\mu(\mathbf{s} : \mathbf{aaa} + \mathbf{s} \cdot \mathbf{a})$$

since $\mathbf{t} : \mathbf{aa} = -2\mu\mathbf{s} : \mathbf{aa}$ also follows from (2.67). Finally, eliminate $\mathbf{t} \cdot \mathbf{a}$ between these last two equations with the help of the third identity above, to obtain the result (2.69).

(Q3) Expand the original expressions (2.72) and (2.73) for the components $\mathbf{t}_g$ and $\mathbf{t}_\perp$, to obtain the respective alternative forms (2.77) and (2.78).

(Q4) Deduce the representations for the successive components $\mathbf{t}_\parallel$, $\mathbf{t}_g$ and $\mathbf{t}_\perp$ in a local orthonormal coordinate system where $\hat{\mathbf{b}} = \hat{\mathbf{e}}_3$ in terms of the entries in the tensor $\mathbf{s}$, recalling that $\mathbf{s}$ is also symmetric and traceless.
(Analogous representations were obtained by Braginskii for the traceless component of his pressure tensor, where the peculiar velocity is identified with the mean species velocity—cf. the Exercises in Sect. 2.6, and for example Ref. [4]. The distinction between fluid and plasma viscosity is also emphasised in the celebrated L.D. Landau & E.M. Lifshitz *Course of Theoretical Physics* (Elsevier)—cf. Volume 10 by L.P. Pitaevskii and E.M. Lifschitz in particular.)

2.10 Parallel Viscosity in an Ion-Electron Plasma

Let us now specifically consider the leading parallel component $\mathbf{t}_\parallel$ in the case of a magnetised fully ionised plasma consisting of electrons and only one ion species (usually protons). There are then useful simplifications due to the much smaller electron mass relative to the ion mass ($m_e \ll m_i$). Collisional coupling between the ion and electron species is negligible, so the collision integral for the ions has the simple form $C_i(m_i\{\mathbf{ww}\}) \simeq -(\vartheta_{ii}/\tau_{ii})\mathbf{t}_i$. Further, the viscosity coefficient μ is much greater for the ions than for the electrons when the electron and ion temperatures are comparable, so the dominant $\nabla \cdot \mathbf{t}$ contribution to render $\nabla \cdot \mathbf{p}$ in the macroscopic equation of motion (2.56) is due to the ion species. The ion velocity $\mathbf{v}_i$ is also approximately the mass-weighted mean velocity $\mathbf{v}$ (i.e. the ion diffusion velocity

$\mathbf{u}_i \simeq 0$),[9] so we may conveniently continue to entirely omit subscripts when discussing the ion species but later include the subscript e to specify the electrons.

The $\hat{\mathbf{b}}\hat{\mathbf{b}}$ projection of the basic equation (2.66) for $\mathbf{t}_s$ produces the equation defining the scalar part $(3/2)t_{\parallel,\parallel} = p_\parallel - p_\perp$ of the parallel viscosity tensor $\mathbf{t}_\parallel$ defined in (2.71) for either the ions or the electrons, since $t_{\parallel,\parallel} \equiv \mathbf{t} : \hat{\mathbf{b}}\hat{\mathbf{b}} = \mathbf{t}_\parallel : \hat{\mathbf{b}}\hat{\mathbf{b}} = -2\mu\,\mathbf{s} : \hat{\mathbf{b}}\hat{\mathbf{b}}$ as we noted in the previous subsection. Let us first consider the usually dominant ion contribution, where since $\mathbf{u} \simeq 0$ we have

$$\frac{dt_{\parallel,\parallel}}{dt} + \frac{\vartheta}{\tau} t_{\parallel,\parallel} + \frac{4}{3} t_{\parallel,\parallel} \nabla \cdot \mathbf{v} + t_{\parallel,\parallel}\, \hat{\mathbf{b}}\hat{\mathbf{b}} : \nabla \mathbf{v} + 2p\,\{\nabla \mathbf{v}\} : \hat{\mathbf{b}}\hat{\mathbf{b}} = 0 \tag{2.79}$$

on ignoring the $\nabla \cdot \mathbf{h}$ term, which corresponds to neglecting thermal effects under the Grad approximation—cf. (2.65). Equation (2.79) could be invoked directly in simulations involving parallel ion viscosity, but let us proceed to extract some approximations for $t_{\parallel,\parallel}$ by considering the various terms in this equation.

In the collisional Braginskii limit (i.e. for sufficiently small τ), balancing the second term with the last term on the left-hand side of (2.79) yields $t_{\parallel,\parallel} = -2\mu\{\nabla \mathbf{v}\} : \hat{\mathbf{b}}\hat{\mathbf{b}}$ where $\mu = p\,\tau/\vartheta$. When ion–ion collisions are rarer (i.e. at larger τ), the other $t_{\parallel,\parallel}$ terms in (2.79) may also be significant. In fast evolving system phases for example—i.e. at large Strouhal numbers $\omega L/U$, where ω denotes the characteristic macroscopic frequency and L and U a characteristic length and velocity—the time derivative term may be important. Indeed, when the first two terms in $t_{\parallel,\parallel}$ jointly balance the last term on the left-hand side of (2.79), the appropriate coefficient is $\mu = p/(\omega + \vartheta/\tau) = p\,\tau/(\omega\tau + \vartheta)$ in $t_{\parallel,\parallel} = -2\mu\{\nabla \mathbf{v}\} : \hat{\mathbf{b}}\hat{\mathbf{b}}$ as defined in Sect. 2.9. Consequently, in either case

$$t_{\parallel,\parallel} = -2\mu\{\nabla \mathbf{v}\} : \hat{\mathbf{b}}\hat{\mathbf{b}} = -2\mu \left[\hat{\mathbf{b}} \cdot \nabla(\mathbf{v} \cdot \hat{\mathbf{b}}) - \mathbf{v} \cdot (\hat{\mathbf{b}} \cdot \nabla \hat{\mathbf{b}}) - \frac{1}{3} \nabla \cdot \mathbf{v} \right], \tag{2.80}$$

where it is notable that the second term on the right-hand side of (2.80) proportional to the magnetic field curvature

$$\boldsymbol{\kappa} = \hat{\mathbf{b}} \cdot \nabla \hat{\mathbf{b}} \tag{2.81}$$

supplements the two velocity gradient terms. Moreover, further curvature terms result on then taking the divergence, to express $\nabla \cdot \mathbf{t}_\parallel$ with $\mathbf{t}_\parallel = (3/2)t_{\parallel,\parallel}\,\{\hat{\mathbf{b}}\hat{\mathbf{b}}\}$ in the macroscopic equation of motion (2.56).

In brief, so far we have proceeded beyond the collisional Braginskii limit for the ions to allow for the contribution $dt_{\parallel,\parallel}/dt$ in (2.79) from the time derivative term in the basic equation (2.66). Reference back to (2.68) shows that representing $\mathbf{t}_\parallel$ with $t_{\parallel,\parallel}$ rendered by (2.80) corresponds to approximating the generalised rate of

---

[9]The definition of all *ion* field quantities following Braginskii can therefore be considered to coincide with ours in an ion-electron plasma. On the other hand, if the electron temperature is so much higher than the ion temperature such that the electron viscosity contribution remains of interest, recall that the Braginskii pressure tensor for the electrons is $\mathbf{p}_e - m_e n_e \mathbf{u}_e \mathbf{u}_e$ in our notation—cf. Exercise (Q1) of Sect. 2.6.

deformation tensor $\mathbf{s}$ with $\{\nabla \mathbf{v}\}$. This velocity gradient approximation $\mathbf{s} \simeq \{\nabla \mathbf{v}\}$, previously familiar in the fluid mechanics context (cf. Sect. 2.2) but that we now find arises more generally from the second term in (2.68), has also been widely adopted for the ion species in plasmas. However, there is a further generalisation due to the residual terms in (2.68) involving $\mathbf{t}$ that we previously noted in Sect. 2.9, which produce additional contributions from the third and fourth terms involving $t_{\|,\|}$ on the left-hand side of Eq. (2.79) for the ions—i.e. we find—cf. Exercise (Q2):

$$t_{\|,\|} = -2p\tau\{\nabla \mathbf{v}\} : \hat{\mathbf{b}}\hat{\mathbf{b}} \left[\vartheta + \tau \left(\omega + \frac{4}{3} \nabla \cdot \mathbf{v} + \hat{\mathbf{b}}\hat{\mathbf{b}} : \nabla \mathbf{v} \right) \right]^{-1}, \tag{2.82}$$

involving additional contributions from the $\nabla \mathbf{v}$ dyadic that could be significant when ion–ion collisions are rare (when the ion–ion collision time τ is relatively large).

In the occasional application where the electron temperature is so very much higher than the ion temperature such that the electron viscosity contribution cannot be ignored, we observe that additional terms from $\{\mathbf{f}_e\, \mathbf{u}_e\}$ due to relative inter-species diffusion arise in the basic equation (2.66) for the electrons. For example, in an ion-electron plasma where the ions are singly charged such that $n_e = n_i = n$ for quasi-neutrality, on retaining only the electron–electron collision term (and ignoring the $\nabla \cdot \mathbf{h}_e$ term) the corresponding $\hat{\mathbf{b}}\hat{\mathbf{b}}$ projection is

$$\begin{aligned} &\frac{dt_{\|,\|e}}{dt} + \frac{\vartheta_{ee}}{\tau_{ee}} t_{\|,\|e} + \frac{4}{3} t_{\|,\|e} \nabla \cdot \mathbf{v}_e + t_{\|,\|e}\, \hat{\mathbf{b}}\hat{\mathbf{b}} : \nabla \mathbf{v}_e \\ &\quad + 2p_e\{\nabla \mathbf{v}_e\} : \hat{\mathbf{b}}\hat{\mathbf{b}} - 2\mathbf{E} \cdot \hat{\mathbf{b}}\, \mathbf{j} \cdot \hat{\mathbf{b}} + \frac{2}{3} \mathbf{j} \cdot (\mathbf{E} + \mathbf{v}_e \times \mathbf{B}) = 0 \end{aligned} \tag{2.83}$$

where the electron velocity $\mathbf{v}_e \simeq \mathbf{v} - \mathbf{j}/(ne)$ and $(e/m_e)(\mathbf{E} + \mathbf{v}_e \times \mathbf{B})$ is a notable contribution in $\mathbf{f}_e$. Thus there are further terms to incorporate in the generalised rate of deformation tensor for the electrons, in addition to the term $2p_e\, \{\nabla \mathbf{v}_e\} : \hat{\mathbf{b}}\hat{\mathbf{b}}$ with form similar to that arising for the ions—viz. those arising from the inter species contribution, the term involving the product of the parallel electric field $E_\| = \mathbf{E} \cdot \hat{\mathbf{b}}$ and the parallel current density $j_\| = \mathbf{j} \cdot \hat{\mathbf{b}}$, and perhaps also the last electromagnetic heating term. Incidentally, the possible inclusion of electron viscosity therefore provides one example where the electric field would need to be retained, whereas usually both $\mathbf{E}$ and $\mathbf{j}$ are readily eliminated in MHD by invoking two electromagnetic equations (cf. Chap. 5).

Finally, if there are significant gradients in the thermal flux for either species such that the basic equation (2.51) for the heat flux vector $\mathbf{R}_s = \mathbf{q}_s - (5/2)p_s\mathbf{u}_s$ should be considered together with the lower level basic equations (2.46)–(2.49), we observe that the $\nabla \cdot \mathbf{h}_s$ term then appropriately re-enters Eqs. (2.79) and (2.83) to produce associated thermal contributions to $\mathbf{t}_\|$. Thus for the Grad approximation in particular, from (2.65) we have $(\nabla \cdot \mathbf{h}) : \hat{\mathbf{b}}\hat{\mathbf{b}} = 4/5\{\nabla \mathbf{q}\} : \hat{\mathbf{b}}\hat{\mathbf{b}}$ and hence

$$t_{\parallel,\parallel} = -2p\tau\{\nabla\mathbf{v} + \frac{2}{5p}\nabla\mathbf{q}\} : \hat{\mathbf{b}}\hat{\mathbf{b}}\left[\vartheta + \tau\left(\omega + \frac{4}{3}\nabla\cdot\mathbf{v} + \hat{\mathbf{b}}\hat{\mathbf{b}} : \nabla\mathbf{v}\right)\right]^{-1} \tag{2.84}$$

as the relevant modification of (2.80) for the ions, with the subscript omitted once again—and there would be a similar modification for the electrons.

Exercises

(Q1) Deduce Eq. (2.79) in $t_{\parallel,\parallel}$ for the ions from (2.66), on neglecting the $\nabla\cdot\mathbf{h}$ term and assuming negligible collisional coupling between the ions and electrons—i.e. from the relevant moment equation (2.49) with the collisional integral $\mathcal{C}(m\{\mathbf{ww}\}) = (\vartheta/\tau)\mathbf{t}$, on omitting the subscript.

(Q2) Retain the two terms involving $\mathbf{t}$ in (2.68) to derive

$$t_{\parallel,\parallel} = -2p\,\tau\{\nabla\mathbf{v}\} : \hat{\mathbf{b}}\hat{\mathbf{b}}\left[\vartheta + \tau\left(\omega + \frac{4}{3}\nabla\cdot\mathbf{v} + \hat{\mathbf{b}}\hat{\mathbf{b}} : \nabla\mathbf{v}\right)\right]^{-1}$$

for the scalar part in the parallel ion viscosity component $\mathbf{t}_{\parallel} = (3/2)t_{\parallel\parallel}\{\hat{\mathbf{b}}\hat{\mathbf{b}}\}$ in a simple ion-electron plasma, consistent with retaining all of the terms in (2.79).

(Q3) Noting $\rho_e = n_e m_e$ and now choosing to omit subscript e, show that

$$-2\,\rho\left[\frac{e}{m}\,(\mathbf{E} + \mathbf{v}\times\mathbf{B})\,\frac{\mathbf{j}}{ne}\right] : \hat{\mathbf{b}}\hat{\mathbf{b}} = -2\,\mathbf{E}\cdot\hat{\mathbf{b}}\,\mathbf{j}\cdot\hat{\mathbf{b}} + \frac{2}{3}\,\mathbf{j}\cdot(\mathbf{E} + \mathbf{v}\times\mathbf{B}),$$

representing some of the additional electromagnetic terms arising in equation (2.83) for the electrons.

(Q4) It is easy to expand $\nabla\cdot\mathbf{t}_{\parallel} = \nabla\cdot[(3/2)t_{\parallel,\parallel}\{\hat{\mathbf{b}}\hat{\mathbf{b}}\}]$ to express the predominant ion viscosity contribution to the macroscopic equation of motion (2.56). It takes much more effort to derive the lower order gyroviscous contribution $\nabla\cdot\mathbf{t}_g$, which leads to a consequence known as "gyroviscous cancellation" in the ion momentum equation and thus in (2.56) for an ion-electron plasma. Although this cancellation in the advective derivative is now mainly of historical interest, show that the gyroviscous component for the ions may be rewritten as

$$\mathbf{t}_g = \frac{\mu}{2|\mathbf{a}|}\left[\hat{\mathbf{b}}\times\mathbf{s}\cdot\left(\mathsf{I} + 3\hat{\mathbf{b}}\hat{\mathbf{b}}\right) - \left(\mathsf{I} + 3\hat{\mathbf{b}}\hat{\mathbf{b}}\right)\cdot\mathbf{s}\times\hat{\mathbf{b}}\right],$$

and then assume the velocity gradient approximation $\mathbf{s} = \{\nabla\mathbf{v}\}$ to obtain

$$\begin{aligned}\nabla\cdot\mathbf{t}_g = &-\rho(\mathbf{v}_*\cdot\nabla)\mathbf{v} - \nabla\chi + 2\,\mathbf{B}\cdot\nabla(B^{-1}\mathbf{A})\\ &-\nabla\times\left[\frac{\mu}{|\mathbf{a}|}\left(\hat{\mathbf{b}}\cdot\nabla\mathbf{v} + \frac{1}{2}(\nabla\cdot\mathbf{v} - 3\,\hat{\mathbf{b}}\hat{\mathbf{b}} : \nabla\mathbf{v})\hat{\mathbf{b}}\right)\right]\end{aligned}$$

where three entities introduced here (with $\omega = \nabla \times \mathbf{v}$ the vorticity vector) are

$$\mathbf{v}_* = -\frac{1}{\rho}\nabla \times \left(\frac{\mu}{|\mathbf{a}|}\hat{\mathbf{b}}\right), \quad \chi = \frac{\mu}{|\mathbf{a}|}\hat{\mathbf{b}}\cdot\omega, \quad \mathbf{A} = \frac{\mu}{|\mathbf{a}|}\hat{\mathbf{b}} \times (3\,\hat{\mathbf{b}}\cdot\nabla\mathbf{v} + \hat{\mathbf{b}}\times\omega) + \chi\hat{\mathbf{b}}.$$

[The first term on the right-hand side of this expression for $\nabla \cdot \mathbf{t}_g$, involving the amended "magnetisation velocity" $\mathbf{v}_*$, produces the legendary cancellation.]

Hint: From the identity $\partial v_j/\partial x_i = \partial v_i/\partial x_j + \epsilon_{ijk}\omega_k$, first verify that

$$\begin{aligned}&\hat{\mathbf{b}} \times [\nabla\mathbf{v} + (\nabla\mathbf{v})^T] \cdot (\mathsf{I} + 3\hat{\mathbf{b}}\hat{\mathbf{b}}) \\ &\quad = 2\,\hat{\mathbf{b}} \times \nabla\mathbf{v} - \hat{\mathbf{b}}\cdot\omega\mathsf{I} + [6\hat{\mathbf{b}} \times (\hat{\mathbf{b}}\cdot\nabla\mathbf{v}) - 2\omega + 3\hat{\mathbf{b}}\cdot\omega\hat{\mathbf{b}}]\,\hat{\mathbf{b}}\,,\end{aligned}$$

and then also consider the identity

$$\frac{\mu}{2|\mathbf{a}|}\,\hat{\mathbf{b}} \times \nabla\mathbf{v} = -\nabla \times \left(\frac{\mu}{2|\mathbf{a}|}\hat{\mathbf{b}}\,\mathbf{v}\right) + \left[\nabla \times \left(\frac{\mu}{2|\mathbf{a}|}\hat{\mathbf{b}}\right)\right]\mathbf{v}.$$

2.11 Heat Flux*

The derivation of an explicit form for the heat flux vector $\mathbf{R}_s$ is more straightforward than that for $\mathbf{t}_s$, although its basic equation may appear more cumbersome. Thus adopting the collision integral representation (2.61) and again suppressing the subscript s where there is no chance of confusion, the relevant basic equation of change (2.51) is

$$\begin{aligned}&\frac{d\mathbf{R}}{dt} + \frac{1}{\tau}\mathbf{R} + \mathbf{R}\nabla\cdot\mathbf{v} + \nabla\cdot\mathsf{L} - \mathbf{f}\cdot\mathbf{t} - \frac{e}{m}\mathbf{R}\times\mathbf{B} + \frac{5}{2}\frac{k}{m}\mathsf{p}\cdot\nabla T \\ &\quad + \frac{5}{2}nk\mathbf{u}\frac{dT}{dt} + \mathbf{R}\cdot\nabla\mathbf{v} + \mathsf{H} : \nabla\mathbf{v} = -\sum_{s\neq j}\frac{1}{\tau_{js}}\mathbf{R}_s - \sum_s \zeta_{js}\rho_s\mathbf{u}_s.\end{aligned} \tag{2.85}$$

Equation (2.85) may be rewritten in a notationally convenient vector form, analogous to the tensor form (2.67) for the traceless component of the pressure tensor $\mathbf{t}$—viz. as

$$\mathbf{R} - \mathbf{R} \times \mathbf{a}_0 = -\lambda\mathbf{d}, \tag{2.86}$$

where the first two terms and the sixth term are represented on the left-hand side, and all the other terms are incorporated in $\mathbf{d}$ on the right-hand side. Thus on writing $d\mathbf{R}/dt = \omega\mathbf{R}$, the thermal conductivity coefficient and the vector

$$\lambda = \frac{5}{2}\frac{k}{m}\frac{p\tau}{\omega\tau + 1} \quad \text{and} \quad \mathbf{a}_0 = \frac{e\mathbf{B}}{m}\frac{\tau}{\omega\tau + 1}$$

introduced here reflect the time derivative and collisional terms. Well-known vector identities then immediately yield the corresponding general explicit form

$$\mathbf{R} = -\frac{\lambda}{1 + |\mathbf{a}|_0^2}(\mathbf{d} + \mathbf{d} \times \mathbf{a}_0 + \mathbf{d} \cdot \mathbf{a}_0\, \mathbf{a}_0), \tag{2.87}$$

which reduces to the classical Fourier law of heat conduction in the limit $\mathbf{a}_0 = 0$.

Once again, the explicit form can be expanded in typical magnetised plasma (where $|\mathbf{a}_0| \gg 1$), except in the neighbourhood of null points in the magnetic field. Thus from (2.87) we obtain

$$\mathbf{R} = \mathbf{R}_{\|} + \mathbf{R}_{g} + \mathbf{R}_{\perp} + \cdots \tag{2.88}$$

in notation analogous to that used for the pressure tensor, where

$$\mathbf{R}_{\|} = -\lambda \mathbf{d} \cdot \hat{\mathbf{b}}\hat{\mathbf{b}}\,, \ \ \mathbf{R}_{g} = -\frac{\lambda}{\mathbf{a}_0}\mathbf{d} \times \hat{\mathbf{b}} \ \text{ and } \ \mathbf{R}_{\perp} = -\frac{\lambda}{\mathbf{a}_0^2}\mathbf{d} \cdot \mathbf{I}_{\perp}. \tag{2.89}$$

Further, in an ion-electron plasma the negligible collisional coupling between the two species again simplifies the relevant collision integral. Thus in this context we have $\mathcal{C}_s(\frac{1}{2}m_s(w^2 - 5kT_s/m_s)\mathbf{w}) = -(\mathbf{R}_s/\tau_{ss} + \zeta_s \rho_s \mathbf{u}_s)$, and we may proceed to consider the leading component $\mathbf{R}_{\|}$ in conjunction with $\mathbf{t}_{\|}$. For the ions $\mathbf{R} \simeq \mathbf{q}$, so on again omitting the subscript and noting that $\hat{\mathbf{b}} \cdot \mathbf{p} = p_{\|}\hat{\mathbf{b}}$ where $p_{\|} = p + t_{\|,\|}$, the $\hat{\mathbf{b}}$ projection of the basic equation of change (2.85) for $q_{\|} = R_{\|} = \mathbf{R} \cdot \hat{\mathbf{b}}$ is

$$\begin{aligned} &\frac{dq_{\|}}{dt} + \frac{1}{\tau}q_{\|} + q_{\|}\nabla\cdot\mathbf{v} + q_{\|}\hat{\mathbf{b}}\hat{\mathbf{b}} : \nabla\mathbf{v} - t_{\|,\|}\mathbf{f} \cdot \hat{\mathbf{b}} \\ &\quad + \frac{5}{2}p_{\|}\frac{k}{m}\hat{\mathbf{b}} \cdot \nabla T + \hat{\mathbf{b}} \cdot (\nabla\cdot\mathsf{L}) + \hat{\mathbf{b}} \cdot \mathsf{H} : \nabla\mathbf{v} = 0. \end{aligned} \tag{2.90}$$

It is conceivable that the last two terms on the left-hand side are negligible, such that the temperature gradient term familiar from the classical Fourier law predominates and therefore determines the heat flux in the direction of the magnetic field. On the other hand, we may invoke Grad's approximation such that

$$\hat{\mathbf{b}} \cdot \nabla\cdot\mathsf{L} = \frac{kT}{m}\nabla\cdot(t_{\|,\|}\,\hat{\mathbf{b}}) + \frac{kT}{m}\,t_{\|,\|}\,\hat{\mathbf{b}}\hat{\mathbf{b}} : \nabla\hat{\mathbf{b}} + t_{\|,\|}\frac{k}{m}\hat{\mathbf{b}} \cdot \nabla T$$

$$\text{and} \quad \hat{\mathbf{b}} \cdot \mathsf{H} : \nabla\mathbf{v} = \frac{2}{5}q_{\|}\,(2\hat{\mathbf{b}}\hat{\mathbf{b}} : \nabla\mathbf{v} + \nabla\cdot\mathbf{v}),$$

from (2.65), associated with $t_{\|,\|}$ given by (2.84). For the electrons, there are again additional terms to include due to inter-species diffusion ($\mathbf{u}_e \neq 0$).

2.12 Magnetic Induction

As indicated in Sect. 2.7, any dynamical macroscopic model includes equations of continuity and motion, supplemented by a barotropic equation of state (i.e. a relation between the pressure p and density ρ) and a constitutive relation for either the fluid or the plasma pressure tensor $\mathbf{p}$ rendered in Sects. 2.2 and 2.9, respectively. Let us also recall that this is a complete (closed) system of equations unless the medium carries an electric current, when the electromagnetic body force term $\mathbf{j} \times \mathbf{B}$ in the equation of motion becomes significant. As discussed further in Sect. 5.3, the pre-Maxwell electromagnetic equation

$$\nabla \times \mathbf{B} = \mu_0 \mathbf{j} \qquad (\mu_0 \text{ a constant}) \tag{2.91}$$

may then be adopted to eliminate $\mathbf{j}$, but another equation involving the magnetic field variable $\mathbf{B}$ is obviously required to complete the system of equations. This equation is obtained from another constitutive relation known as *Ohm's law* for a conducting fluid, or a *generalised Ohm's law* in the case of plasma.

The usual form of Ohm's law for an electrically conducting fluid is

$$\mathbf{E} + \mathbf{v} \times \mathbf{B} = \frac{1}{\sigma}\,\mathbf{j}, \tag{2.92}$$

where σ here denotes the *conductivity coefficient*, inversely related to the *resistivity coefficient* $\eta \equiv 1/(\mu_0\sigma)$. The terms on the left-hand side of (2.92) are often referred to as ideal, as they remain in the limit of zero resistivity, when σ is infinite. Both the current density $\mathbf{j}$ and the electric field $\mathbf{E}$ may be eliminated using (2.91) and the additional electromagnetic equation

$$\nabla \times \mathbf{E} = -\frac{\partial \mathbf{B}}{\partial t}, \tag{2.93}$$

respectively (cf. Chap. 5). Equation (2.92) yields the equation of *magnetic induction* as

$$\frac{\partial \mathbf{B}}{\partial t} = \nabla \times (\mathbf{v} \times \mathbf{B}) - \nabla \times (\eta \nabla \times \mathbf{B}), \tag{2.94}$$

although another term due to the Hall effect may also arise if a generalised Ohm's law is used [cf. (2.100)].

For a plasma, a generalised Ohm's law is often identified as the second level of another hierarchy of moment equations, obtained by multiplying the Boltzmann equation (2.41) by e_s/m_s beforehand. Thus there is the charge conservation equation

$$\frac{\partial q}{\partial t} + \nabla \cdot \mathbf{j} = 0, \tag{2.95}$$

at the first level, where q denotes the charge density (cf. also Chap. 5); and at the second level,

$$\mathbf{E}+\mathbf{v}_e \times \mathbf{B} = \frac{1}{\sigma}\,\mathbf{j} - \frac{1}{n_e e}\nabla p_e + \cdots \tag{2.96}$$

where the subscript e again denotes electron field quantities.[10] In passing, we note that on introducing the unit vector $\hat{\mathbf{b}} = \mathbf{B}/|\mathbf{B}|$ in the direction of the magnetic field, Eq. (2.96) implies that the component of the electron velocity perpendicular to the magnetic field is

$$\mathbf{v}_{e\perp} = \frac{\mathbf{E}\times\mathbf{B}}{B^2} - \frac{1}{n_e e B}\,\hat{\mathbf{b}}\times\nabla p_e + \cdots . \tag{2.97}$$

The terms on the right-hand side of (2.97) may be identified from quite elementary physical considerations as "polarisation", "diamagnetic", etc., contributions—although plasma physicists may also use such terminology to identify analogous contributions in the perpendicular velocity of the much more massive positive ions in a plasma, derived from the corresponding ion momentum equation.

In a neutral plasma containing an equal number of single-charge ions and electrons ($n_i = n_e = n$), on noting the electron velocity is $\mathbf{v}_e \simeq \mathbf{v} - \mathbf{j}/(ne)$ we may rewrite Eq. (2.96) as

$$\mathbf{E}+\mathbf{v}\times\mathbf{B} = \frac{1}{\sigma}\,\mathbf{j} + \frac{1}{ne}\,\mathbf{j}\times\mathbf{B} - \frac{1}{ne}\nabla p_e + \cdots , \tag{2.98}$$

where the second term $(1/ne)\,\mathbf{j}\times\mathbf{B}$ on the right-hand side is due to the *Hall effect* corresponding to the relative inter-species. The finite conductivity or *resistive* term arising when the conductivity coefficient σ is finite, and also the Hall term or any other term retained on the right-hand side of (2.98), may be referred to as *non-ideal*.

The current density $\mathbf{j}$ and electric field $\mathbf{E}$ appearing in (2.98) can be eliminated using the electromagnetic Eqs. (2.91) and (2.93). The electron pressure term in (2.96) or (2.98) has often been neglected—indeed, it has been argued that when the electrons are largely isothermal ($T_e \simeq$ constant) we have

$$\nabla\times\left(\frac{1}{ne}\nabla p_e\right) \simeq -\frac{kT_e}{n^2 e}\nabla n\times\nabla n = 0, \tag{2.99}$$

on writing $p_e = nkT_e$ (where the subscript is again suppressed on the electron number density and k is the Boltzmann constant). Retaining the other four terms in (2.98) then produces the magnetic induction equation

[10]Most authors have associated the generalised Ohm's law with the electron momentum equation in some way. Another viewpoint that produces the terms of interest as in (2.96), but with an alternative identification of the conductivity coefficient σ, was proposed by H.S. Green (*Physics of Fluids* **2**, 341–349, 1959).

$$\frac{\partial \mathbf{B}}{\partial t} = \nabla \times (\mathbf{v} \times \mathbf{B}) - \nabla \times (\eta \nabla \times \mathbf{B}) - \nabla \times \left(\frac{1}{\mu_0 n e} (\nabla \times \mathbf{B}) \times \mathbf{B} \right), \tag{2.100}$$

with the resistivity coefficient $\eta = 1/(\mu_0 \sigma)$ and $1/(\mu_0 n e)$ the Hall coefficient. However, the electron pressure term would necessarily be retained if we were to account for electron temperature gradients.

On incorporating $\mathsf{p} = p\,\mathsf{I}$ and invoking (2.91), Eqs. (2.55), (2.56), (2.59) and the appropriate equation of magnetic induction constitute a complete system in the variables $\{\rho, \mathbf{v}, p, \mathbf{B}\}$. The ideal MHD model corresponds to omitting all non-ideal terms from the equation of magnetic induction, as discussed further in Sect. 5.7 in the plasma context. Resistive MHD corresponds to the retention of the term $\nabla \times (\eta \nabla \times \mathbf{B})$, and when the Hall term is retained we have a Hall MHD model (ideal or resistive)—cf. Sect. 5.5 on conducting liquids, and Sect. 5.14 and Sect. 5.15 in the plasma context. All of these models may be modified further, such as when the plasma viscosity is important—using the pressure tensor discussed in Sect. 2.2 for a conducting fluid, or as in Sect. 2.9 for a plasma.

Bibliography

1. R. Balescu, *Transport Processes in Plasmas, vol. 1, Classical Transport Theory, vol. 2, Neoclassical Transport* (North Holland, 1989). (Provides a unified treatment of plasma transport, with underlying concepts from kinetic theory and physics discussed, making the two volumes largely self-contained)
2. C. Borgnakke, R.E. Sonntag, *Fundamentals of Thermodynamics*, 8th edn. (Wiley, New York, 2013). (Discusses the principles of thermodynamics, with notable applications to diverse fields and problem solving)
3. S. Chapman, T.G. Cowling, *The Mathematical Theory of Non-Uniform Gases*, 3rd edn. (Cambridge University Press, Cambridge, 1991). (Classic book providing a detailed development of the Maxwell-Boltzmann approach to kinetic theory and the resulting macroscopic equations, with ionised gases in electric and magnetic fields considered in the later chapters)
4. H. Goedbloed, S. Poedts, *Principles of Magnetohydrodynamics* (Cambridge University Press, Cambridge, 2004). (Graduate textbook emphasising the physical background and detailed development of the ideal MHD model, in the context of thermonuclear fusion research and plasma astrophysics)
5. R.D. Hazeltine, J.D. Meiss, *Plasma Confinement* (Dover, New York, 2003). (Graduate textbook that presents fundamental theory for the magnetic confinement of plasma at or near thermonuclear conditions)
6. Yu.L Klimontovich, *Kinetic Theory of Non-Ideal Gases and Non-Ideal Plasmas* (Pergamon Press, Oxford, 1982). (Specialist exposition by a respected researcher in kinetic theory, treating strongly coupled non-ideal gases in the first part and fully ionised non-ideal plasmas in the second, followed by a third part on quantum aspects)
7. H. Lamb, *Hydrodynamics*, 6th edn. (Cambridge University Press, Cambridge, 1993). (The remarkable classic text, noted for its excellent detailed presentation of the fundamentals of fluid mechanics and various solution methods)
8. W. Prager, *An Introduction to Mechanics of Continua* (Dover, New York, 2004). (Originally published in 1961, this now classic textbook systematically addresses the foundations of fluid mechanics, plasticity and elasticity)

9. I.P. Shkarovsky, T.W. Johnston, M.P. Bachynski, *The Particle Kinetics of Plasmas* (Addison-Wesley, Reading, 1966). (A thorough treatment of the kinetic theory of plasmas, with some interesting material and results not readily accessible elsewhere)
10. L. Spitzer, *Physics of Fully Ionized Gases*, 2nd edn. (Dover, New York, 2006). (Second edition of a slim volume on plasmas and fusion research by one of the founders of the field, emphasising the macroscopic equations but with an appendix on kinetic theory)

Chapter 3
Basic Fluid Dynamics

Although the classical ideal fluid model entirely neglects fluid viscosity, it nevertheless describes some features in certain realistic flows or flow regions, and it is often applicable to wave motion as discussed in the next chapter. The inherent nonlinearity of this ideal model was addressed in remarkable ways by many famous mathematicians, who developed various concepts and results that still remain important. First integrals of the inviscid equation of motion (known as Bernoulli equations), aspects of vorticity, and potential theory for incompressible irrotational flow are landmarks of the classical ideal theory. An ordering procedure establishes that the incompressibility assumption applies in any subsonic flow, and confirms the relevant Bernoulli equation for the pressure variation in the ideal model. We then observe that the shear viscosity (whether large or small) must be included to account for the drag and enhanced vorticity in flow past an obstacle, and that perturbation or numerical methods are usually required since exact viscous solutions are rare. The chapter concludes with an optional (starred) section as an introduction to some simplified equations of motion in dynamical meteorology and oceanography, with some references for further reading. Some other notable fluid mechanics textbooks and several related sources (on asymptotic and perturbation methods, potential theory and hydrodynamic stability) are listed in the bibliography for this chapter.

3.1 Fundamental Model

As discussed in Chap. 2, fluid motion is traditionally described by no more than five macroscopic field variables—viz. the fluid density $\rho(\mathbf{r}, t)$, the fluid velocity $\mathbf{v}(\mathbf{r}, t)$, and the fluid pressure $p(\mathbf{r}, t)$. Thus a complete set of macroscopic equations in the Eulerian description consists of the continuity equation

$$\frac{d\rho}{dt} + \rho\nabla\cdot\mathbf{v} = 0, \tag{3.1}$$

R.J. Hosking and R.L. Dewar, *Fundamental Fluid Mechanics and Magnetohydrodynamics*, DOI 10.1007/978-981-287-600-3_3

the equation of motion including gravity

$$\rho \frac{d\mathbf{v}}{dt} + \nabla \cdot \mathsf{p} = \rho \mathbf{g}, \tag{3.2}$$

and the adiabatic equation of state

$$\frac{d}{dt}(p\rho^{-\gamma}) = 0 \quad \text{or} \quad \frac{dp}{dt} + \gamma p \nabla \cdot \mathbf{v} = 0, \tag{3.3}$$

with the Newtonian fluid pressure tensor

$$\mathsf{p} = p\,\mathsf{I} - 2\mu\,\{\nabla \mathbf{v}\} \tag{3.4}$$

where the shear viscosity is explicitly retained [9]. Incidentally, we recall from Sect. 2.2 that the field p is interpreted as the total hydrostatic pressure when the volume viscosity term is implicitly incorporated.

It is notable that the fluid mechanics model is inherently nonlinear, since the material derivative

$$\frac{d}{dt} = \frac{\partial}{\partial t} + \mathbf{v} \cdot \nabla$$

produces advective terms throughout, including the advective term $\mathbf{v} \cdot \nabla \mathbf{v}$ in the equation of motion which only vanishes in special flow configurations. However, when the shear viscosity term may be omitted from (3.2) such that $\mathsf{p} = p\,\mathsf{I}$ and hence $\nabla \cdot \mathsf{p} = \nabla p$, the mathematical difficulty presented by the nonlinearity may sometimes be avoided—viz. when a pressure equation can be combined with potential theory. The ideal fluid model and vorticity are first considered in Sects. 3.2 and 3.3 respectively, before our asymptotic ordering procedure is introduced in Sect. 3.4 to establish the incompressible approximation that applies in any subsonic flow. After a brief discussion of consequent potential theory for irrotational flow in Sects. 3.5 and 3.6, in Sect. 3.7 we not only recover the ideal equations of motion and vorticity conservation but also confirm the relevant Bernoulli equation for the pressure variation, under the subsonic asymptotic ordering procedure of Sect. 3.4.

On the other hand, other approaches are needed when the shear viscosity must be retained, including analytical techniques that depend upon whether the shear viscosity coefficient μ is small or relatively large. In Sect. 3.8, we first mention d'Alembert's Paradox and the "no slip" boundary condition, before obtaining the Navier–Stokes equation of motion on assuming a larger viscosity coefficient under our asymptotic ordering procedure. The Reynolds number Re defined in Sect. 3.12 is often used to characterise the importance of the viscous term relative to the advective term in this equation, and the necessary retention of the viscous term in any boundary layer leads us to consider possible perturbation solutions (matched asymptotic expansions) when Re $\gg$ 1. After discussing vortices in Sect. 3.9, we turn to Stokes flow past a sphere (slow Re $\ll$ 1 fluid motion) in Sect. 3.10, briefly mention some classical exact viscous flow solutions in Sect. 3.11, and then summarise various characteristic numbers (including the Reynolds number) in further briefly considering singular perturbation theory in Sect. 3.12. The Oseen correction for slow flow past a sphere in Sect. 3.13 then illustrates the use of matched asymptotic expansions, via the small parameter Re $\ll 1$; and dynamic meteorology and oceanography models are then considered very briefly in the starred Sect. 3.14, for those who may be interested in an introduction to these environmental scientific areas.

3.2 Ideal Fluid Model

When the shear viscosity contribution is negligible, Eq. (3.2) reduces to the ideal (inviscid) equation of motion

$$\rho\frac{d\mathbf{v}}{dt}+\nabla p=\rho\mathbf{g} \tag{3.5}$$

in the Eulerian form [8]. The evolution of the fluid density ρ is of course governed by the continuity equation (3.1); and the evolution of the scalar fluid pressure p by the adiabatic equation (3.3), corresponding to neglecting heat flow between fluid elements (i.e. thermal conduction), which is justified if timescales are not too long. As also foreshadowed in Sect. 2.7, this system of equations in $\{\rho, \mathbf{v}, p\}$ constituting the *ideal fluid* (or *inviscid* fluid) model is mathematically complete. Its solution is often subject to the kinematic (non-cavitation) boundary condition

$$\hat{\mathbf{n}}\cdot\mathbf{v}=\hat{\mathbf{n}}\cdot\mathbf{V}, \tag{3.6}$$

applicable when the fluid is confined by an impenetrable boundary moving at velocity $\mathbf{V}$ and with unit normal $\hat{\mathbf{n}}$ (usually directed away from the fluid). Note that the tangential component of $\mathbf{v}$ is not constrained, however, for there is no tangential surface force acting on the fluid at the boundary when the shear viscosity is neglected.

Energy transport during the motion of an ideal fluid may be considered by taking the dot product of (3.5) with $\mathbf{v}$, whence

$$\rho\frac{d}{dt}\left(\frac{1}{2}v^2\right)+\mathbf{v}\cdot\nabla p=\rho\mathbf{g}\cdot\mathbf{v}$$

or

$$\frac{\partial}{\partial t}\left(\frac{1}{2}\rho v^2\right)+\nabla\cdot\left(\frac{1}{2}\rho v^2\mathbf{v}\right)-p\nabla\cdot\mathbf{v}=\rho\mathbf{g}\cdot\mathbf{v}-\nabla\cdot(p\mathbf{v}), \tag{3.7}$$

on invoking the commutation result

$$\rho\frac{df}{dt}=\frac{\partial}{\partial t}(\rho f)+\nabla\cdot(\rho\mathbf{v}f)$$

with $f \equiv v^2/2$. Equation (3.7) has been arranged such that the right-hand side represents the rate of work done by the body and surface forces (the gravity and hydrostatic pressure, respectively).

Although (3.7) is not in conservation form due to the third term $-p\nabla\cdot\mathbf{v}$ on the left-hand side, we can proceed to derive a total energy conservation equation as follows. From (3.1) and (3.3)

$$\frac{\rho}{p}\frac{d}{dt}\left(\frac{p}{\rho}\right) = \frac{1}{p}\frac{dp}{dt} - \frac{1}{\rho}\frac{d\rho}{dt} = -(\gamma-1)\nabla\cdot\mathbf{v},$$

such that (provided $\gamma \neq 1$) we obtain

$$-p\nabla\cdot\mathbf{v} = \frac{\rho}{\gamma-1}\frac{d}{dt}\left(\frac{p}{\rho}\right) = \frac{1}{\gamma-1}\left(\frac{\partial p}{\partial t} + \nabla\cdot(p\mathbf{v})\right), \tag{3.8}$$

on invoking the commutation result with $f \equiv p/\rho$—i.e.

$$\rho\frac{d}{dt}\left(\frac{p}{\rho}\right) = \frac{\partial p}{\partial t} + \nabla\cdot(p\mathbf{v}).$$

Thus we replace $-p\nabla\cdot\mathbf{v}$ in (3.7) using (3.8), to obtain the conservation equation for the total energy—viz.

$$\frac{\partial U}{\partial t} + \nabla\cdot\mathbf{S} = \rho\mathbf{g}\cdot\mathbf{v}, \tag{3.9}$$

where the total (kinetic plus adiabatic) energy density is

$$U = \frac{1}{2}\rho v^2 + \frac{p}{\gamma-1} \tag{3.10}$$

and the energy flux vector (including the work done by the hydrostatic pressure) is

$$\mathbf{S} = \left(\frac{1}{2}\rho v^2 + \frac{\gamma}{\gamma-1}p\right)\mathbf{v}. \tag{3.11}$$

Thus the energy flux in the ideal fluid model is purely advective, consistent with negligible thermal conduction under the adiabatic assumption reducing (2.58) to (3.3).

Occasionally, it is more realistic to go to the opposite limit of very strong thermal conduction, and assume the temperature is constant everywhere on the longer timescale of interest. This is the isothermal assumption, where the equation of state is

$$\frac{d}{dt}\left(\frac{p}{\rho}\right) = 0, \tag{3.12}$$

which formally corresponds to setting $\gamma = 1$ in (3.3)—but where of course the adiabatic energy conservation equation (3.9) is singular.

Let us now consider Bernoulli equations, beginning with the case of steady flow relative to some observer, when the dependent variables are functions of position $\mathbf{r}$ but not explicitly functions of time t (all ∂_t terms are zero). Writing $\mathbf{g}$ in terms of the gravitational potential V

$$\mathbf{g} = -\nabla V, \tag{3.13}$$

and noting that (3.1) is $\nabla \cdot (\rho \mathbf{v}) = 0$ for steady flow, reduces Eq. (3.9) to the form

$$\mathbf{v} \cdot \nabla \left(\frac{1}{2} v^2 + \frac{\gamma}{\gamma - 1} \frac{p}{\rho} + V \right) = 0. \tag{3.14}$$

In order to integrate this first-order partial differential equation, let us consider a streamline (or flowline)—i.e. a curve on which $\mathbf{v}$ is always tangential. In the case of steady flow, the streamlines are identical to the Lagrangian fluid trajectories mentioned in Sect. 2.4. Thus on any streamline defined by $d\mathbf{r} = \lambda \mathbf{v}$ (with parameter λ), Eq. (3.14) becomes

$$d\mathbf{r} \cdot \nabla \left(\frac{1}{2} v^2 + \frac{\gamma}{\gamma - 1} \frac{p}{\rho} + V \right) = 0, \tag{3.15}$$

with first integral a form of Bernoulli equation for steady flow—viz.

$$\frac{1}{2} v^2 + \frac{\gamma}{\gamma - 1} \frac{p}{\rho} + V = \text{const} \tag{3.16}$$

on any streamline.[1] Ignoring V (gravity) in (3.16), we have Bernoulli's Principle—viz. that the faster the flow the lower the pressure, which is exploited to produce the lift on an aircraft wing for example. Since $\rho = \text{const.}\ p^{1/\gamma}$ for any adiabatic fluid element, we have

$$\int \frac{dp}{\rho} = \frac{\gamma}{\gamma - 1} \frac{p}{\rho} \tag{3.17}$$

and hence from (3.16) that

$$\frac{1}{2} v^2 + \int \frac{dp}{\rho} + V = \text{const} \tag{3.18}$$

on any streamline, a more general form of Bernoulli equation for the steady flow of any barotropic fluid (including the isothermal case $\gamma = 1$). Note that the constant on the right-hand side of (3.16) or (3.18) is typically different for each streamline, unless the flow is irrotational (cf. the next section).

---

[1] This procedure is a simple example of the solution of a quasilinear partial differential equation by the method of characteristics.

It is notable that Bernoulli equations may also be derived in the following way, which identifies them as first integrals of the Euler equation. Substituting (3.13) and the vector identity

$$\mathbf{v} \cdot \nabla \mathbf{v} = \nabla \left(\frac{1}{2} v^2 \right) - \mathbf{v} \times (\nabla \times \mathbf{v}) \tag{3.19}$$

into the ideal equation of motion (3.5) yields

$$\frac{\partial \mathbf{v}}{\partial t} + \nabla \left(\frac{1}{2} v^2 + V \right) + \frac{\nabla p}{\rho} = \mathbf{v} \times (\nabla \times \mathbf{v}).$$

For a barotropic fluid (so ρ is a function of p only), or more generally in isentropic flow, we have $\nabla p/\rho = \nabla h$ where $h = \int dp/\rho$ is called the (specific) enthalpy so that

$$\frac{\partial \mathbf{v}}{\partial t} + \nabla \left(\frac{1}{2} v^2 + h + V \right) = \mathbf{v} \times (\nabla \times \mathbf{v}). \tag{3.20}$$

For steady flow, the dot product of (3.20) with $\mathbf{v}$ yields the generalised form of (3.14) corresponding to (3.17), leading to (3.18).

We have observed that the Bernoulli equation (3.18) applies on a streamline in any steady ideal barotropic flow, and it evidently determines the pressure if the flow field is somehow found. The form of Bernoulli equation applicable in any ideal flow that is irrotational ($\nabla \times \mathbf{v} = 0$) but not necessarily completely steady ($\partial \mathbf{v}/\partial t \neq 0$ everywhere) is met in Exercise (Q2) of the next section, where the important concept of vorticity is introduced. As then discussed in Sects. 3.5 and 3.6, an incompressible irrotational flow field may be obtained from classical potential theory. The ordering process in Sect. 3.4, justifying the incompressibility assumption whenever the flow is *subsonic*, is extended in Sect. 3.7 to produce the corresponding incompressible irrotational form of Bernoulli equation that determines the pressure variation. As we noted at the end of the previous section, we then proceed to include shear viscosity that (however small) typically produces rotational flow (vorticity) past any sufficiently "blunt" object, with flow separation and vortex sheets in its wake and associated viscous drag.

Exercises

(Q1) Steam escapes from a boiler through a conical pipe with diameter d_1 at the boiler and diameter d_2 at the open end. If v_1, v_2 denote the corresponding magnitudes of the velocities of the steam (both uniform over the respective pipe cross-sections) at the ends, and if the streamlines diverge from the vertex of the cone, assuming an isothermal equation of state $p = k\rho$ (k a constant) and neglecting gravity show that

$$\frac{v_1}{v_2} = \frac{{d_2}^2}{{d_1}^2} \exp \frac{{v_1}^2 - {v_2}^2}{2k}.$$

(Q2) Derive the energy principle for the barotropic flow under gravity of a fluid volume V bounded by closed fluid surface S—viz. that the material rate of change

$$\frac{d}{dt}(\mathcal{K}+\mathcal{V}) = \int_V p\nabla\cdot\mathbf{v}d\tau - \int_S p\mathbf{v}\cdot d\mathbf{S},$$

where the total kinetic energy of the fluid volume V is

$$\mathcal{K} \equiv \frac{1}{2}\int \rho v^2 d\tau$$

and its total potential energy is

$$\mathcal{V} \equiv \int \rho V d\tau.$$

Interpret each of the terms on the right-hand side of the energy principle.

3.3 Vorticity and Irrotational Flow

Recall that any dyadic may be written as the sum of a symmetric and antisymmetric part. Thus the velocity gradient $\nabla\mathbf{v}$ may be written as

$$\nabla\mathbf{v} = \boldsymbol{\epsilon} + \boldsymbol{\Omega},$$

where $\quad \boldsymbol{\epsilon} = \frac{1}{2}[\nabla\mathbf{v} + (\nabla\mathbf{v})^T] \quad$ and $\quad \boldsymbol{\Omega} = \frac{1}{2}[\nabla\mathbf{v} - (\nabla\mathbf{v})^T]$

denote symmetric and antisymmetric dyadics, respectively. Let us also recall from Sect. 2.2 that the rate of deformation tensor $\{\nabla\mathbf{v}\}$ incorporates the symmetric part ϵ, which is sometimes called the rate of strain tensor.

The antisymmetric part $\boldsymbol{\Omega}$ is associated with fluid rotation, so it may be called the rotation (or vorticity [13]) tensor. Thus the vorticity

$$\boldsymbol{\omega} \equiv \nabla\times\mathbf{v} \tag{3.21}$$

has the ith component (where $i \in \{1, 2, 3\}$)

$$\omega_i = [\nabla\times\mathbf{v}]_i = \epsilon_{ijk}\frac{\partial v^k}{\partial x^j} = \epsilon_{ijk}\frac{1}{2}\left(\frac{\partial v^k}{\partial x^j} - \frac{\partial v^j}{\partial x^k}\right) = \epsilon_{ijk}[\boldsymbol{\Omega}]_{jk}, \tag{3.22}$$

such that the local rotation of a fluid line element $d\mathbf{r}$ is

$$\frac{1}{2}\boldsymbol{\omega} \times d\mathbf{r} = (d\mathbf{r}) \cdot \boldsymbol{\Omega}, \tag{3.23}$$

since the vorticity is twice its angular velocity. One may also recall Stokes Theorem (1.57), to observe that the circulation K around any closed curve corresponds to the vorticity field—i.e.

$$K \equiv \oint_C \mathbf{v} \cdot d\mathbf{r} = \int_S d\mathbf{S} \cdot \nabla \times \mathbf{v} = \int_S \boldsymbol{\omega} \cdot d\mathbf{S}. \tag{3.24}$$

The circulation is conserved in ideal (inviscid) flow as outlined immediately below, arises in constraints when solving the Laplace equation for the velocity potential (cf. Sect. 3.5), and is directly related to the lift on a body in a flow (cf. Sect. 3.6).

The surface integral in the vorticity on the right-hand side of (3.24) may be equated to $2\pi\kappa$, where κ is the strength of the vortex tube defined by vortex lines in the vorticity field, analogous to streamlines in the velocity field—i.e. such that $d\mathbf{r} = \lambda\boldsymbol{\omega}$, where λ is a parameter—emanating from the closed curve C. The solenoidal property $\nabla \cdot \boldsymbol{\omega} = 0$ and the Divergence Theorem (1.60) imply that the strength is characteristic of the vortex tube, since

$$\int_{S_1} \boldsymbol{\omega} \cdot d\mathbf{S} = \int_{S_2} \boldsymbol{\omega} \cdot d\mathbf{S}$$

where S_1 and S_2 are two surfaces of cross-section anywhere along the tube.

Now the curl of the ideal equation of motion (3.20) gives the vorticity equation

$$\frac{\partial \boldsymbol{\omega}}{\partial t} = \nabla \times (\mathbf{v} \times \boldsymbol{\omega}) \tag{3.25}$$

or

$$\frac{d\boldsymbol{\omega}}{dt} = \boldsymbol{\omega} \cdot \nabla\mathbf{v} - \boldsymbol{\omega}\nabla \cdot \mathbf{v} = \boldsymbol{\omega} \cdot (\nabla\mathbf{v} - \mathbf{I}\nabla \cdot \mathbf{v}), \tag{3.26}$$

where the material rate of change of the vorticity is dependent upon the velocity and vorticity fields alone. In the next section, we find that $\nabla \cdot \mathbf{v} = 0$ in subsonic flow, so this ideal equation suggests that the vorticity is often conserved (assuming there are no regions of large velocity gradient)—in particular, if the vorticity of any fluid element is zero then it remains zero as the fluid element is advected in the flow. The well-known related result (Kelvin's circulation theorem) is that the circulation over any material closed curve remains constant in any ideal constant density or barotropic flow under conservative body forces. Thus from (2.40), (3.5) and (3.13) the material rate of change of the circulation

$$\frac{dK}{dt} = \frac{d}{dt}\oint_C \mathbf{v} \cdot d\mathbf{r} = \oint_C \frac{d\mathbf{v}}{dt} \cdot d\mathbf{r} = -\oint_C d\mathbf{r} \cdot \left(\frac{\nabla p}{\rho} + \nabla V\right) = 0, \tag{3.27}$$

whether or not $\omega = 0$ on the closed curve C.[2] Moreover, Eq. (3.26) was essentially first obtained by Helmholtz. Thus substituting in (3.26) for $\nabla \cdot \mathbf{v}$ from the continuity equation (3.1) yields

$$\frac{d}{dt}\left(\frac{\omega}{\rho}\right) = \frac{\omega}{\rho} \cdot \nabla \mathbf{v}, \tag{3.28}$$

so the material rate of change of ω/ρ depends upon its instantaneous value and the local velocity gradient.

The above results in the classical ideal model were largely obtained with water or air in mind, and the physical importance of the small but finite viscosity in boundary layers was only appreciated much later—cf. Sect. 3.8. Although the fluid viscosity introduces a dissipative term into (3.26), significant vorticity is generated in the neighbourhood of the boundary, where the velocity gradient is very large when the viscosity is small.[3] Nevertheless, there are circumstances where the ideal fluid assumption is almost universally appropriate such that the bulk of the flow remains irrotational (i.e. $\omega = 0$ almost everywhere)—e.g. the flow field over a smooth streamlined object, where the flow is irrotational upstream of the object and remains largely irrotational downstream. An important mathematical observation is that the irrotational (zero vorticity) flow condition $\nabla \times \mathbf{v} = 0$ is both necessary and sufficient for the existence of a velocity potential ϕ such that

$$\mathbf{v} = -\nabla \phi, \tag{3.29}$$

when we may proceed to exploit potential theory if the flow is subsonic as discussed below.

Exercises

(Q1) Use a Cartesian representation to derive (3.23) from (3.22).

(Q2) For incompressible irrotational but not necessarily steady flow, from (3.20) derive the equation

$$-\frac{\partial \phi}{\partial t} + \frac{1}{2}|\nabla \phi|^2 + h + V = 0, \tag{3.30}$$

where enthalpy $h = \int dp/\rho$ and we may also adopt (3.17) as before.

---

[2] The consequence that a flow irrotational at any time remains irrotational was originally suggested by Lagrange and proven by Cauchy.

[3] The terminology "Hydrodynamics" and "Aerodynamics" reflects the history of the subject—i.e. with reference to incompressible fluids ("liquids") and compressible fluids, respectively. However, the incompressible approximation is also applicable to aubsonic aerodynamic flow (as shown in the next section) and a "magnetohydrodynamic" (MHD) model may be compressible, so the historical distinction has become blurred. Incidentally, some might prefer to substitute "Dynamics" for "Mechanics" in the title to this chapter, but we trust the qualifier "Basic" serves to indicate the nature of the topics presented!

Hint: Note that the velocity potential is defined within an arbitrary function of time t, and in particular that $\mathbf{v} = -\nabla\phi' = -\nabla\phi$ where $\phi' = \phi - \int^t \phi(u)du$.

3.4 Subsonic Flow and the Incompressible Approximation

It can often be assumed that the motion is incompressible such that

$$\nabla \cdot \mathbf{v} = 0, \tag{3.31}$$

corresponding to no change in volume (no expansion nor contraction) anywhere in the fluid (cf. Sect. 2.5). This assumption is widely made for liquids such as water, sometimes with passing reference to the limit of (3.3) when $\gamma \rightarrow \infty$. However, it turns out to be a good approximation for highly compressible fluids such as air too, provided the flow speed $|\mathbf{v}|$ is everywhere much less than the sound speed $c_s \equiv \sqrt{\gamma p/\rho}$. We can reach this conclusion as an example of a simplification under an asymptotic ordering scheme, which is often quite useful in applied mathematics. Thus rather than merely introducing (3.31) ad hoc, let us treat subsonic flow as a case study in "asymptotology".

Consider an asymptotic analysis in terms of the small parameter

$$\varepsilon = |\mathbf{v}|/c_s \ll 1.$$

The theory of asymptotic expansions [10] deals with behaviour in a limit, in this case $\varepsilon \rightarrow 0$. The "big oh" asymptotic ordering notation $f = \mathrm{O}(\varepsilon^n)$ means that f scales like ε^n (i.e. $\varepsilon^{-n} f \rightarrow$ const) as $\varepsilon \rightarrow 0$ whereas the "little oh" notation $f = \mathrm{o}(\varepsilon^n)$ means f approaches zero *faster* than ε^n (i.e. $\varepsilon^{-n} f \rightarrow 0$) as $\varepsilon \rightarrow 0$.

All physical quantities are formally expanded in powers of ε—i.e.

$$\begin{aligned}\rho &= \rho^{(0)} + \varepsilon\rho^{(1)} + \varepsilon^2\rho^{(2)} + \mathrm{O}(\varepsilon^3)\\ p &= p^{(0)} + \varepsilon p^{(1)} + \varepsilon^2 p^{(2)} + \mathrm{O}(\varepsilon^3)\\ \mathbf{v} &= \varepsilon\mathbf{v}^{(1)} + \varepsilon^2\mathbf{v}^{(2)} + \mathrm{O}(\varepsilon^3),\end{aligned} \tag{3.32}$$

where the velocity is notably one order smaller than the other two quantities, which define the larger sound speed c_S. The inertia term $\rho\mathbf{v} \cdot \nabla\mathbf{v}$ in the equation of motion (3.2) therefore appears at second order, an issue that is taken up in Sect. 3.7 and again later in this chapter. Let us also assume the gravity $\mathbf{g}$ is sufficiently weak such that buoyancy forces are not dominant—i.e. adopt

$$\mathbf{g} = \varepsilon^2 g^{(2)} + \mathrm{O}(\varepsilon^3). \tag{3.33}$$

Finally, let us assume there are no sound waves excited in the fluid so that time variations are sufficiently slow, represented by defining a "slow" time variable $t_{(1)} \equiv \varepsilon t = O(1)$ such that

$$\frac{\partial}{\partial t} = \varepsilon \frac{\partial}{\partial t_{(1)}}. \tag{3.34}$$

Equations (3.33) and (3.34) are substituted into the fundamental equations (3.1)–(3.3), and like powers of ε equated. At order ε^0, only the equation of motion (3.2) contributes—i.e.

$$\nabla p^{(0)} = 0, \tag{3.35}$$

which on trivial integration yields

$$p^{(0)} = \text{const} \tag{3.36}$$

(and assumed to be time independent). At order ε,

$$(3.1)^{(1)} \implies \frac{\partial \rho^{(0)}}{\partial t_{(1)}} = -\nabla \cdot \left(\rho^{(0)} \mathbf{v}^{(1)}\right) \tag{3.37}$$

$$(3.2)^{(1)} \implies \nabla p^{(1)} = \nabla \cdot (2\mu\{\nabla \mathbf{v}^{(1)}\}) \tag{3.38}$$

$$(3.3)^{(1)} \implies \left(\frac{\partial}{\partial t_{(1)}} + \mathbf{v}^{(1)} \cdot \nabla\right) p^{(0)} = -\gamma p^0 \nabla \cdot \mathbf{v}^{(1)}$$

$$\implies \nabla \cdot \mathbf{v}^{(1)} = 0 \tag{3.39}$$

on using (3.36). Thus to first order there is the continuity equation (3.37) in the lowest order quantities $\rho^{(0)}$ and $\mathbf{v}^{(1)}$—but the equation of motion reduces to (3.38) that expresses pressure balance against any substantial viscous stress, and in (3.39) the incompressibility condition (3.31) emerges from the adiabatic equation as a constraint arising from a solvability condition (cf. [10], pp. 358–360). This is quite typical of such an asymptotic procedure, where the general rule is to introduce an ordering that builds in a constraint. (It may be said that "sound waves have been eliminated".)

In passing, we note from Exercise (Q3) in Sect. 1.8 that

$$\frac{\oint_{\Delta S} \mathbf{v} \cdot d\mathbf{S}}{\Delta V} \to \nabla \cdot \mathbf{v}$$

when the volume element ΔV enclosed by the surface ΔS shrinks towards a point, which is sometimes viewed as defining divergence. The ratio on the left-hand side represents the outflow per unit spatial volume when $\mathbf{v}$ denotes the fluid velocity vector, and if fluid is neither created nor destroyed then $\nabla \cdot \mathbf{v} = 0$—i.e. we have the incompressibility constraint (3.31) wherever there is neither divergent flow from any source ($\nabla \cdot \mathbf{v} > 0$) nor convergent flow to any sink ($\nabla \cdot \mathbf{v} < 0$). Combining (3.31) with (3.29) for subsonic irrotational flow is discussed in the next section.

Exercises

(Q1) A stream of water flows steadily through a horizontal pipe at a rate of k m^3 per second, and the water emerges at atmospheric pressure. The cross section of the pipe has area $A\,\mathrm{m}^2$, everywhere except at a short radial contraction of cross-sectional area $B\,\mathrm{m}^2$ ($B < A$), where there is a tap leading to a vertical open tube connected to the pipe. Assuming the water has a constant density but ignoring gravity, show that the eventual height of water in the tube above the pipe reaches up to $k^2(A^2 - B^2)/(2gA^2B^2)$ metres when the tap is opened.

(Q2) At time $t = 0$ a vacuous (negligible pressure) bubble arises at the centre of a spherical volume of water. If the pressure p_0 at the water boundary of radius $S(t)$ is kept constant, and if $R(t)$ denotes the radius of the bubble at time t, show that

$$S^3 = R^3 + b^3 - a^3 \qquad \text{and} \qquad \dot{R}^2 = \frac{2p_0}{3\rho}\frac{a^3 - R^3}{R^4}\left(\frac{1}{R} - \frac{1}{S}\right)^{-1},$$

where $\dot{R}(0) = \dot{S}(0) = 0, R(0) = a, S(0) = b$ and ρ is the constant water density.

3.5 Potential Flow

When the incompressibility constraint (3.31) is combined with the irrotational flow result (3.29), we obtain the Laplace equation for the velocity potential

$$\nabla^2\phi = 0. \tag{3.40}$$

Thus for subsonic irrotational flow the original nonlinear hyperbolic equation of motion is replaced by this linear elliptic equation of potential theory we mentioned in Sect. 1.11, significantly simplifying calculations [7]. Indeed, in this context Neumann boundary conditions also apply. In particular, from (3.6) the boundary condition at a stationary impenetrable wall is

$$\frac{\partial\phi}{\partial n} \equiv \hat{\mathbf{n}}\cdot\nabla\phi = 0. \tag{3.41}$$

It is well-known but nevertheless quite remarkable that the Laplace equation (3.40) subject to the relevant boundary conditions supplemented by a finite number of constraints (the circulation constants) determines subsonic irrotational flow—there is no need to know the pressure, although one may solve for the pressure after finding the fluid velocity $\mathbf{v}$ (cf. Sect. 3.7). A brief discussion of existence and uniqueness of the solutions to (3.40) and (3.41) follows.

First, consider the case of flow in a simply connected region V, which is completely enclosed by a fixed impenetrable boundary S. The unique solution to (3.40) and (3.41) is ϕ = constant, so $\mathbf{v}^{(1)} \equiv 0$. This follows immediately from

$$\int_V |\nabla\phi|^2 \, d\tau = \int_S [\phi\nabla\phi] \cdot d\mathbf{S} = \int_S \phi \frac{\partial\phi}{\partial n} \, dS = 0, \tag{3.42}$$

for any harmonic function ϕ—cf. the first Green identity (1.62).

Now consider flow in a *doubly* connected region V, which means that V can be filled with two distinct families of closed curves. All the closed curves within each family are topologically equivalent, since they can be continuously deformed into each other while remaining in V. However, the curves in one family $\{C_R\}$ can be deformed to a point (they are reducible), but not the curves in the other family $\{C_I\}$ (they are irreducible). Because of the assumed irrotational nature of the flow, it follows that the circulations

$$K_R \equiv \oint_{C_R} \mathbf{v} \cdot d\mathbf{r} \quad \text{and} \quad K_I \equiv \oint_{C_I} \mathbf{v} \cdot d\mathbf{r} \tag{3.43}$$

for any curve in each of the respective families are topological invariants (in fact, $K_R \equiv 0$). Consequently, the only solution of (3.40) and (3.41) such that $K_I = 0$ is the trivial solution ϕ = const.

This is proven by first noting that $K_I = 0$ implies that $\phi(\mathbf{r}) = -\int_{\mathbf{r}_0}^{\mathbf{r}} \mathbf{v} \cdot d\mathbf{r}$ is a single-valued function of $\mathbf{r}$, returning to the same value even if the line integral completely encompasses a C_I contour. Suppose V is now cut to make a simply connected domain V'. Since ϕ is single-valued, the contributions to the surface integral in (3.42) from either side of the cut cancel because $\partial\phi/\partial n$ changes sign, hence $|\nabla\phi| \equiv 0$. Therefore, any two solutions with the same non-zero value of K_I can differ only by a constant, since their difference is a solution with $K_I = 0$. Note that solutions with $K_I \neq 0$ are *not* single-valued. If there is flow across the boundary S, then $\partial\phi/\partial n \neq 0$. However, applying the Divergence Theorem (1.60) to $\nabla^2\phi = 0$ demonstrates that the constraint

$$\int_S [\nabla\phi] \cdot d\mathbf{S} = \int_S \frac{\partial\phi}{\partial n} dS = 0 \tag{3.44}$$

must be satisfied for a solution to exist. (Since $\mathbf{v} = -\nabla\phi$, this is merely a statement of mass conservation.) Given K_I, uniqueness within an additive constant follows as before. The analysis generalises to n-ply connected domains, where $n - 1$ circulation integrals are required for uniqueness.

3.6 Solving the Laplace Equation

Various analytical and numerical methods have been devised to solve the Laplace equation. In this section, we restrict the discussion to two approaches for flows that do not vary in one direction (the z-direction) in which there is no flow ($v_z = 0$)—i.e. when the coordinate z is ignorable and the motion is two-dimensional ("planar"). The first is the use of a complex potential, and in the second the solution is represented by a series expansion in terms of a basis set of harmonic functions. The classic text by Lamb [8] is an excellent starting point for further reading on ideal (inviscid) incompressible irrotational flows, in both two and three dimensions.

When a complex potential can be introduced, the theory of complex functions may be used, including images or conformal transformations. Typically, planes of symmetry are exploited in the method of images. Thus the image of any line source or line vortex in a flow bounded by an infinite plane is a reflection through that plane that is then notionally removed, analogous to the adoption of the complex potential in the solution to the flow past a semicircular cylinder here. The method of images may also be exploited in three dimensions, such as for point sources or sinks. Conformal transformations preserve the Laplace equation.

Thus for planar steady flow, in Cartesian coordinates we set out to construct the complex potential

$$f(\zeta) = \phi(x, y) + i\psi(x, y),$$

an analytic function of the complex variable $\zeta = x + iy$ under continuous first derivatives of the real and imaginary parts ϕ and ψ satisfying the Cauchy–Riemann equations

$$\frac{\partial \phi}{\partial x} = \frac{\partial \psi}{\partial y}, \qquad \frac{\partial \psi}{\partial x} = -\frac{\partial \phi}{\partial y}$$

such that $\nabla^2 \phi = 0$ and $\nabla^2 \psi = 0$. The variable ϕ is of course the velocity potential, and the stream function ψ defines the flux everywhere in the two-dimensional flow. Correspondingly, $\psi =$ const along a streamline $d\mathbf{r} = \lambda \mathbf{v}$ (i.e. $dx/u = dy/v$ where $\mathbf{v} = u\,\hat{\mathbf{i}} + v\,\hat{\mathbf{j}}$). Put another way, the streamlines lie within the level surfaces of ψ, because the total differential

$$d\psi(x, y) = \frac{\partial \psi}{\partial x} dx + \frac{\partial \psi}{\partial y} dy = v dx - u dy = 0;$$

and the equipotential lines $\phi =$ const are perpendicular to the streamlines since

$$\nabla \phi \cdot \nabla \psi = \frac{\partial \phi}{\partial x}\frac{\partial \psi}{\partial x} + \frac{\partial \phi}{\partial y}\frac{\partial \psi}{\partial y} = 0.$$

In plane polar coordinates where $f(z) = \phi(r, \theta) + i\psi(r, \theta)$, the Cauchy–Riemann equations are

$$\frac{\partial \phi}{\partial r} = \frac{1}{r}\frac{\partial \psi}{\partial \theta}, \qquad \frac{\partial \psi}{\partial r} = -\frac{1}{r}\frac{\partial \phi}{\partial \theta}.$$

The flow field may be obtained from the first derivative of the complex potential function, or the conjugate—cf. Exercise (Q1) below. Note also that the linearity of $\nabla^2\phi = 0$ and $\nabla^2\psi = 0$ enables superposition of specific complex potential functions representing particular contributions to the total flow, such as sources or sinks in an otherwise uniform flow. Indeed, even if the flow is not irrotational (when the potential function does not exist), a stream function may still be defined such that

$$\mathbf{v} = \hat{\mathbf{k}} \times \nabla\psi, \tag{3.45}$$

which implies not only $v_z = 0$ and $\nabla \cdot \mathbf{v} = -\hat{\mathbf{k}} \cdot \nabla \times \nabla\psi = 0$ (planar subsonic flow) but also $\mathbf{v} \cdot \nabla\psi = 0$—i.e. $\psi = $ const along a streamline.

There are various theorems on the complex potential for planar flows in the literature, including the following that we choose to quote here (cf. [11] for further relevant discussion):

- (Milne–Thomson Circle Theorem) When a solid circular cylinder $|\zeta| = a$ is introduced into a flow where there are no rigid boundaries, with a complex potential $f(z)$ where any singularity is in $\zeta > a$, the complex potential becomes $f(\zeta) + f^*(a^2/\zeta)$, f^* denoting the complex conjugate of f;
- (Blasius Theorem) if $f(\zeta)$ is the complex potential in a flow past a solid cylinder of any shape, then neglecting external forces (a) the thrust per unit length on the cylinder $X\hat{\mathbf{i}} + Y\hat{\mathbf{j}}$ is given by

$$X - iY = \frac{1}{2}\, i\rho \oint_C \left(\frac{df}{d\zeta}\right)^2 d\zeta,$$

  and (b) the moment M about the origin of the thrust per unit length is the real part of

$$-\frac{1}{2}\rho \oint_C \zeta \left(\frac{df}{d\zeta}\right)^2 d\zeta,$$

  where ρ is the fluid density and the integrals are taken around the cylinder contour; and
- (Theorem of Kutta and Joukovskii) when a solid cylinder of any shape is placed in a uniform stream of speed U, the resultant thrust is perpendicular to the stream (the "lift") with magnitude $\rho U K$, where ρ is the fluid density and K is the circulation around the cylinder.

The lift on an aerofoil can be determined using ideal (inviscid) theory, but a discussion of the drag requires the introduction of shear viscosity (cf. Sect. 3.8).

For series expansion solutions in potential theory, the appropriate basis set is determined by the geometry of the problem—in particular, by the boundaries. For the two-dimensional Laplace equation in plane polar coordinates, two countably infinite sets of single-valued basis functions ("polar harmonics") may also be obtained by separation of variables:

$$u_m(r,\theta) \equiv r^{|m|} e^{im\theta} \; ; \; m = 0, \pm 1, \pm 2, \ldots, \tag{3.46}$$

regular at $r = 0$ (including the trivial constant solution when $m = 0$); and

$$v_m(r,\theta) \equiv r^{-|m|} e^{im\theta} \; ; \; m = \pm 1, \pm 2, \ldots, \tag{3.47}$$

regular at $r = \infty$. There is also an $m = 0$ solution that is singular at both 0 and ∞—viz. the solution (corresponding to a source at infinity and a sink at $r = 0$)

$$v_0(r,\theta) \equiv \ln r; \tag{3.48}$$

and finally, $\nabla^2\theta = 0$ so that θ can be used as a particular multi-valued solution (corresponding to flows with circulation about $r = 0$). Thus the series expansion for any planar flow is

$$\phi = \sum_{m=-\infty}^{\infty} A_m r^{|m|} e^{im\theta} + B_0 \ln r + \sum_{m=-\infty}^{\infty} B_m r^{-|m|} e^{im\theta} - \frac{K_1}{2\pi}\theta, \tag{3.49}$$

where A_m and B_m are coefficients to be determined from the boundary conditions (at $r = 0$ and ∞), and K_1 is the circulation about $r = 0$. (Note that the $m = 0$ term can be omitted under either of the two summation symbols in this result.)

For example, suppose that $\partial\phi/\partial n$ is specified by its Fourier expansion on concentric circular cylindrical surfaces at $r = a$ and $r = b$:

$$\left.\frac{\partial\phi}{\partial r}\right|_{r=a} = \sum_{m=-\infty}^{\infty} \alpha_m e^{im\theta} \tag{3.50}$$

$$\left.\frac{\partial\phi}{\partial r}\right|_{r=b} = \sum_{m=-\infty}^{\infty} \beta_m e^{im\theta}. \tag{3.51}$$

Comparing the partial derivative with respect to r of (3.43) at $r = a$ and b, and equating coefficients of $e^{im\theta}$, yields

$$m = 0: \quad B_0/a = \alpha_0, \tag{3.52}$$

$$B_0/b = \beta_0; \tag{3.53}$$

$$m \neq 0: \quad |m| \left(A_m\, a^{|m|-1} - B_m\, a^{-|m|-1} \right) = \alpha_m, \tag{3.54}$$

$$|m| \left(A_m\, b^{|m|-1} - B_m\, b^{-|m|-1} \right) = \beta_m. \tag{3.55}$$

The relations $B_0 = a\alpha_0$ and $B_0 = b\beta_0$ for $m = 0$ are compatible only if the existence condition (3.44) is satisfied. For $m \neq 0$,

$$A_m = |m|^{-1}\left(b^{2|m|} - a^{2|m|}\right)^{-1}\left(b^{|m|+1}\beta_m - a^{|m|+1}\alpha_m\right), \tag{3.56}$$

$$B_m = |m|^{-1}\left(b^{2|m|} - a^{2|m|}\right)^{-1}\left(a^{2|m|}b^{|m|+1}\beta_m - b^{2|m|}a^{|m|+1}\alpha_m\right). \tag{3.57}$$

Thus the solution to the Neumann problem has been found explicitly, and the convergence of (3.49) is assured. Note that this series solution depends upon the boundaries being symmetric about $r = 0$. However, for any planar flow the method of conformal mapping may be used to convert a non-symmetric into a suitably symmetric domain.

Exercises

(Q1) Show that $u - iv = -df/d\zeta$, where $\mathbf{v} = u\,\hat{\mathbf{i}} + v\,\hat{\mathbf{j}}$ is the two-dimensional fluid velocity; and that $|\mathbf{v}|^2 = (df/d\zeta)(df/d\zeta)^*$, with the star superscript denoting the complex conjugate. Then deduce the flow field corresponding to each of the following complex potential functions:

(a) $f(\zeta) = -U\mathrm{e}^{-i\alpha}\zeta$ (U, α constants); (b) $f(\zeta) = -m\ln\zeta$ (m constant);
(c) $f(\zeta) = i\kappa\ln\zeta$ (κ constant).

(Q2) Consider steady ideal planar subsonic flow over

$$y = \begin{cases} 0, & \text{for } x \le -a,\ x \ge a \\ \sqrt{a^2 - x^2}, & \text{for } |x| < a \end{cases},$$

the surface of a semicircular cylinder of radius a. Assuming the flow originates as a uniform stream from left to right so the complex potential is $f(\zeta) = -U\zeta$ at infinity, ignoring gravity find the resultant vertical and horizontal forces on the cylinder.

(Q3) The Laplace equation $\nabla^2\phi = 0$ is to be solved in the region V between two circular cylindrical surfaces, with axes of symmetry parallel but distinct. In particular, assume that ϕ is independent of the axial coordinate z and consider the area of cross-section defined between an outer cylindrical surface of radius $r = 1$ and an inner cylindrical surface where $|r - 2/5| = 2/5$. Consider the bilinear conformal transformation

$$w = \frac{\zeta - \alpha}{\alpha^*\zeta - 1}, \quad |\alpha| < 1$$

to map the area of cross-section to an annular region $1/2 \leq |w| \leq 1$, and hence obtain the velocity potential

$$\phi(x, y) = c_1 \ln\left(\frac{2\zeta - 1}{\zeta - 2}\right) + c_0$$

where $\zeta = x + iy$ and c_0, c_1 are constants.

3.7 Pressure Variation in Ideal Subsonic Flow

As we have found, it is often important to know the pressure variation (e.g. to calculate the lift on an aerofoil), and in this section it is considered more carefully in ideal subsonic flow. From the ordering in Sect. 3.4 that led to the incompressible approximation, we recall that $p^{(0)}$ is constant; and we now observe that $p^{(1)}$ is also constant, if the shear viscosity μ is small enough to render the right-hand side of (3.38) zero. Thus the pressure variation is $O(\varepsilon^2)$, and appears in the next (second) order equation obtained from (3.2)—viz.

$$\rho^{(0)}\left(\frac{\partial \mathbf{v}^{(1)}}{\partial t_{(1)}} + \mathbf{v}^{(1)} \cdot \nabla \mathbf{v}^{(1)}\right) + \nabla p^{(2)} = \rho^{(0)}\mathbf{g}^{(2)} \tag{3.58}$$

if $\mu = o(\epsilon)$, recalling the "little oh" asymptotic ordering notation introduced above (3.32). It is notable that all of the quantities in (3.58) are leading order under the subsonic asymptotic analysis, with the exception of the pressure variation in the flow—and as the pressure variation is the fluid pressure within a constant, this is consistent with the ideal equation of motion (3.5).

For a barotropic equation of state (when p is a function of ρ alone), constant $p^{(0)}$ implies that $\rho^{(0)}$ = constant. Then dividing through by $\rho^{(0)}$ and adopting $\mathbf{g}^{(2)} = -\nabla V^{(2)}$ yields the equation

$$\frac{\partial \mathbf{v}^{(1)}}{\partial t_{(1)}} + \nabla\left(\frac{1}{2}|\mathbf{v}^{(1)}|^2 + \frac{p^{(2)}}{\rho^{(0)}} + V^{(2)}\right) = \mathbf{v}^{(1)} \times \left(\nabla \times \mathbf{v}^{(1)}\right)$$

in the form of (3.20), from the identity $\mathbf{v} \cdot \nabla \mathbf{v} = \nabla\left(\frac{1}{2}|\mathbf{v}|^2\right) - \mathbf{v} \times (\nabla \times \mathbf{v})$ introduced before. This result is to be solved for $p^{(2)}/\rho^{(0)}$, but the equation $\nabla[p^{(2)}/\rho^{(0)}] = \mathbf{f}$ cannot have a solution unless $\nabla \times \mathbf{f} = 0$. Hence we have

$$\frac{\partial(\nabla \times \mathbf{v}^{(1)})}{\partial t_{(1)}} = \nabla \times [\mathbf{v}^{(1)} \times (\nabla \times \mathbf{v}^{(1)})]$$

—i.e. the conservation of vorticity equation (3.25) to lowest order, which has already been used in Sect. 3.4 to justify irrotational flow and hence the existence of a scalar potential for $\mathbf{v}^{(1)}$. The derivation of information about a lower order quantity from the solvability condition for a higher order quantity, in this case about $\mathbf{v}^{(1)}$ from $p^{(2)}$, is a common feature of asymptotic expansion procedures [10]. It is often said that one annihilates the higher order information, and in this case the annihilator is the operator $\nabla \times$.

Thus introducing the irrotational flow velocity potential, the equation for $p^{(2)}$ becomes

$$\nabla\left(-\frac{\partial\phi^{(1)}}{\partial t_{(1)}}+\frac{1}{2}|\nabla\phi^{(1)}|^2+\frac{p^{(2)}}{\rho^{(0)}}+V^{(2)}\right)=0;$$

which on integrating yields

$$-\frac{\partial\phi^{(1)}}{\partial t_{(1)}}+\frac{1}{2}|\nabla\phi^{(1)}|^2+\frac{p^{(2)}}{\rho^{(0)}}+V^{(2)}=F(t) \tag{3.59}$$

throughout the flow, where the arbitrary function of time $F(t)$ may be removed by redefining the velocity potential to be $\phi^{(1)}-\int^t F(u)du$ as in Exercise (Q2) of Sect. 3.3. The pressure variation due to the flow is therefore defined by the leading order approximations to the density and velocity fields—i.e.

$$p^{(2)}=\rho^{(0)}\frac{\partial\phi^{(1)}}{\partial t_{(1)}}-\frac{1}{2}\rho^{(0)}|\nabla\phi^{(1)}|^2-\rho^{(0)}V^{(2)}, \tag{3.60}$$

where of course the nonlinear contribution on the right-hand side is due to the advection term $\mathbf{v}\cdot\nabla\mathbf{v}$ in the consistent second-order rendition (3.58) of the equation of motion. The ordering indices are also usually omitted from (3.59) and (3.60), and sometimes the pressure variation $p^{(2)}$ is written as δp (rather than p) to emphasise that it is the pressure perturbation due to the flow.

Equation (3.60) is the appropriate equation for the pressure variation throughout an ideal but not necessarily steady subsonic flow. While (3.60) is the reduced form of the Bernoulli equation (3.30) derived in Exercise (Q2) of Sect. 3.3 in the limit $\gamma\to\infty$, our asymptotic analysis has demonstrated the validity of (3.60) for arbitrary γ provided the flow is subsonic, as was the case with (3.31) in Sect. 3.4. Pressure variations are immediately transmitted throughout the fluid because the sound speed is considered indefinitely fast on the subsonic time-scale—and notions of cause and effect are altered. Rather than pressure variations "causing" fluid accelerations and thus changes in fluid velocity, the fluid velocity is completely determined by the boundary conditions through the Laplace equation, and the pressure variations are then "caused" by the variations in fluid velocity as specified by Eq. (3.60).

In the following sections, the shear viscosity is retained in the equation of motion, such that the vorticity is no longer conserved throughout the flow under Kelvin's circulation theorem—cf. Eq. (3.27). We begin with a brief discussion of the production of vorticity in regions where velocity gradients are large, although the shear viscosity coefficient is small.

Exercises

(Q1) Consider a planar irrotational flow field described by the velocity potential

$$\phi(r,\theta) = a\,|\mathbf{v}_0|\,\theta - |\mathbf{v}_\infty|\left(r + \frac{a^2}{r}\right)\cos\theta$$

in the domain $r > a$, where (r, θ) denote polar coordinates and a, $|\mathbf{v}_0|$ and $|\mathbf{v}_\infty|$ are constants.

(a) Calculate the flow velocity $\mathbf{v} = -\nabla\phi$, and consider the streamlines for (i) $\mathbf{v}_0 = 0$ and (ii) $\mathbf{v}_0 = \mathbf{v}_\infty$. Is the flow compatible with the presence of a solid body in the region $r \leq a$?

(b) What is the vorticity $\boldsymbol{\omega}$ of the flow? Calculate the circulation about two circuits in the plane—viz. (i) the circle $r = 2a$, $0 \leq \theta < 2\pi$ with centre at the origin; and (ii) the circle of radius a with centre at the point $r = 3a$, $\theta = 0$.

(c) Does the flow satisfy the incompressibility constraint? Assuming that the density is constant (ρ_0 say) far upstream, what is the density elsewhere in the flow?

(d) If the fluid is assumed to be ideal, show that:
(i) if $\mathbf{v}_0 \neq 0$, there is a lift (a force in the vertical $\theta = \pi/2$ direction); but
(ii) there is no drag (no force in the horizontal $\theta = 0$ direction).

(Q2) Consider steady subsonic flow past a sphere of radius a. Assuming that the flow originates as a uniform stream with constant velocity $\mathbf{U}$ and ignoring gravity, show that the ideal (inviscid) theory once again predicts there is no drag on the sphere.

3.8 Boundary Layers

The zero drag predicted on an obstacle in a flow, such as on a cylinder or sphere as considered in the previous Exercises and earlier in Sect. 3.6, is d'Alembert's Paradox from the eighteenth century that was not resolved for about 250 years! It is now well-known that this absurd theoretical result is due to the inviscid assumption underlying the Bernoulli equation—and in particular, that the viscosity causes a fluid to adhere to any boundary such that there is a "no slip" boundary condition (see below). The shear viscosity generates vorticity in any flow past a "blunt" object as mentioned before, so the irrotational flow assumption ($\boldsymbol{\omega} = \nabla \times \mathbf{v} = 0$) must be abandoned, at least in regions adjacent to and behind the object in the flow. Even if the shear viscosity coefficient is small, there is typically a wake generated downstream, behind a boundary layer (itself a layer of vorticity) adhering to the surface of the object. Indeed, the boundary layer may detach at some point along the surface of the object, producing a wake with very strong velocity shear (a vortex sheet, described further in Sect. 3.9). The unrealistic arbitrary transverse fluid velocity $\hat{\mathbf{n}} \times \mathbf{v}$ at any boundary in the ideal fluid model is incompatible with the more physical (viscous fluid) boundary condition $\mathbf{v} = \mathbf{V}$, such that we also have the "no slip" condition

$$\hat{\mathbf{n}} \times \mathbf{v} = \hat{\mathbf{n}} \times \mathbf{V} \tag{3.61}$$

in addition to the non-cavitation condition (3.6) at any boundary. Mathematically, the loss of any boundary condition is a classical sign of singular behaviour, typically associated with a reduction of order in the governing differential equation. All fluid viscosity terms have been ordered out in the preceding analysis in this chapter, not only in (3.38) at first order but also in (3.58) at second order, with the shear viscosity coefficient assumed to be $o(\epsilon)$.

Let us therefore consider the restoration of fluid viscosity in the subsonic model, beginning with the modification of (3.58). Formally, when the shear viscosity coefficient μ is $O(\epsilon)$ the ideal equation of motion (3.58) is replaced by

$$\rho\left(\frac{\partial \mathbf{v}}{\partial t} + \mathbf{v}\cdot\nabla\mathbf{v}\right) + \nabla p = \rho\mathbf{g} + \mu\nabla^2\mathbf{v}, \tag{3.62}$$

where the ordering indices are again omitted. Once again, the quantities in the equation are leading order under the subsonic asymptotic analysis, except for the pressure variation that is nevertheless the total fluid pressure within a constant as before—cf. Sect. 3.7. Thus this equation agrees with the fundamental equation of motion (3.2) with the Newtonian pressure tensor (3.4) when the fluid is uniform (ρ and μ are both constant) and the flow is subsonic ($\nabla\cdot\mathbf{v} = 0$). A secondary asymptotic expansion in powers of any smaller shear viscosity coefficient $\mu = o(\epsilon)$—i.e. writing $\rho = \rho^{(0)} + \mu\rho^{(1)} + \cdots$ and similar expressions for p and $\mathbf{v}$ in (3.62)—reproduces the previous ideal (inviscid) equation of motion (3.58), assuming that $\mathbf{r}$ and t do *not* scale with μ (i.e. both $\partial/\partial t$ and ∇ are $O(1)$ operators). Equation (3.62) is commonly called the Navier–Stokes equation, sometimes even when the fluid has variable density ρ.

Following Prandtl, the apparent conflicts in the limit $\mu \to 0$ mentioned in the first paragraph above were resolved by recognising that for small but finite μ there is a narrow boundary layer in the region of fluid near any solid boundary (a region with a very short scale length in the direction normal to its surface).[4] Although the magnitude of the viscous term $\mu\nabla^2\mathbf{v}$ relative to the inertia term $\rho\mathbf{v}\cdot\nabla\mathbf{v}$ may indeed be very small elsewhere in the flow such that the Reynolds number $\text{Re} \gg 1$ (cf. Sect. 3.12), in the boundary layer the fluid velocity shear and hence $\nabla^2\mathbf{v}$ is so large that the viscous term $\mu\nabla^2\mathbf{v}$ remains comparable with the other terms in

[4] Ludwig Prandtl's revolutionary contribution to fluid mechanics was presented at the Third International Mathematics Congress held in Heidelberg, Germany in 1904. His subsequent article was published in the Congress proceedings—cf. *Early Developments of Modern Aerodynamics*, J.A.K. Ackroyd, B.P. Axell and A.I. Rubin (eds.), Butterworth-Heinemann (Oxford, 2001), p. 77 for an English translation. His contribution was a major step in convincing the engineering community that mathematical and later computational fluid mechanics can realistically describe fluid motion, replacing some scepticism of the classical ideal theory. There is a well-written historical article by John D. Anderson Jr. in *Physics Today*, pp. 42–48, December 2005.

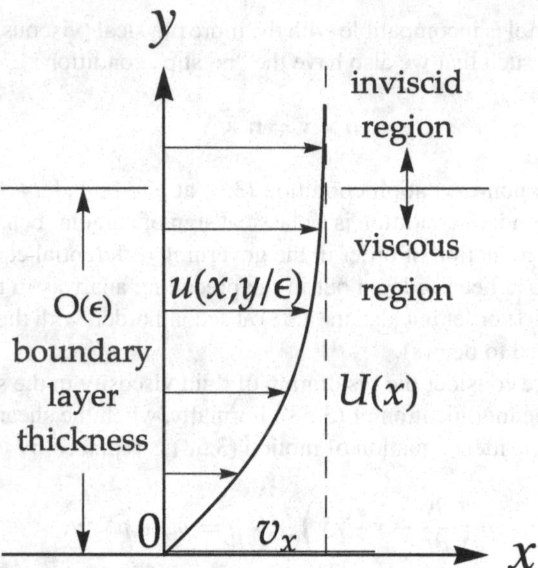

Fig. 3.1 Planar boundary layer flow near a solid surface (*horizontal wall*) with no-slip boundary condition using a stretched length scale such that the rapid, y/ϵ, variation in the normal direction is resolved, but on which the streamwise x variation is negligible

(3.62). Omission of the viscosity term $\mu\nabla^2\mathbf{v}$, involving higher order (second order) spatial derivatives, in the ideal fluid model represents the singular mathematical feature anticipated above. Thus the flow may include a substantial essentially ideal (inviscid) region outside the boundary layer, where the irrotational assumption is generally adequate. In a formal asymptotic analysis taking $\mu = \mathrm{O}(\varepsilon^3)$, the solution involving viscosity in the boundary layer, often referred to as the "inner" region, may be "matched" to the solution in the inviscid "outer" region—as in the simple planar flow $\mathbf{v} = u(x, y)\,\hat{\mathbf{i}}$ shown schematically in Fig. 3.1 for example.

Some of the mathematical features can be illustrated by a one-dimensional model problem:

$$\varepsilon y'' - y' = 0,\ y(0) = 0,\ y(1) = 1.$$

This is a singular boundary value problem, because the $\varepsilon \to 0$ limit of the equation with $y'' = \mathrm{O}(1)$ is $y' = 0$, which cannot satisfy the boundary conditions—i.e. no solution exists to the problem if ε is strictly zero and the boundary conditions are as given. However, for arbitrarily small ε the exact solution is

$$y(x) = \frac{e^{x/\varepsilon} - 1}{e^{1/\varepsilon} - 1}.$$

Thus when $\varepsilon \to 0+$ there are two regions—viz. the outer region, where $y^{(0)} = 0$ is a good solution of the "ideal" equation $y' = 0$; and the narrow boundary layer or inner region where the solution varies rapidly, so a different asymptotic ordering is appropriate. A similar simple boundary value problem to illustrate this behaviour is given in Ref. [12], where related perturbation methods are applied to a remarkably wide variety of applications. Reference may also be made to Refs. [1, 3], for further aspects of boundary layer theory.

3.9 Vortices

A vortex is a region of fluid in which the vorticity $\boldsymbol{\omega}$ is non-zero. Let us recall from Sect. 3.3 that vortex lines have tangents parallel to $\boldsymbol{\omega}$, the direction of the swirl in the flow, perpendicular to the flow direction associated with the otherwise analogous streamlines. The vortex line density, the number of lines per unit area passing through an infinitesimal surface with normal parallel to $\boldsymbol{\omega}$, is of course proportional to $|\boldsymbol{\omega}|$. We recall that (3.24) associated the circulation $K \equiv \int_C \mathbf{v} \cdot d\mathbf{r}$ around a closed material curve C with the corresponding strength κ of a vortex tube.

We also recall that that the strength is a characteristic of any vortex tube, since it is constant along the tube. This immediately implies that in the absence of dissipation a vortex tube can not terminate within the fluid, for any geometric continuation of the tube into a region where $\boldsymbol{\omega} \equiv 0$ produces a contradiction. Thus a vortex tube either has a closed form in the fluid (e.g. a "smoke ring"), or ends at a fluid boundary. Moreover, the strength of any vortex tube completely surrounded by an inviscid irrotational flow region is conserved according to the Kelvin theorem (3.27), since the curve C may be any topologically equivalent circuit in that region.

All of this advances the important concept that vortex lines move with the fluid (Helmholtz theorem)—or equivalently, that vortex tubes move with the fluid—implicit in the ideal vorticity equation (3.26). Whereas previously (3.26) was invoked to justify the persistence of irrotational motion in an ideal fluid and therefore the introduction of a velocity potential, the emphasis here has shifted to the implication that any fluid element with vorticity must subsequently possess vorticity, and hence that the fluid in a vortex line (or vortex tube) must subsequently form a vortex line (or vortex tube). The respective magnitude and direction of the fluid vorticity varies with the length and direction of the corresponding vortex line. Some aspects of whirlwinds or tornados may be recognised in several of the above remarks!

We have recognised that additional vorticity can be generated due to the shear viscosity—e.g. in the case of formerly irrotational flow past an obstacle, from where new vortex lines and vortex tubes originate. The curl of the Navier–Stokes equation (3.62) yields

$$\frac{d\boldsymbol{\omega}}{dt} = \boldsymbol{\omega} \cdot (\nabla \mathbf{v} - \mathbf{I} \nabla \cdot \mathbf{v}) + \nu \nabla^2 \boldsymbol{\omega} = \boldsymbol{\omega} \cdot \nabla \mathbf{v} - \boldsymbol{\omega} \nabla \cdot \mathbf{v} + \nu \nabla^2 \boldsymbol{\omega}. \tag{3.63}$$

The coefficient $\nu = \mu/\rho$ is often called the kinematic viscosity, or alternatively the diffusivity because the viscous term $\nu \nabla^2 \boldsymbol{\omega}$ produces vorticity diffusion. Since the coefficient ν is often small, the vorticity diffusion term may be small in regions where the vorticity gradient remains small—i.e. there can be slow dissipation of an ideal (inviscid) solution, such as occurs for example with a "smoke ring". However, let us recall that the viscous term $\mu\nabla^2\mathbf{v}$ must be retained in (3.62) to describe boundary layers, where the velocity gradient is large. The corresponding viscous term $\nu\nabla^2\boldsymbol{\omega}$ in (3.63), omitted from (3.26), must be retained to predict the vortex wake generated in flow past an obstacle. As already mentioned in Sect. 3.8, a boundary layer may be viewed as a layer of vorticity. Due to the "no slip" boundary condition, vorticity is produced at the obstacle and diffused outward by the surface viscous forces, and then swept along by the flow. Mathematically speaking, the major interest in the inviscid limit $\nu \to 0$ (equivalent to $\mu \to 0$) is therefore again when it is singular, with an associated loss of boundary conditions.

Sometimes the vorticity is confined to a set of vortex tubes, each of which has a diameter much less than the distances separating them. Such a vortex tube may be idealised as having negligible thickness but finite strength (cf. the Dirac delta function) and called a *line vortex*. Ideal flow solutions in this context have singularities along the line vortices, and the circulation around them must take the values they prescribe. Another useful concept is the *vortex sheet* mentioned in the previous section, where $\boldsymbol{\omega}$ is non-zero only within a sheet of negligible width, which can also be thought of as a distributed line vortex. If an infinitesimal rectangular circuit γ is drawn through a vortex sheet, coincident with a level surface of a curvilinear coordinate χ, from Stokes Theorem the tangential velocity has a discontinuity. Specifically, $[\![\mathbf{v}]\!] \cdot \boldsymbol{\tau} = \mathbf{K} \cdot \hat{\mathbf{n}} \times \boldsymbol{\tau}$ for any arbitrary unit tangent vector $\boldsymbol{\tau}$, and hence

$$[\![\mathbf{v}]\!] = \mathbf{K} \times \hat{\mathbf{n}}, \tag{3.64}$$

where $\hat{\mathbf{n}}$ is the unit normal to the sheet and $\mathbf{K}$ is the surface vorticity. Thus

$$\boldsymbol{\omega} = \mathbf{K}\,\delta(\chi), \tag{3.65}$$

where χ is a function such that its level surface $\chi = 0$ coincides with the sheet, and $\hat{\mathbf{n}} \cdot \nabla\chi = 1$ on the sheet. In terms of χ, the jump in $\mathbf{v}$ is defined by

$$[\![\mathbf{v}]\!] \equiv \lim_{\delta\to 0} \left(\mathbf{v}(\chi = \delta) - \mathbf{v}(\chi = -\delta)\right), \tag{3.66}$$

and the jump condition $\hat{\mathbf{n}} \cdot [\![\mathbf{v}]\!] = 0$ from $\nabla \cdot \mathbf{v} = 0$ is used.

In cylindrical coordinates (r, θ, z), for incompressible axisymmetric flow with a swirl the velocity and vorticity fields are

$$\mathbf{v} = -\frac{1}{r}\frac{\partial\psi}{\partial z}\hat{\mathbf{e}}_r + v_\theta\hat{\mathbf{e}}_\theta + \frac{1}{r}\frac{\partial\psi}{\partial r}\hat{\mathbf{e}}_z, \tag{3.67}$$

$$\boldsymbol{\omega} = -\frac{\partial v_\theta}{\partial z}\hat{\mathbf{e}}_r - \left[\left(\frac{1}{r}\frac{\partial^2}{\partial r^2} - \frac{1}{r}\frac{\partial}{\partial r} + \frac{\partial^2}{\partial z^2}\right)\psi\right]\hat{\mathbf{e}}_\theta + \frac{1}{r}\frac{\partial v_\theta}{\partial r}\hat{\mathbf{e}}_z, \tag{3.68}$$

where $\psi(r, z, t)$ is the relevant Stokes stream function and $v_\theta(r, z, t)$ is the swirl velocity component (cf. also the next section). The Helmholtz equation (3.28) has azimuthal and meridional components

$$\frac{d}{dt}\left(\frac{\omega_\theta}{r}\right) = \frac{1}{r^2}\frac{\partial v_\theta^2}{\partial z}, \tag{3.69}$$

$$\frac{d(rv_\theta)}{dt} = 0, \tag{3.70}$$

respectively. If the shape of the vorticity is preserved such that the flow is steady in some moving reference frame, then from (3.70) we have $rv_\theta = C(\psi)$ where $C(\psi)$ is an arbitrary function. Since $d\psi = rv_z dr - rv_r dz$, the right-hand side of (3.69) can be rewritten as

$$\frac{1}{r^2}\frac{\partial}{\partial z}\left(\frac{C^2}{r^2}\right) = -v_r\frac{1}{r^3}\frac{dC^2}{d\psi} = \frac{1}{2}\frac{d}{dt}\left(\frac{1}{r^2}\frac{dC^2}{d\psi}\right) \tag{3.71}$$

because $d\psi/dt = 0$ and $dr/dt = v_r$. Consequently, (3.69) can be integrated to give

$$\frac{\omega_\theta}{r} = \frac{dH}{d\psi} + \frac{C}{r^2}\frac{dC}{d\psi} \tag{3.72}$$

where H is an arbitrary function of ψ in (3.20) written as $-\nabla H = \mathbf{v} \times \boldsymbol{\omega}$ for steady flow, on noting that

$$\nabla H = rv_z\frac{dH}{d\psi}\hat{\mathbf{e}}_r - rv_r\frac{dH}{d\psi}, \tag{3.73}$$

$$\mathbf{v} \times \boldsymbol{\omega} = \left(\frac{C}{r^2}\frac{\partial C}{\partial r} - v_z\omega_\theta\right)\hat{\mathbf{e}}_r + \left(\frac{C}{r^2}\frac{\partial C}{\partial z} + v_r\omega_\theta\right)\hat{\mathbf{e}}_z. \tag{3.74}$$

Finally, on substituting the expression for ω_θ into the θ-component of the vorticity we have

$$\left(\frac{\partial^2}{\partial r^2} - \frac{1}{r}\frac{\partial}{\partial r} + \frac{\partial^2}{\partial z^2}\right)\psi + r^2\frac{dH}{d\psi} + C\frac{dC}{d\psi} = 0. \tag{3.75}$$

This equation has the same form as the Grad-Shafranov equation (5.68) in Sect. 5.10, expressed as

$$\left(\frac{\partial^2}{\partial R^2} - \frac{1}{R}\frac{\partial}{\partial R} + \frac{\partial^2}{\partial Z^2}\right)\psi + \mu_0 R^2\frac{dp}{d\psi} + F\frac{dF}{d\psi} = 0, \tag{3.76}$$

corresponding to $\boldsymbol{\omega} \times \mathbf{v} = -\nabla H$ and $\boldsymbol{\omega} = \nabla \times \mathbf{v}$ having the respective MHD counterparts $\mathbf{j} \times \mathbf{B} = -\nabla p$ and $\mu_0 \mathbf{j} = \nabla \times \mathbf{B}$.

A particular historical case is Hill's spherical vortex that corresponds to $H = -A\psi$ and $C = 0$, where the vorticity is confined to the interior of a uniformly translating sphere of fluid of radius a. In this axisymmetric flow, the vortex lines are circles about an axis through the sphere and the streamlines lie in the meridional planes. Thus with the origin instantaneously coinciding with the centre of the sphere of radius a, we have

$$\omega_\theta = \begin{cases} Ar, \text{ if } r^2 + z^2 < a^2, \\ 0, \text{ if } r^2 + z^2 > a^2. \end{cases} \tag{3.77}$$

and the governing elliptic partial differential equation

$$\frac{\partial^2 \psi}{\partial r^2} - \frac{1}{r}\frac{\partial \psi}{\partial r} + \frac{\partial^2 \psi}{\partial z^2} = -Ar^2. \tag{3.78}$$

Since ψ and $\partial\psi/\partial z$ are continuous at the surface $\sqrt{r^2 + z^2} = a$ (velocity continuity), the solution for the stream function is

$$\psi = \begin{cases} -A(r^4 + r^2z^2 + \frac{5}{3}r^2a^2)/10, & r^2 + z^2 < a^2, \\ Aa^5r^2/15(r^2 + z^2)^{3/2}, & r^2 + z^2 > a^2, \end{cases} \tag{3.79}$$

and the corresponding velocity potential for the irrotational motion when $r^2+z^2 > a^2$ is $\phi = -Aa^5z/15(r^2 + z^2)^{3/2}$—i.e. the ideal incompressible flow outside the radius a is equivalent to that due to a rigid sphere moving with a velocity of magnitude $2Aa^2/15$. Further discussion on vortex dynamics is given in the excellent book by Saffman [13].

Exercises

(Q1) Two line vortices of the same strength and direction of circulation initially occupy the lines $x = \pm a,\ y = 0,\ -\infty < z < \infty$ in an infinite liquid. Determine the flow.

(Q2) (Karman vortex street) Line vortices of equal strength lie in two parallel rows $x = \pm na (n = 0, \pm 1, \pm 2, ...),\ y = \pm b,\ -\infty < z < \infty$ in an infinite liquid at time $t = 0$, where the vortices in the upper row $y = b$ all have anticlockwise circulation and those in the lower row all have clockwise circulation. Given the result

$$\sum_{n=0}^{\infty} \frac{\alpha}{\alpha^2 + (2n+1)^2} = \frac{\pi}{4} \tanh\left(\frac{\pi\alpha}{2}\right),$$

show that each line vortex moves in the x-direction with uniform speed

$$\frac{K}{2a} \tanh\left(\frac{2\pi b}{a}\right),$$

where K is the circulation of each line vortex.

(Q3) If a uniform vortex sheet in the plane $y = 0$ is assumed to consist of a continuous distribution of line vortices, aligned in the z-direction and with strength κ per unit length of the x-axis, show that the corresponding fluid velocity at any point $P(x, y, z)$ is $\mathbf{v}(x, y) = u(x, y)\hat{\mathbf{i}} + v(x, y)\hat{\mathbf{j}}$ where

$$u = -\kappa \int_{-\infty}^{\infty} \frac{y}{y^2 + (x - \xi)^2}\, d\xi, \quad v = \kappa \int_{-\infty}^{\infty} \frac{x - \xi}{y^2 + (x - \xi)^2}\, d\xi.$$

3.10 Stokes Flow Past a Sphere

The formal ordering initiated in Sect. 3.4 indicates that slow flow of a more strongly viscous uniform fluid may be described by Eqs. (3.38) and (3.39). Thus if the constant shear viscosity coefficient μ is assumed to be O(1) rather than O(ϵ) as in Sect. 3.8, the resulting viscous model is

$$\nabla p = \mu \nabla^2 \mathbf{v} = -\mu \nabla \times (\nabla \times \mathbf{v}), \tag{3.80}$$

$$\nabla \cdot \mathbf{v} = 0. \tag{3.81}$$

Stokes carried out one of the earliest viscous flow analyses, in essentially solving Eqs. (3.80) and (3.81) subject to the non-cavitation condition (3.6) and the "no slip" condition (3.61) for slow flow past a sphere.

In an otherwise uniform flow, with constant velocity $\mathbf{U}$ far away from the sphere, a system of spherical coordinates (r, θ, ϕ) may be oriented such that $\theta = 0$ in the direction of $\mathbf{U}$, whence $\mathbf{v} = v_r(r, \theta)\hat{\mathbf{e}}_r + v_\theta(r, \theta)\hat{\mathbf{e}}_\theta$—i.e. the flow is entirely axisymmetric, independent of ϕ. Stokes introduced a stream function to satisfy the incompressibility condition

$$\nabla \cdot \mathbf{v} = \frac{1}{r^2}\frac{\partial}{\partial r}(r^2 v_r) + \frac{1}{r \sin\theta}\frac{\partial}{\partial \theta}(\sin\theta\, v_\theta) = 0,$$

which defines the exact differential

$$d\psi(r, \theta) = (r^2 \sin\theta) v_r d\theta - (r \sin\theta) v_\theta dr$$

such that

$$v_r = \frac{1}{r^2 \sin\theta}\frac{\partial \psi}{\partial \theta}, \quad v_\theta = -\frac{1}{r \sin\theta}\frac{\partial \psi}{\partial r}, \tag{3.82}$$

or in vector form

$$\mathbf{v} = \nabla \psi \times \nabla \phi, \tag{3.83}$$

the analogue for spherical coordinates of the planar stream function representation (3.45). Indeed, apart from the appearance of the angle around the symmetry axis, the representation (3.83) is of course coordinate independent and therefore also applies to the cylindrical coordinates used in Sect. 3.9.

Thus Stokes had effectively shown it remained to solve

$$\nabla^2(\nabla \times \mathbf{v}) = \nabla \times \left(\frac{1}{\mu}\nabla p\right) = \frac{1}{\mu}\nabla \times \nabla p = 0, \tag{3.84}$$

the curl of the equation of motion (3.80), or

$$\left[\frac{\partial^2}{\partial r^2} + \frac{\sin\theta}{r^2}\frac{\partial}{\partial\theta}\left(\frac{1}{\sin\theta}\frac{\partial}{\partial\theta}\right)\right]^2 \psi = 0 \tag{3.85}$$

on introducing the stream function. This partial differential equation admits solutions of form $\psi(r,\theta) = f(r)\sin^2\theta$, where $f(r)$ satisfies the ordinary differential equation:

$$\frac{d^4 f}{dr^4} - \frac{4}{r^2}\frac{d^2 f}{dr^2} + \frac{8}{r^3}\frac{df}{dr} - \frac{8}{r}f = 0. \tag{3.86}$$

Substituting the trial form $f(r) = r^\alpha$ produces the indicial equation

$$\alpha(\alpha-1)(\alpha-2)(\alpha-3) - 4\alpha(\alpha-1) + 8\alpha - 8 = (\alpha-4)(\alpha-2)(\alpha-1)(\alpha+1) = 0,$$

so that the general solution of (3.86) is

$$f(r) = c_1 r^4 + c_2 r^2 + c_3 r + \frac{c_4}{r} \text{ (where the } c_i \text{ are constants).}$$

For the uniform incoming flow, the boundary condition from (3.82) is

$$\psi(r,\theta) \sim \frac{1}{2}r^2 \sin\theta \text{ as } r \to \infty$$

and hence $c_1 = 0$ and $c_2 = 1/2$. The variable r may be scaled in terms of the sphere's radius, so that "no slip" boundary conditions at the sphere $\psi(1,\theta) = \psi_r(1,\theta) = 0$ (corresponding to $v_\theta(1,\theta) = 0$) yield

$$\frac{1}{2} + c_3 + c_4 = 0 \quad \text{and} \quad 1 + c_3 - c_4 = 0,$$

whence $c_1 = -3/4$, $c_4 = 1/4$. Thus the Stokes solution for the stream function in the flow past a stationary sphere is

$$\psi(r,\theta) = \frac{1}{4}\left(2r^2 - 3r + \frac{1}{r}\right)\sin^2\theta; \tag{3.87}$$

or in "dimensional form", for a uniform flow of speed U past a sphere of radius a:

$$\psi(r,\theta) = \frac{1}{4}U\left(2r^2 - 3ar + \frac{a^3}{r}\right)\sin^2\theta. \tag{3.88}$$

Note that the stream function is defined within an arbitrary constant factor, because (3.84) is an homogeneous equation. It follows that the viscous force per unit area on the sphere includes a component $(3\mu/2a)\mathbf{U}$, producing the viscous drag of magnitude $6\pi\mu aU$ found by Stokes.

Exercises

(Q1) Deduce (3.85) from (3.80), and then also directly from (3.84).

(Q2) Derive the Stokes drag $D = 6\pi\mu aU$ from (3.88), noting that the viscous force per unit area follows from

$$\hat{\mathbf{e_r}} \cdot \mathbf{P} = \hat{\mathbf{e_r}} \cdot \left(-p\mathbf{I} + \mu[\nabla\mathbf{v} + (\nabla\mathbf{v})^T]\right),$$

since $\hat{\mathbf{e_r}}$ is the unit normal $\hat{\mathbf{n}}$ at any point on the sphere.

(Q3) (Hele-Shaw cell) A viscous liquid flows slowly between two parallel plates a distance $2a$ apart, which is much less than the length and width of the plates. The pressure distribution along the edge of the liquid layer is arbitrary, but varies slowly with respect to the scale of the separation. Adopt Cartesian axes Oxyz with $z = 0$ the mid-plane of the layer, and make the orderings $\mu = \mathrm{O}(1)$ and $a/L = \mathrm{O}(\varepsilon)$, $\varepsilon \to 0$, where L is a characteristic horizontal length. Assume that p and $\mathbf{v}$ are functions of $\{X, Y, z\}$ where $X \equiv \varepsilon x$ and $Y \equiv \varepsilon y$, and expand p and $\mathbf{v}$ in the formal asymptotic series

$$p = p^{(0)} + \varepsilon p^{(1)} + \cdots$$
$$\mathbf{v} = \varepsilon\mathbf{v}^{(1)} + \varepsilon^2\mathbf{v}^{(2)} + \cdots$$

From (3.80) and (3.81), show that $p^{(0)}$ and $p^{(1)}$ are functions of X and Y only, and that

$$v_z^{(1)} = v_z^{(2)} = 0, \quad \mathbf{v}^{(1)} = -\frac{(a^2 - z^2)}{2\mu}\nabla_\perp p^{(0)} \quad \text{where } \nabla_\perp \equiv \hat{\mathbf{i}}\frac{\partial}{\partial X} + \hat{\mathbf{k}}\frac{\partial}{\partial Y}.$$

Also show that $p^{(0)}$ obeys the planar Laplace equation

$$\nabla_\perp^2 p^{(0)} = 0.$$

3.11 Exact Viscous Flow Solutions

Exact viscous flow solutions often depend upon special mathematical or geometric features, and include unidirectional shear flows where the nonlinear inertia term $\mathbf{v} \cdot \nabla \mathbf{v}$ in the Navier–Stokes equation (3.62) is zero—e.g. flows between infinite parallel plates or through a straight pipe. Other exact viscous solutions include steady plane flow between divergent plates, axial or azimuthal flow between concentric circular cylinders, flow due to steady rotation of an infinite disc, plane or axisymmetric flow against a flat plate, or flow due to the impulsive or sinusoidal motion of an infinite flat plate in its own plane.

However, we recall that a stream function ψ can always be introduced for subsonic planar flow. Thus if z again denotes an ignorable coordinate, Eq. (3.45) enforces both $v_z = 0$ and $\nabla \cdot \mathbf{v} = -\hat{\mathbf{k}} \cdot \nabla \times \nabla \psi = 0$. Moreover,

$$\mathbf{v} \cdot \nabla \psi = 0. \tag{3.89}$$

so that ψ is constant on a streamline—i.e. the streamlines lie within the level surfaces of ψ.

Exercises

(Q1) A uniform fluid of density ρ and shear viscosity coefficient μ is contained between two fixed parallel plates, which are distance d apart and at an angle α to the horizontal. Assuming steady subsonic unidirectional flow under gravity, show that the flow volume per unit width passing between the plates is

$$Q = \frac{(G + \rho g \sin \alpha) d^3}{12\mu},$$

where G denotes the negative downward pressure gradient.

(Q2) A pressure gradient ($-G$ say) is suddenly applied to a uniform fluid of density ρ and viscosity μ in a long circular cylindrical pipe of radius $r = a$. Show that the magnitude of the time-dependent unidirectional velocity along the pipe is given by

$$v(r,t) = \frac{G}{4\mu}(a^2 - r^2) - \frac{2Ga^2}{\mu} \sum_{n=1}^{\infty} \frac{J_0(\lambda_n r/a)}{\lambda_n^3 J_1(\lambda_n)} \exp\left(-\lambda_n^2 \mu t/(\rho a^2)\right),$$

where J_0 and J_1 denote Bessel functions of the first kind, and the λ_n are the successive positive roots of $J_0(\lambda) = 0$.

(Q3) (Couette Flow): A uniform liquid with shear viscosity coefficient μ is contained in the annular region between an inner cylinder of radius a, slowly rotating with angular velocity ω_0 about its axis of symmetry, and an outer fixed coaxial cylinder of radius $b > a$. Assuming no slip boundary conditions and steady flow, show that the angular velocity $\omega = \mathbf{r} \times \mathbf{v}/r^2$ of the liquid at radius r is

$$\omega = \frac{a^2}{r^2}\frac{b^2 - r^2}{b^2 - a^2}\,\omega_0,$$

and that

$$-4\pi\mu\frac{a^2b^2}{b^2 - a^2}\omega_0$$

is the torque per unit length exerted by the liquid on the central cylinder.[5]

3.12 Singular Perturbation Theory

Since exact solutions of the Navier–Stokes equation (3.62) are so exceptional, numerical or perturbation solutions are often sought. However, a modern perspective places both boundary layer theory introduced in Sect. 3.8 and the Stokes solution discussed in Sect. 3.10 in the context of singular asymptotic expansions in perturbation theory [2]—and this provides another insight into boundary layer theory, and a suitable preparation for further discussion of slow viscous flow past a sphere in the next section. Indeed, the relevant asymptotic analysis for both can be developed with reference to a dimensionless form of the Navier–Stokes equation (3.62) as follows.

If L and T denote a characteristic length and a characteristic time for the flow, then $\mathbf{r}^* = \mathbf{r}/L$ and $t^* = t/T$ are dimensionless independent variables. Let us also introduce the dimensionless dependent variables $\rho^* = \rho/ <\rho>$, $p^* = p/ <p>$ and $\mathbf{v}^* = \mathbf{v}/U$, where $<\rho>$, $<p>$ and U denote characteristic reference quantities. On dropping the asterisks, the consequent dimensionless form of (3.62) is

$$\frac{L}{TU}\rho\frac{\partial \mathbf{v}}{\partial t} + \rho\mathbf{v}\cdot\nabla\mathbf{v} + \frac{<p>}{<\rho>\,U^2}\nabla p = \frac{gL}{U^2}\rho\hat{\mathbf{g}} + \frac{\mu}{<\rho>\,UL}\nabla^2\mathbf{v}. \tag{3.90}$$

There are several important *characteristic numbers* in (3.90), associated with famous names in fluid mechanics—viz.

• a dimensionless frequency $L/(TU)$, sometimes called the *Strouhal number*, such that $L/(TU) \ll 1$ defines steady flow;

• a pressure to kinetic energy ratio $< p >/(< \rho > U^2)$—written $(\gamma \mathrm{M}^2)^{-1}$ for an adiabatic fluid where $\mathrm{M} \equiv U/c_\mathrm{s}$ is the *Mach number*, the ratio of the flow speed to the sound speed;

[5]Couette flow is a classical area of investigation in hydrodynamic stability, where theoretical and experimental investigations of ideal and viscous flow were first carried out by Rayleigh and G.I. Taylor [4, 6]. In particular, Taylor showed that viscous Couette flow is stable only if the angular speed $|\,\omega_0|$ of the inner cylinder is sufficiently small, and there is a similar stability condition when the outer cylinder rotates and the inner cylinder is fixed.

• a potential energy to kinetic energy ratio gL/U^2, the *Froude number*, which is usually denoted by Fr; and
• a viscous dissipation to kinetic energy ratio $\mu/(<\rho>UL) = (\text{Re})^{-1}$, where Re is the *Reynolds number*.

Each of these dimensionless numbers is a measure of the importance of the associated term relative to the nonlinear inertia term $\rho\mathbf{v}\cdot\nabla\mathbf{v}$ in a particular flow. Flows that have identical dimensionless numbers and boundary conditions are called similar, and this concept is exploited in physical modelling (e.g. in wind tunnels), although typically not all similarity conditions may be met simultaneously—i.e. they are not usually entirely compatible.

As mentioned in Sect. 3.8, boundary layer theory corresponds to the Reynolds number $\text{Re} \gg 1$, when the viscous term in (3.90) with coefficient $\text{Re}^{-1} \ll 1$ is neglected except where $\nabla^2\mathbf{v}$ is large. Thus a perturbation expansion in terms of the small parameter $\text{Re}^{-1} \ll 1$ renders the dimensionless form of the inviscid equation (3.5) to leading order, except in the boundary layer where the length must be scaled such that the viscous term is retained. As mentioned in Sect. 3.8, an inner boundary layer solution satisfying "no slip" boundary conditions may then be matched to an outer ideal (inviscid) solution, to obtain perturbation solutions in high Reynolds number flows.

In contrast, the steady slow flow discussed in Sect. 3.10 corresponds to a low kinetic energy to pressure ratio $<\rho> U^2/p > \ll 1$ and a low Reynolds number $\text{Re} \ll 1$, such that the pressure gradient and viscous terms in (3.90) are dominant and comparable to leading order. However, at some distance from the sphere considered the omitted inertia term $\rho\mathbf{v}\cdot\nabla\mathbf{v}$ becomes comparable to these terms. In brief, a perturbation expansion in terms of the small parameter $\text{Re} \ll 1$ introduces the inertia term at first order. Consequently, although the zeroth-order Stokes solution is valid throughout the flow, the straightforward perturbation solution to first order is singular (not "uniformly valid")—and this led Oseen to a development we discuss in the following section, from the modern perspective of matched asymptotic expansions.

3.13 The Oseen Correction

Let us first recall that the Navier–Stokes equation (3.62) is obtained on introducing the fluid viscosity at the second order under the the formal procedure begun in Sect. 3.4. On the other hand, the fluid viscosity term is essentially treated as first order in adopting Eq. (3.80) to investigate slow viscous flow, where the fluid inertia term appearing in (3.62) is omitted. If the fluid inertia is retained, the equation replacing (3.80) is

$$\rho\mathbf{v}\cdot\nabla\mathbf{v} + \nabla p = \mu\nabla^2\mathbf{v} = -\mu\nabla\times(\nabla\times\mathbf{v}), \tag{3.91}$$

with its corresponding dimensionless form obtained from (3.90) when the Strouhal and Froude numbers are small—a notion refining the assumption of steady flow and omission of the gravity term. Equation (3.85) in the Stokes stream function $\psi(r,\theta)$ is then replaced by

$$\mathcal{L}^2\psi = \frac{\text{Re}}{r^2\sin\theta}\left(\psi_\theta\frac{\partial}{\partial r} - \psi_r\frac{\partial}{\partial\theta} + 2\cot\theta\,\psi_r - \frac{2\psi_\theta}{r}\right)\mathcal{L}\psi \tag{3.92}$$

where for convenience below the subscripts r and θ on the right-hand side denote the respective first partial derivatives of ψ, the linear second-order differential operator

$$\mathcal{L} \equiv \frac{\partial^2}{\partial r^2} + \frac{\sin\theta}{r^2}\frac{\partial}{\partial\theta}\left(\frac{1}{\sin\theta}\frac{\partial}{\partial\theta}\right),$$

and where the radius of the sphere a and the speed U of the original uniform flow are appropriate characteristic parameters to use in defining the Reynolds number Re. The right-hand side of (3.92) vanishes in the limit $\text{Re} \to 0$, when it reduces to (3.85) solved by Stokes for slow flow past a sphere.

Let us first develop a perturbation solution of (3.92) for $\text{Re} \ll 1$ in the form

$$\psi(r,\theta;\text{Re}) = \sum_{n=0}^{\infty}(\text{Re})^n\,\psi_n(r,\theta). \tag{3.93}$$

On substituting (3.93) into (3.92), Eq. (3.85) in ψ_0 is obtained at the leading (zeroth) order as one would expect; and application of the zeroth-order boundary conditions $\psi_0(1,\theta) = 0$ and $\psi_{0r}(1,\theta) = 0$ at the sphere (of dimensionless unit radius) and

$$\psi_0(r,\theta) \sim \frac{1}{2}r^2\sin\theta \text{ as } r \to \infty$$

then of course produces the Stokes solution (3.87) for $\psi_0(r,\theta)$ as before. At the next order in the small perturbation parameter Re, the equation

$$\mathcal{L}^2\psi_1 = \frac{1}{r^2\sin\theta}\left(\psi_{0\theta}\frac{\partial}{\partial r} - \psi_{0r}\frac{\partial}{\partial\theta} + 2\cot\theta\,\psi_{0r} - \frac{2}{r}\psi_{0\theta}\right)\mathcal{L}\psi_0 \tag{3.94}$$

yields the first-order contribution $\psi_1(r,\theta)$. By substituting the Stokes solution (3.87) into the right-hand side of (3.94), and then noting that this equation has a solution of the form $\psi_1(r,\theta) = g(r)\sin^2\theta\cos\theta$, there again remains an ordinary differential equation to solve—viz.

$$\frac{d^4g}{dr^4} - \frac{12}{r^2}\frac{d^2g}{dr^2} + \frac{24}{r}\frac{dg}{dr} = -\frac{9}{4}\left(\frac{2}{r^2} - \frac{3}{r^3} + \frac{1}{r^5}\right), \tag{3.95}$$

which has the solution

$$g(r) = \frac{d_{-2}}{r^2} + d_0 + d_3 r^3 + d_5 r^5 - \frac{3}{16}r^2 + \frac{9}{32}r + \frac{3}{32}\frac{1}{r}$$

where the $\{d_i\}$ are constants. The first-order boundary conditions $\psi_1(r,\theta) = o(r^2)$ as $r \to \infty$, $\psi_1(1,\theta) = 0$ and $\psi_{1r}(1,\theta) = 0$ (at the sphere) respectively require that $d_5 = d_3 = 0$ and $d_2 = d_0 = -3/32$, whence

$$\psi_1(r,\theta) = -\frac{3}{32}\left(2r^2 - 3r + 1 - \frac{1}{r} + \frac{1}{r^2}\right)\sin^2\theta\cos\theta. \tag{3.96}$$

However, the term $-(3/16)\,r^2$ in (3.96) means that this particular solution does *not* satisfy the condition at infinity; and there is no complementary solution ψ_{1C} of (3.94) that can be added (3.96), to render a first-order solution $o(r^2)$ as $r \to \infty$. Thus the perturbation solution (3.93) to first-order

$$\psi(r,\theta) = \frac{1}{4}\left(2r^2 - 3r + \frac{1}{r}\right)\sin^2\theta$$
$$-\frac{3}{32}\mathrm{Re}\left(2r^2 - 3r + 1 - \frac{1}{r} + \frac{1}{r^2}\right)\sin^2\theta\,\cos\theta + \mathrm{O}(\mathrm{Re}^2) \tag{3.97}$$

satisfies the "no slip" boundary conditions at the sphere but not the condition $\psi(r,\theta) \sim (1/2)r^2\sin\theta$ when $r \to \infty$. This solution was found by Whitehead, who pointed out its nonuniformity in what became known as "Whitehead's Paradox", and its resolution by Oseen implicitly recognised the singular nature of the boundary value problem.

The perturbation solution (3.97) is an asymptotic expansion in the limit $\mathrm{Re} \to 0$, but it is evidently not valid when the term $(3/16)\mathrm{Re}\, r^2\sin^2\theta\cos\theta$ becomes greater than $(3/4)r\sin^2\theta$—i.e. for $r > \mathrm{Re}^{-1}$. Indeed, Oseen had noticed that terms $\mathrm{O}(\mathrm{Re}\, r^{-2})$ are ignored on the right-hand side of (3.92) in rendering (3.85) in the limit $\mathrm{Re} \to 0$, whereas the term

$$\frac{\partial^2}{\partial r^2}\left[\frac{\sin\theta}{r^2}\frac{\partial}{\partial\theta}\left(\frac{1}{\sin\theta}\frac{\partial}{\partial\theta}\right)\right]\psi$$

on the left-hand side of (3.92) is $\mathrm{O}(r^{-3})$. Thus rather than the limit $\mathrm{Re} \to 0$ with r fixed, Oseen considered the limit $\mathrm{Re} \to 0$ with a new variable $r^{(-1)} = \mathrm{Re}\, r$ *fixed*—i.e. the independent variable is "contracted" (since $\mathrm{Re} \ll 1$), enhancing the focus on the distant flow field away from the sphere.[6] Introducing this new independent variable into (3.92) leads to

[6]In contrast, the term "stretched" is used when magnifying the thin inner viscous region in boundary layer theory (when $\mathrm{Re} \gg 1$).

$$\mathcal{L}^2\psi = \frac{\mathrm{Re}^2}{(r^{(-1)})^2 \sin\theta}\left(\frac{\partial\psi}{\partial\theta}\frac{\partial}{\partial r^{(-1)}} - \frac{\partial\psi}{\partial r^{(-1)}}\frac{\partial}{\partial\theta} + 2\cot\theta\,\frac{\partial\psi}{\partial r^{(-1)}} - \frac{2}{r^{(-1)}}\frac{\partial\psi}{\partial\theta}\right)\mathcal{L}\psi, \tag{3.98}$$

where here and henceforth

$$\mathcal{L} \equiv \frac{\partial^2}{\partial (r^{(-1)})^2} + \frac{\sin\theta}{(r^{(-1)})^2}\frac{\partial}{\partial\theta}\left(\frac{1}{\sin\theta}\frac{\partial}{\partial\theta}\right),$$

and we may proceed to obtain an outer solution from (3.98) that matches the inner solution (3.97) in the neighbourhood of $r = \mathrm{Re}^{-1}$.

Oseen's expansion may be obtained by first rewriting the uniformly valid (i.e. valid $\forall\, r$) Stokes solution, the leading term in (3.97), in the variable $r^{(-1)}$—i.e.

$$\frac{1}{4}\left(2\frac{(r^{(-1)})^2}{(\mathrm{Re})^2} - 3\frac{r^{(-1)}}{\mathrm{Re}} + \frac{\mathrm{Re}}{r^{(-1)}}\right)\sin^2\theta$$

$$= \frac{1}{2}\frac{1}{(\mathrm{Re})^2}(r^{(-1)})^2 \sin^2\theta - \frac{3}{4}\frac{1}{\mathrm{Re}} r^{(-1)} \sin^2\theta + \mathrm{O}(\mathrm{Re})$$

suggests an expansion of the form

$$\psi(r^{(-1)}, \theta) = \frac{1}{2}\frac{1}{(\mathrm{Re})^2}(r^{(-1)})^2 \sin^2\theta + \frac{1}{\mathrm{Re}}\psi_1(r^{(-1)}, \theta) + \psi_2(r^{(-1)}, \theta) + \ldots. \tag{3.99}$$

Then noting that $(1/2)(r^{(-1)})^2 \sin^2\theta \equiv \psi^0$ say in the leading term satisfies $\mathcal{L}\psi = 0$, and that

$$\frac{1}{(r^{(-1)})^2\sin\theta}\frac{\partial\psi^0}{\partial\theta} = \frac{\cos\theta}{(\mathrm{Re})^2}, \quad \frac{1}{(r^{(-1)})^2\sin\theta}\frac{\partial\psi^0}{\partial r^{(-1)}} = \frac{\sin\theta}{r^{(-1)}(\mathrm{Re})^2}$$

$$\text{and}\quad 2\cot\theta\frac{\partial\psi^0}{\partial r^{(-1)}} = \frac{2}{r^{(1)}}\frac{\partial\psi^0}{\partial\theta},$$

substituting (3.99) into (3.98) produces the Oseen equation

$$\left(\mathcal{L} - \cos\theta\frac{\partial}{\partial r^{(-1)}} + \frac{\sin\theta}{r^{(-1)}}\frac{\partial}{\partial\theta}\right)\mathcal{L}\,\psi_1(r^{(-1)}, \theta) = 0. \tag{3.100}$$

On setting $\mathcal{L}\,\psi_1 = \phi\exp[(r^{(-1)}/2)\cos\theta]$, Eq. (3.100) yields $(\mathcal{L}-1/4)\phi = 0$, with a solution in the form $\phi = f(r^{(-1)})\sin^2\theta$ where

$$\frac{d^2 f}{dr^{(-1)2}} - \left(\frac{2}{(r^{(-1)})^2} + \frac{1}{4}\right) f = 0. \tag{3.101}$$

Now the solution of Eq. (3.101) that does not grow exponentially is

$$f(r^{(-1)}) = \mathrm{C}\left(1 + \frac{2}{r^{(-1)}}\right) \exp\left(-\frac{1}{2} r^{(-1)}\right) \tag{3.102}$$

where C is a constant, so that $\mathcal{L}\psi_1 = \mathrm{C}(1 + 2/r^{(-1)}) \exp[-(r^{(-1)}/2)(1 - \cos\theta)]$ has the particular solution

$$\psi_{1P} = \frac{3}{2}(1 + \cos\theta) \exp[-(r^{(-1)}/2)(1 - \cos\theta)], \tag{3.103}$$

where the value $3/2$ has been chosen in anticipation of matching the Oseen expansion with the dominant terms of (3.97) as follows. Thus upon adding the complementary function $\psi_{1C} = -(3/2)(1 + \cos\theta)$ obviously satisfying $\mathcal{L}\psi_1 = 0$, we obtain the Oseen asymptotic expansion

$$\psi(r^{(-1)}, \theta) = \frac{1}{2}\frac{1}{\mathrm{Re}^2} (r^{(-1)})^2 \sin^2\theta$$

$$- \frac{3}{2}\frac{1}{\mathrm{Re}} (1 + \cos\theta) \left(1 - \exp[-(r^{(-1)}/2)(1 - \cos\theta)]\right) + \mathrm{O}(1) \tag{3.104}$$

that produces

$$\psi(r, \theta) = \frac{1}{2} r^2 \sin^2\theta - \frac{3}{4} r \sin^2\theta + \mathrm{Re}\frac{3}{16} r^2 \sin^2\theta\,(1 - \cos\theta) + \ldots$$

when rewritten in terms of r and expanding the exponential. Thus (3.104) matches the asymptotic expansion (3.97) near $r = \mathrm{Re}^{-1} \gg 1$, and satisfies the boundary condition $\psi \sim (1/2)(r^{(-1)}/\mathrm{Re})^2 \sin^2\theta = (1/2)\, r^2 \sin^2\theta$ as $r \to \infty$ as sought.

Matching an inner asymptotic expansion written in terms of the outer variable to an outer asymptotic expansion written in terms of the inner variable, which determines the constant C and complementary function ψ_{1C} in this case, has become a well-known analytic technique [2].

Exercises

(Q1) Obtain the modified dimensionless equation (3.92) for slow steady viscous flow from (3.91).

(Q2) Derive (3.101) from the Oseen equation (3.100). Then verify (3.102) and the consequent particular solution (3.103).

(Q3) Darcy's law for flow through porous media, first deduced experimentally and now sometimes viewed as a statistical development from the Navier–Stokes equation, has led to various potential flow problems. For example, when water seeps through homogeneous soil to drain away into small gaps between horizontal hollow circular cylindrical pipes of equal length placed in a row, the flow may be represented by a velocity potential $\phi(r, z)$ satisfying an axisymmetric boundary value problem in dimensionless independent variables—viz.

$$\frac{1}{r}\frac{\partial}{\partial r}\left(\frac{\partial \phi}{\partial r}\right) + \frac{\partial^2 \phi}{\partial z^2} = 0 \qquad (\lambda_1 < r < \lambda_2, \quad |z| < 1)$$

subject to $\phi(\lambda_1, z) = V_0$ (a constant) for $|z| < \epsilon$ but $\phi_r(\lambda_1, z) = 0$ for $\epsilon < |z| < 1$ where $\epsilon \ll 1$ is the dimensionless small gap width, $\phi(\lambda_2, z) = 0$ and $\phi_z(r, \pm 1) = 0$. (The dimensionless λ_1 and λ_2 are the respective ratios of the pipe radius and the seepage distance to the horizontal reference length.) One may obtain an outer asymptotic solution some distance away from the drainpipe

$$\begin{aligned}\phi_{\text{outer}}(r, z) = {} & \frac{1}{2}\lambda_1 \ln(\lambda_2/r) + \frac{1}{\pi}\sum_{n=1}^{\infty}\frac{1}{n}R_n(r)\cos(n\pi z) - \frac{1}{\pi}\sqrt{\frac{\lambda_1}{r}}\ln|1 - x| \\ & - \frac{1}{8\pi^2}\sqrt{\frac{\lambda}{r}}\left(\frac{1}{r} + \frac{3}{\lambda_1}\right)\Re[x + (1 - x)\ln(1 - x)] + \mathrm{O}(\epsilon^2),\end{aligned}$$

where $x = \exp[-\pi(r - \lambda_1) - iz]$ and the $R_n(r)$ are known functions; and an inner asymptotic solution near any gap $\phi_{\text{inner}}(X, Z) \simeq V_0 - (1/\pi)\, u(X, Z) + \mathrm{O}(\epsilon)$, in terms of stretched variables $X \equiv (r - \lambda_1)/\epsilon = \sinh u \sin v$, $Z \equiv z/\epsilon = \cosh u \cos v$ involving convenient elliptic cylindrical coordinates (u, v). Show that these solutions match provided

$$V_0 = \frac{1}{2}\ln\left(\frac{\lambda_2}{\lambda_1}\right) + \frac{1}{\pi}\ln\left(\frac{2}{\pi\epsilon}\right) - \frac{1}{2\pi^2\lambda_1} + \frac{1}{\pi}\sum_{n=1}^{\infty}\frac{1}{n}R_n(\lambda_1).$$

3.14 Dynamical Meteorology and Oceanography*

We close this chapter with our brief presentation of the dynamical equations of meteorology and oceanography, as a point of departure for those interested in either of these two scientific disciplines. The predominant new feature is the Coriolis acceleration due to the rotation of the Earth, in the reformulated equation of motion in a local terrestrial frame of reference—i.e. in a moving reference frame with origin at a point on the surface of the Earth. Although it also revolves around the Sun, the Earth's rotation about its axis in about 24 h (at angular velocity $\omega_0 \simeq 7.3 \times 10^{-5}$ rad/s) produces the dominant centripetal acceleration for geophysical phenomena.

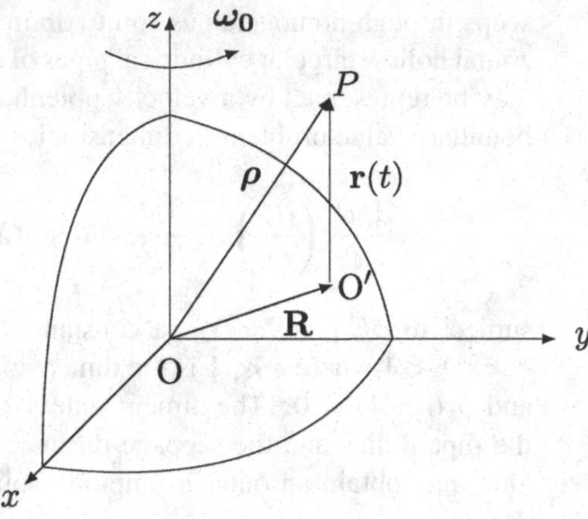

Fig. 3.2 The related inertial and terrestrial frames

Let $\mathbf{r}$ and $\mathbf{v}$ denote the position and velocity of the fluid particle P with reference to the moving terrestrial frame with origin O′. The *true* acceleration in the equation of motion (3.2) is relative to the "fixed" inertial frame with origin O denoted by the subscript F (cf. the last paragraph in Sect. 2.4)—t.e.

$$\left.\frac{d\mathbf{v}}{dt}\right|_F = \left.\frac{d}{dt}\right|_F \left.\frac{d}{dt}\right|_F \rho = \left.\frac{d^2\mathbf{R}}{dt^2}\right|_F + \frac{d\mathbf{v}}{dt} + \frac{d\omega_0}{dt} \times \mathbf{r} + 2\,\omega_0 \times \mathbf{v} + \omega_0 \times (\omega_0 \times \mathbf{r})$$

on expanding (2.29) with $\rho = \mathbf{R}+\mathbf{r}$ in lieu of $\mathbf{r}$, where ρ and $\mathbf{R}$ denote position vectors relative to the inertial frame Oxyz—cf. Fig. 3.2. It is conventional to choose Cartesian axes in the terrestrial frame, with axis O′z' vertically upward, O′x' eastwards and O′y' northwards.

Now the Earth's rotation varies slowly on the geophysical timescales of interest, so we may take $d\omega_0/dt \simeq 0$; and the acceleration of the origin O′ of the terrestrial frame relative to to the origin O of the inertial frame is

$$\left.\frac{d^2\mathbf{R}}{dt^2}\right|_F = \omega_0 \times (\omega_0 \times \mathbf{R}).$$

Further, assuming $|\mathbf{r}| \ll \mathbf{R}$ and noting that gravitational acceleration corresponds to a combination of Newton's law of attraction and the predominant centripetal acceleration

$$\mathbf{g} \equiv -\frac{GM_{\text{Earth}}}{\rho^2}\hat{\rho} - \omega_0 \times (\omega_0 \times \mathbf{R}),$$

we obtain the *equation of motion in the terrestrial frame*

$$\frac{d\mathbf{v}}{dt} + 2\,\omega_0 \times \mathbf{v} + \frac{1}{\rho}\nabla p = \mathbf{g} + \mathbf{F}', \tag{3.105}$$

where $2\boldsymbol{\omega}_0 \times \mathbf{v}$ is known as the Coriolis acceleration and $\mathbf{F}'$ includes specific body forces other than gravity. The Coriolis acceleration is usually resolved into its components in the terrestrial frame. If ϕ denotes the latitude (the angle of the radial line OO' in degrees from the equatorial plane xOy), the components of the Earth's angular velocity are

$$\boldsymbol{\omega}_0 = (\boldsymbol{\omega}_0 \cdot \hat{\mathbf{i}})\,\hat{\mathbf{i}} + (\boldsymbol{\omega}_0 \cdot \hat{\mathbf{j}})\,\hat{\mathbf{j}} + (\boldsymbol{\omega}_0 \cdot \hat{\mathbf{k}})\,\hat{\mathbf{k}} = \omega_0\,(\cos\phi\,\hat{\mathbf{j}} + \sin\phi\,\hat{\mathbf{k}}),$$

so the Coriolis acceleration is

$$2\boldsymbol{\omega}_0 \times \mathbf{v} = 2\omega_0(w\cos\phi - v\sin\phi)\,\hat{\mathbf{i}} + 2\omega_0 u\sin\phi\,\hat{\mathbf{j}} - 2\omega_0 u\cos\phi\,\hat{\mathbf{k}}$$

where $\mathbf{v} = u\,\hat{\mathbf{i}} + v\,\hat{\mathbf{j}} + w\,\hat{\mathbf{k}}$.

Approximations can be made for large-scale motion in the atmosphere or the sea. First of all, both the vertical acceleration and Coriolis acceleration are usually much smaller than the gravity g, so the vertical component of the equation of motion (3.105) is approximately

$$\frac{1}{\rho}\frac{\partial p}{\partial z} + g = 0, \tag{3.106}$$

known as the *hydrostatic equation*. The two horizontal velocity components u and v then satisfy the remaining two scalar equations obtained from the equation of motion (3.105). Thus since the vertical velocity component w is negligible ($|w| \ll |u|, |v|$), the equations for essentially planar motion in the atmosphere or ocean are

$$\begin{aligned} \frac{\partial u}{\partial t} - (2\omega_0\sin\phi)\,v + \frac{1}{\rho}\frac{\partial p}{\partial x} &= F'_x \\ \frac{\partial v}{\partial t} + (2\omega_0\sin\phi)\,u + \frac{1}{\rho}\frac{\partial p}{\partial y} &= F'_y, \end{aligned} \tag{3.107}$$

where ϕ again denotes the latitude angle and it is common to write $f = 2\omega_0\sin\phi$.

When the Rossby number $\mathrm{Ro} = U/fL$ is small (U again denotes a characteristic velocity and L a characteristic length), the motion is often treated as steady. The *geostrophic equations*

$$\begin{aligned} -fv + \frac{1}{\rho}\frac{\partial p}{\partial x} &= 0 \\ fu + \frac{1}{\rho}\frac{\partial p}{\partial y} &= 0 \end{aligned} \tag{3.108}$$

are then suitable for upper atmospheric motion, assuming friction is negligible. On the other hand, friction is important in the atmospheric boundary layer, when certain non-zero inputs for F'_x and F'_y in (3.107) are required. Friction is also important at the top or bottom of the ocean, and equations that have been used to describe *ocean circulation* are

$$-fV + \frac{1}{\rho}\frac{\partial P}{\partial x} = X_{\text{top}} - X_{\text{bottom}}$$
$$fU + \frac{1}{\rho}\frac{\partial P}{\partial y} = Y_{\text{top}} - Y_{\text{bottom}}, \tag{3.109}$$

obtained by integrating (3.107) through the depth of the ocean, where

$$U \equiv \int_0^h u\,dz,\ V \equiv \int_0^h v\,dz,\ P \equiv \int_0^h p\,dz$$

and $\mathbf{F}' = (\partial X/\partial z, \partial Y/\partial z)$.

Large-scale motion is emphasised in A.E. Gill, *Atmosphere-Ocean Dynamics* (Academic Press, 1982), and we also recommend J.R. Holton and G.J. Hakim, *An Introduction to Dynamic Meteorology* (5th Edition, Academic Press, 2012) and J. Pedlosky, *Geophysical Fluid Dynamics* (2nd Edition, Springer, 1987) as further reading. Incidentally, the meteorologist Lorenz was the first to note that nonlinear terms in a deterministic model can lead to a lack of predictability associated with chaos—cf. Drazin [5], for an excellent discussion of nonlinear dynamical systems.

Bibliography

1. D.J. Acheson, *Elementary Fluid Dynamics* (Clarendon Press, Oxford, 2000). (The introductory textbook on theory and applications recommended in the Preface)
2. C.M. Bender, S.A. Orszag, *Advanced Mathematical Methods for Scientists and Engineers I: Asymptotic Methods and Perturbation Theory* (Springer, New York, 1999). (Practical and self-contained presentation, applied to both differential and difference equations)
3. G.K. Batchelor, *An Introduction to Fluid Dynamics* (Cambridge University Press, Cambridge, 2000). (Highly recommended textbook on the dynamics of real fluids)
4. S. Chandrasekhar, *Hydrodynamic and Hydromagnetic Stability* (Dover, New York, 1981). (The widely referenced treatise mentioned in the Preface, with clear and detailed linear analysis of both thermal and dynamic instabilities)
5. P.G. Drazin, *Nonlinear Systems* (Cambridge University Press, Cambridge, 1992). (Graduate level text treating nonlinear difference and differential equations, with applications carefully chosen to explore the fascinating topics of bifurcation and chaos in particular)
6. P.G. Drazin, W.H. Reid, *Hydrodynamic Stability*, 2nd edn. (Cambridge University Press, Cambridge, 2004). (On linear and nonlinear hydrodynamic stability analysis and the transition to turbulence, including a notable contribution on parallel shear flow)
7. O.D. Kellogg, *Foundations of Potential Theory* (Dover, New York, 1953). (Classic systematic treatment of potential functions and applications)
8. H. Lamb, *Hydrodynamics*, 6th edn. (Cambridge University Press, Cambridge, 1993). (First referenced in Chapter 2)
9. L.D. Landau, E.M. Lifshitz, *Fluid Mechanics*, 2nd edn. (Pergamon Press, New York, 1987). (Volume 6 in the celebrated series of graduate textbooks on theoretical physics, in this case

presenting the fundamentals and applications of fluid mechanics, and also a chapter on computational fluid mechanics (available online))

10. P.D. Miller, Applied asymptotic analysis, in *Graduate Studies In Mathematics*, vol. 75, ed. by D. Saltman, et al. (American Mathematical Society, Providence, 2006). (Textbook providing a survey of asymptotic methods, especially in the context of wave propagation)
11. L.M. Milne-Thomson, *Theoretical Hydrodynamics* (Dover, New York, 1996). (Mainly discusses the ideal fluid model, and most notably complex variable theory for two-dimensional flows)
12. A.H. Nayfeh, *Perturbation Methods* (Wiley, New York, 2008). (An orderly survey of the application of perturbation methods in many fields (available online))
13. P.G. Saffman, *Vortex Dynamics* (Cambridge University Press, Cambridge, 1992). (Recommended text on important aspects of fluid motion that are primarily controlled by vorticity)

Chapter 4
Waves in Fluids

Sound and water waves are familiar longitudinal and transverse disturbances relative to the direction of propagation in a fluid, respectively. Sound waves arise in a compressible fluid, but water (gravity) waves are well described in the subsonic incompressible approximation. Our main emphasis in this chapter is on linear wave theory, where the disturbances from an equilibrium or steady state are assumed small and solutions may be obtained by superposition as Fourier series or integral forms. Our analysis is extended to superposed fluids, where hydrodynamic instability may occur—and the other topics chosen either consolidate earlier concepts or are relevant to developments in the following two chapters. The additional bibliographic entries at the end of this chapter provide relevant supplementary reading.

4.1 Introduction

Waves are often propagated through fluids or solids, when some equilibrium or quasi-steady state is disturbed. Waves in fluids result from restoring forces such as pressure or gravity that tend to return displaced fluid elements to their original relative positions, causing bounded oscillatory perturbations of the field variables (the velocity, pressure or density). If the wave source does not persist, the wave energy may propagate away from the local vicinity or the waves may be damped out due to viscous effects (when the fluid kinetic energy dissipates as heat), so the system returns to its original state. On the other hand, if the perturbations grow such that the system departs from its original equilibrium or quasi-steady state, the system is unstable [1, 3].

A wave may have a fairly regular oscillation frequency and wavelength between adjacent crests. It may be a simple infinite wave train represented by the form

$$\zeta(x, t) = a \cos(kx - \omega t), \tag{4.1}$$

R.J. Hosking and R.L. Dewar, *Fundamental Fluid Mechanics and Magnetohydrodynamics*, DOI 10.1007/978-981-287-600-3_4

where x is a space coordinate and t is the time. Equation (4.1) is a special case of the wave form

$$\zeta(\mathbf{r}, t) = a \exp[i(\mathbf{k} \cdot \mathbf{r} - \omega t)] \tag{4.2}$$

where $\mathbf{r}$ is the position vector, representing wave propagation in the direction of the *wave vector* $\mathbf{k}$. (It is usually implicit that the real part of such an exponential form is intended.) The *amplitude a* need not be constant, in what is known as a *Fourier component* in the context of linear wave theory. A Fourier component is evidently characterised by the amplitude, the *wave number k* or *wavelength* $2\pi/k$, and the *frequency* ω or *period* $2\pi/\omega$. Linear theory is appropriate for small-amplitude waves, where all products (quadratic and higher degree terms) in the time-dependent variables are neglected and Fourier components may be superposed to provide the complete signal as a linear *wave form*—cf. Exercise (Q1). Waves are said to be *dispersive* when the *phase speed* $c = \omega/k$ is not constant but depends upon k, such that the components propagate at different speeds and a wave form represented by a superposition of Fourier components changes shape (suffers *dispersion*). However, a wave form may appear to be more random, whether linear or nonlinear—when it is sometimes called a signal. Moreover, it often distorts or changes in magnitude or velocity, but yet remains recognisable [14].

Small-amplitude waves are not always described using the traditional Fourier components (4.1) or (4.2). Consider the wave form

$$\zeta = a \cos S, \tag{4.3}$$

where the amplitude a and the *phase function S* are both functions of x and t. This form includes (4.1) when $S = kx - \omega t$, but with inhomogeneous media in mind one may define a local wave number $k(x, t)$ and local frequency $\omega(x, t)$ by

$$k(x, t) = \frac{\partial S}{\partial x} \quad \text{and} \quad \omega(x, t) = -\frac{\partial S}{\partial t}, \tag{4.4}$$

to more generally identify k as the phase density (radians/unit length) and ω the phase flux (radians/unit time) satisfying the conservation equation

$$\frac{\partial k}{\partial t} + \frac{\partial \omega}{\partial x} = 0. \tag{4.5}$$

Indeed, if the local frequency is related to the wave number k by a *dispersion relation* $\omega = \Omega(k, x, t)$, (4.5) produces the first-order hyperbolic partial differential equation

$$\frac{\partial k}{\partial t} + c_{\mathrm{g}}(k, x, t)\frac{\partial k}{\partial x} = -\frac{\partial \Omega}{\partial x} \tag{4.6}$$

—i.e. a wave equation for the local wave number k involving the *group speed* $c_{\mathrm{g}}(k, x, t) = \partial\Omega/\partial k$, which is also associated with energy transport [14]. In particular, if the frequencies and wave numbers of components of form (4.1) are confined to quite narrow bands, then the envelope of the *wave packet* given by their superposition changes slowly and propagates at the group speed.

The above discussion is extended in the next section on the short wave approximation, subsequently used in our presentation on sound waves in Sects. 4.3 and 4.4. Water waves are then analysed in Sect. 4.5, modifications due to either a floating flexible plate or surface tension are considered in Sect. 4.6, and two classical hydrodynamical instabilities are identified in Sect. 4.7. Local averaging in Sect. 4.8 can provide a general formulation that is applied in Sect. 4.9; the oscillation-centre description in Sect. 4.10 is related to variational stability analysis discussed in Chap. 6; and Sects. 4.11 and 4.12 on shock waves and shock structure preface some important developments in Chap. 5.

Exercises

(Q1) Consider a wave component $a(k)\exp[i(kx-\omega t)]$ with wave number in the near neighbourhood of a particular value k_0, so the dispersion relation is approximated by the truncated Taylor series

$$\omega(k) \simeq \omega(k_0) + \left.\frac{d\omega}{dk}\right|_{k=k_0} (k-k_0) = \omega(k_0) + c_g(k_0)\,(k-k_0).$$

Noting that superposition defines the wave form

$$\zeta(x,t) = \int_{\mathcal{D}} a(k)\exp[i(kx-\omega t)]\,dk$$

over its domain $\mathcal{D}$, show that:

$$\zeta(x,t) \simeq F(x - c_g(k_0)t)\exp[i(k_0 x - \omega(k_0))t],$$

with $F(\xi) = \int_{\mathcal{D}} a(k)\exp[i(k-k_0)\xi]\,dk$; and, if the amplitudes have a Gaussian distribution $a(k) = a_0\exp[-\lambda(k-k_0)^2]$ where $\lambda \gg 1$ (corresponding to wave numbers localised about k_0), show that the travelling wave form is the real part of

$$\zeta(x,t) \simeq a_0\sqrt{\frac{\pi}{\lambda}}\,\exp\left[-\frac{(x-c_g(k_0))^2}{4\lambda}\right]\exp[i(k_0 x - \omega(k_0))t].$$

(Q2) Denoting the phase velocity by $\mathbf{c} = (\omega/k)\hat{\mathbf{e}}_k$, show the group velocity $\mathbf{c}_g$ for a wave with dispersion relation $\omega = kf(k_x/k, k_y/k, k_z/k)$ and f an arbitrary function is the hypotenuse of a right-angled triangle where $\mathbf{c}$ is the base.

4.2 Short-Wavelength Approximation

In this section, we describe a well-known asymptotic approximation method in wave theory to be invoked in the following two sections, applicable when the wavelength may be regarded as small. This short-wavelength assumption may be introduced through the *eikonal ansatz*

$$q_1 = \bar{q}(\mathbf{r}, t, \varepsilon) \exp\left[i\frac{S(\mathbf{r}, t)}{\varepsilon}\right] \tag{4.7}$$

suggested on extending (4.3) as discussed in detail below, and the method is sometimes called the JWKB or WKB approximation following work by Jeffreys, Wentzel, Kramers and Brillouin in the early years of quantum mechanics [7, p. 310]—although it was introduced much earlier by others including Liouville and Green, and is alternatively called the Liouville–Green method. The basic notion is to seek solutions of the form (4.7), where the amplitude $\bar{q}$ is slowly varying but the exponential component (primarily oscillatory) varies rapidly.

As in the previous section, the *eikonal function* (or phase function) S is a generalisation of the wave phase to take into account the variation with position and time in an inhomogeneous medium, but now as a function of $\mathbf{r}$ and t relevant to $\mathbf{k}\cdot\mathbf{r} - \omega t$. The generalised wave vector and frequency corresponding to (4.4) are

$$\mathbf{k} \equiv \nabla S, \tag{4.8}$$

$$\omega \equiv -\frac{\partial S}{\partial t}, \tag{4.9}$$

from which we obtain

$$\nabla \times \mathbf{k} = 0 \left[\text{ or equivalently, } \nabla\mathbf{k} = (\nabla\mathbf{k})^T\right], \tag{4.10}$$

$$\frac{\partial \mathbf{k}}{\partial t} + \nabla\omega = 0 \left[\text{ analogous to (4.5)}\right]. \tag{4.11}$$

In our context, (4.7) represents an array of linearised perturbations of the physical variables involved in the wave motion, $\varepsilon \to 0$ is a formal asymptotic ordering parameter to express the smallness of the wavelength with respect to typical length and time scales of the background system that are *slowly varying* compared with the waves, and $\bar{q}$ is the array of slowly varying wave amplitudes. *The "dummy" parameter ε is included to get the ordering correct, but then set to* 1 *at the end of the calculation.* The background scale length and characteristic time are considered to be O(1), and the amplitude functions $\bar{q}(\mathbf{r}, t, \varepsilon)$ are also assumed to vary on these scales. To develop the formal short-wavelength expansion, the amplitude quantities are expanded in ε—i.e.

$$\bar{q} = \bar{q}^{(0)} + \varepsilon\bar{q}^{(1)} + \mathrm{O}(\varepsilon^2), \tag{4.12}$$

and the linearised equations of motion are required to be satisfied order by order in ε. As the coefficients $\bar{q}^{(n)}$ may be complex, we have not taken S to be a function of ε, because we assume that small phase corrections can be handled through the amplitude expansion. Incidentally, since $\mathbf{k}$ and ω are given by (4.8) and (4.9), in general the eikonal $S(\mathbf{r}, t)$ satisfies a partial differential equation of Hamilton–Jacobi form with Ω acting as the Hamiltonian—viz.

$$\frac{\partial S}{\partial t} + \Omega(\nabla S, \mathbf{r}, t) = 0, \tag{4.13}$$

which can be solved by the method of characteristics.

Let us now extend the analysis of the previous section to consider a dispersion relation of the form

$$\omega = \Omega(\mathbf{k}, \mathbf{r}, t), \tag{4.14}$$

relating the generalised frequency ω and wave vector $\mathbf{k}$. On substituting (4.14) into (4.11), we obtain

$$\frac{\partial \mathbf{k}}{\partial t} + \frac{\partial \Omega}{\partial \mathbf{k}} \cdot (\nabla \mathbf{k})^T + \frac{\partial \Omega}{\partial \mathbf{r}} = 0, \tag{4.15}$$

where $\partial/\partial\mathbf{k}$ is now the vector differential operator with $\mathbf{r}$ held fixed, and $\partial/\partial\mathbf{r}$ has the same form as ∇ but differs from it because the partial derivatives are taken with $\mathbf{k}$ fixed. From (4.10), Eq. (4.15) may be rewritten as the generalisation of (4.6)—viz.

$$\left(\frac{\partial}{\partial t} + \mathbf{c}_g \cdot \nabla\right)\mathbf{k} = -\frac{\partial \Omega}{\partial \mathbf{r}}, \tag{4.16}$$

involving the *group velocity*

$$\mathbf{c}_g = \frac{\partial \Omega}{\partial \mathbf{k}}. \tag{4.17}$$

Further, we may also write

$$\frac{d\mathbf{k}}{dt} = -\frac{\partial \Omega}{\partial \mathbf{r}} \tag{4.18}$$

on defining an advective derivative $d/dt \equiv \partial/\partial t + \mathbf{c}_g \cdot \nabla$ in this context, so the group velocity defines the propagation of wave phase information—and for linear waves, amplitude information.

To complete the solution by characteristics, let us adopt the Lagrangian viewpoint where the group velocity $\mathbf{c}_g$ may be interpreted as the velocity of a fictitious particle with position $\mathbf{r}$ and velocity $\mathbf{c}_g$, according to

$$\frac{d\mathbf{r}}{dt} = \frac{\partial \Omega}{\partial \mathbf{k}}. \tag{4.19}$$

(In the case of sound waves considered in the next sections, the particle is often called a *phonon*.) Equations (4.18) and (4.19) have the form of the Hamiltonian equations of motion with **k** the "momentum" conjugate to **r**, and they map the values of **k** and **r** at time t_0 to their values at time t. Thus $\nabla\mathbf{k}$ remains symmetric ($\nabla \times \mathbf{k} = 0$ is preserved), so in principle S can be constructed at any later time by integrating $\nabla S = \mathbf{k}$, because the identity $\Delta\mathbf{k}_1 \cdot \Delta\mathbf{r}_2 - \Delta\mathbf{k}_2 \cdot \Delta\mathbf{r}_1 = 0$ is preserved by the Hamiltonian flow (cf. the Exercise below). The trajectories of the fictitious particles are called *rays*, and the Hamiltonian equations the *ray equations*.

A more practical way to calculate S is by integrating along the ray trajectories. From the definitions (4.8) and (4.9), the dispersion relation for an arbitrary wave (4.14) and the definition of group velocity (4.17) yield

$$\frac{dS}{dt} = \mathbf{k} \cdot \frac{\partial \Omega}{\partial \mathbf{k}} - \Omega. \tag{4.20}$$

This equation is analogous to that for the *action* $\int L\,dt$ in classical mechanics, where $L \equiv \mathbf{p} \cdot \partial H/\partial \mathbf{p} - H$ is the *Lagrangian function*. The dispersion relation (4.31) for sound waves derived in the next section is homogeneous and linear in **k** such that they are non-dispersive, so the right-hand side of (4.20) vanishes and the phase S is constant on a ray trajectory. However, this is not so for dispersive waves (where the dispersion relations are nonlinear in **k**).

Exercise

(Q1) Consider ray trajectories starting from the points $\mathbf{r}, \mathbf{r} + \Delta\mathbf{r}_1, \mathbf{r} + \Delta\mathbf{r}_2$, and define $\Delta\mathbf{k}_1 \equiv \Delta\mathbf{r}_1 \cdot \nabla\mathbf{k}$, $\Delta\mathbf{k}_2 \equiv \Delta\mathbf{r}_2 \cdot \nabla\mathbf{k}$. Initial conditions $\{(\mathbf{r},\ \mathbf{k})|t\}$ are not arbitrary, since **k** must be derivable from an eikonal such that $\nabla \times \mathbf{k} = 0$ or $\nabla\mathbf{k} = (\nabla\mathbf{k})^T$. Show that $\Delta\mathbf{k}_1 \cdot \Delta\mathbf{r}_2 - \Delta\mathbf{k}_2 \cdot \Delta\mathbf{r}_1 = 0$, and that this result is propagated in time by the ray equation.

4.3 Sound Waves on a Slowly Varying Background Flow

We consider an ideal barotropic fluid (i.e. the density ρ is a function of the fluid pressure p) to discuss sound wave propagation. The flow is described by (3.5) and (3.3) conveniently rewritten as

$$\frac{d\mathbf{v}}{dt} = -\nabla \int \frac{dp}{\rho} \tag{4.21}$$

$$\frac{d \ln p}{dt} = -\gamma \nabla \cdot \mathbf{v}, \tag{4.22}$$

and we consider the behaviour of small-amplitude short-wavelength perturbations—i.e. short with respect to the scale lengths for variations in the background density and fluid velocity. (We shall comment on a consequence of omitting the continuity Eq. (3.1) at the end of this section.)

Thus, substituting $\mathbf{v} = \mathbf{v}_0 + \mathbf{v}_1$ and $p = p_0 + p_1$, where $\mathbf{v}_1$ and p_1 denote small perturbations of the background ("quasi-equilibrium") values $\mathbf{v}_0$ and p_0, into (4.21) and (4.22) and linearising (retaining only first-order terms):

$$\frac{d\mathbf{v}_1}{dt} + \mathbf{v}_1 \cdot \nabla \mathbf{v} = -\nabla \left(\frac{p_1}{\rho} \right) \tag{4.23}$$

$$\frac{d}{dt}\left(\frac{p_1}{p} \right) + \frac{\mathbf{v}_1 \cdot \nabla p}{p} = -\gamma \nabla \cdot \mathbf{v}_1 . \tag{4.24}$$

Note that $\ln p = \int dp/p$, and for convenience the zero subscript on background quantities has been omitted.

Using the eikonal ansatz in (4.7)–(4.12) with $\bar{q}$ denoting $\{\bar{\mathbf{v}}, \bar{p}\}$, at $O(\varepsilon^{-1})$ Eqs. (4.23) and (4.24) yield

$$\omega' \bar{\mathbf{v}}^{(0)} = \frac{\mathbf{k}\, \bar{p}^{(0)}}{\rho} \tag{4.25}$$

$$\omega' \bar{p}^{(0)} = \gamma p \, \mathbf{k} \cdot \bar{\mathbf{v}}^{(0)}, \tag{4.26}$$

where $\omega' \equiv \omega - \mathbf{k} \cdot \mathbf{v}$ is the Doppler shifted frequency seen by an observer moving with the local background fluid velocity. Consequently, substituting (4.26) into (4.25) yields

$$\left(\omega'^2 \mathbf{I} - c_s^2 \mathbf{k}\mathbf{k} \right) \cdot \bar{\mathbf{v}}^{(0)} = 0, \tag{4.27}$$

where

$$c_s \equiv \sqrt{\frac{\gamma p}{\rho}} \tag{4.28}$$

is the sound speed. The dyadic dotting $\bar{\mathbf{v}}^{(0)}$ can be diagonalised by using as basis vectors $\hat{\mathbf{e}}_1 \equiv \mathbf{k}/k$, a unit vector in the $\mathbf{k}$ (longitudinal) direction, and two mutually orthogonal unit vectors $\hat{\mathbf{e}}_2$ and $\hat{\mathbf{e}}_3$ orthogonal (transverse) to $\mathbf{k}$—i.e.

$$\left[\left(\omega'^2 - k^2 c_s^2 \right) \hat{\mathbf{e}}_1 \hat{\mathbf{e}}_1 + \omega'^2 \left(\hat{\mathbf{e}}_2 \hat{\mathbf{e}}_2 + \hat{\mathbf{e}}_3 \hat{\mathbf{e}}_3 \right) \right] \cdot \bar{\mathbf{v}}^{(0)} = 0. \tag{4.29}$$

Equation (4.27) has nontrivial (i.e. nonzero) solutions provided the determinant of the dyadic representation vanishes, giving the dispersion relation

$$\omega'^4(\omega'^2 - k^2c_s^2) = 0. \tag{4.30}$$

The solution $\omega'^2 = k^2c_s^2$ corresponds to sound waves in the compressible fluid [8]. The dispersion relation for sound waves thus has two branches in the form assumed in (4.14), with

$$\Omega(\mathbf{k}, \mathbf{r}) \equiv \mathbf{k} \cdot \mathbf{v} \pm |\mathbf{k}|c_s, \tag{4.31}$$

linear in $\mathbf{k}$ so that sound waves are *non-dispersive* (the phase speed is independent of the wave number, and coincides with the group speed).

The solution $\omega'^4 = 0$ corresponds to transverse perturbations (i.e. $\bar{\mathbf{v}}^{(0)}$ perpendicular to $\mathbf{k}$), which have zero frequency in the fluid reference frame because any transverse restoring force such as gravity has been neglected in the equation of motion (4.21). If we had not ignored the continuity equation (3.1) in the above derivation, we would have found *another* zero-frequency branch corresponding to wave-like modulations of density and temperature compensating each other in such a way as to leave the pressure unaffected. This branch is known as the *entropy wave* [4, p. 189].

Exercises

(Q1) If $x' = x - Vt,\ y' = y,\ z' = z, t' = t$ define the transformation relating the space coordinates (x', y', z') and time t' in a Cartesian reference frame moving along the x-axis with velocity $\mathbf{V} = V\hat{\mathbf{i}}$ relative to another Cartesian frame with space coordinates (x, y, z) and time t, show that $\omega' = \omega - \mathbf{k} \cdot \mathbf{v}$ is an invariant. (The familiar transformation is called Galilean, and the form $\omega' = \omega - \mathbf{k} \cdot \mathbf{v}$ is said to be Galilean invariant—cf. also Sect. 5.4.)

(Q2) Find the frequency of the lowest mode of sound vibration in a long, thin tube of length ℓ: (a) connected end-to-end to form a toroidal cavity; and (b) with closed ends. Assume the pressure, density and ratio of specific heats of the air (taken to be an ideal fluid) in the tube to be given.
Hint: Assume purely axial flow, and first obtain the wave equation for the velocity perturbation.

Q3) Invoking (3.13) for the gravitational potential V, the self-gravitation equation $\nabla \cdot \mathbf{g} = -4\pi G\rho$ introduced in Sect. 2.3 becomes $\nabla^2 V = 4\pi G\rho$. Hence obtain the dispersion relation $\omega^2 = k^2c_s^2 - 4\pi G\rho$ in a uniform compressible self-gravitating fluid, on assuming "plane wave" disturbances of form $C \exp[i(\mathbf{k} \cdot \mathbf{r} - \omega t)]$ where C is a constant. Then determine when the self-gravitation renders the corresponding sound waves unstable.

4.4 The Sonic Wake from a Point Source

By examining the solvability condition for $\mathbf{k} \times \bar{\mathbf{v}}^{(1)}$ in the asymptotic expansion (4.12), it can be shown that the wave amplitude variation is determined by the equation

$$\frac{\partial N}{\partial t} + \nabla \cdot (N\mathbf{c}_g) = 0\,, \tag{4.32}$$

where

$$N \equiv \frac{2\rho|\bar{\mathbf{v}}^{(1)}|^2}{\omega'} = \frac{2|\bar{p}^{(0)}|^2}{\rho {c_s}^2 \omega'}. \tag{4.33}$$

The quantity N is known as the wave action density and (4.32) is a conservation equation for *wave action*, where the associated flux vector is $N\mathbf{c}_g$. An elegant alternative derivation of (4.32) may be obtained by using Whitham's averaged Lagrangian method [14], showing that (4.33) is a very general result, by no means limited to sound waves. Although the result (4.32) is purely classical, an appealing picture for sound waves is that $N/\hbar$ (where $\hbar$ is the Planck constant) is the number density of the phonons that move at the group velocity $\mathbf{c}_g$. Their number is conserved because of the slow variation of the background flow, corresponding to constancy of the occupation number in the *adiabatic approximation* of quantum mechanics, and because dissipation (e.g. due to viscosity) has been neglected they are not absorbed.

Let us now proceed to illustrate the general ray trajectory equations set out in Sect. 4.2, using sound waves as a simple example representative of non-dispersive waves. A sound source such as a jet plane may be treated as quite localised on an atmospheric scale, so let us consider a uniform fluid moving with velocity $v_0\hat{\mathbf{e}}_1$ past a monochromatic *point* source of sound waves located at $\mathbf{r} = 0$, to model a single Fourier component of the noise from a jet plane viewed in a Galilean frame of reference in which it is at rest. From (4.31), on choosing the positive branch the group velocity is

$$\mathbf{c}_g = v_0\hat{\mathbf{e}}_1 + c_s\hat{\mathbf{e}}_k, \tag{4.34}$$

where $\hat{\mathbf{e}}_k \equiv \mathbf{k}/|\mathbf{k}|$ as before. Since $\partial\Omega/\partial\mathbf{r} = 0$, from (4.18) we have that $\mathbf{k}$ is constant on a ray trajectory. The ray trajectories emanating from $\mathbf{r} = 0$ at t_0 are thus $\mathbf{r} = \mathbf{c}_g(t - t_0)$, or from (4.34)

$$\mathbf{r} = (v_0\hat{\mathbf{e}}_1 + c_s\hat{\mathbf{e}}_k)(t - t_0). \tag{4.35}$$

The set of such trajectories for fixed $t - t_0$ but arbitrary $\hat{\mathbf{e}}_k$ (in any arbitrary direction) is a circle centred on $\mathbf{r}_0 = v_0(t - t_0)\hat{\mathbf{e}}_1$ with radius $c_s(t - t_0)$. Because $S(\mathbf{r}, t)$ is constant on a ray trajectory for sound waves, they are circles of constant phase or *wavefronts*. There are two fundamentally different cases—viz. subsonic and supersonic flow.

As illustrated in Fig. 4.1a, for subsonic flow ($v_0 < c_s$) the circular wavefronts do not intersect and can fill all space. For a monochromatic source of frequency ω we have $S(0, t_0) = -\omega t_0$, and thus the solution for $S(\mathbf{r}, t)$ is found by solving (4.35) for t_0 as a function of $\mathbf{r}$ and t. Writing (4.35) as

$$c_s(t - t_0)\hat{\mathbf{e}}_k = \mathbf{r} - v_0(t - t_0)\hat{\mathbf{e}}_1, \tag{4.36}$$

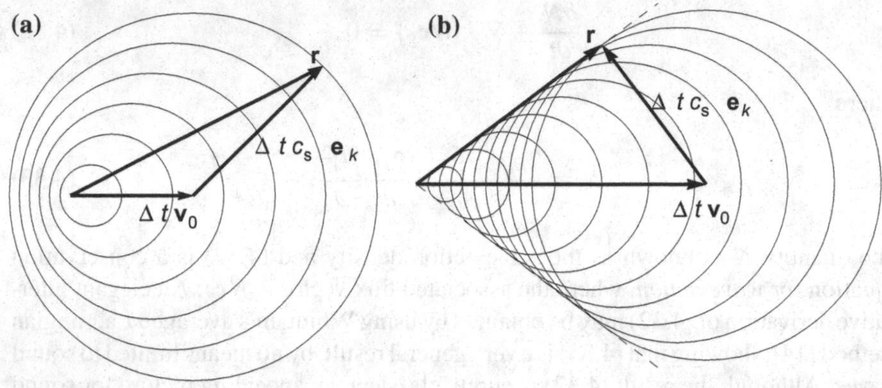

Fig. 4.1 Circular phase fronts of a selected frequency component in the sound wake behind an object moving to the left: **a** subsonic case; **b** supersonic case, showing a point **r** on the Mach cone (*dashed*)

and squaring to remove the arbitrary direction of $\hat{\mathbf{e}}_k$, we obtain the quadratic equation

$$\left(c_s^2 - v_0^2\right)(t - t_0)^2 + 2v_0(\hat{\mathbf{e}}_1 \cdot \mathbf{r})(t - t_0) - |\mathbf{r}|^2 = 0 \tag{4.37}$$

and hence

$$t - t_0 = \frac{-v_0(\hat{\mathbf{e}}_1 \cdot \mathbf{r}) \pm \left[v_0^2(\hat{\mathbf{e}}_1 \cdot \mathbf{r})^2 + \left(c_s^2 - v_0^2\right)|\mathbf{r}|^2\right]^{\frac{1}{2}}}{c_s^2 - v_0^2},$$

so on choosing the causal root $t - t_0 > 0$ the eikonal function is

$$S(\mathbf{r}, t) = -\omega t + \omega|\mathbf{r}| \frac{(c_s^2 - v_0^2 \sin^2\theta)^{\frac{1}{2}} - v_0 \cos\theta}{c_s^2 v_0^2} \tag{4.38}$$

in spherical polar coordinates. Given $(t - t_0)$, we can then also solve (4.35) to obtain

$$\mathbf{c}_g = \frac{\mathbf{r}}{t - t_0} = \frac{\left(c_s^2 - v_0^2\right)\hat{\mathbf{e}}_r}{(c_s^2 - v_0^2 \sin^2\theta)^{\frac{1}{2}} - v_0 \cos\theta} \tag{4.39}$$

where $\hat{\mathbf{e}}_r \equiv \mathbf{r}/|\mathbf{r}|$, so that $\mathbf{c}_g$ is purely radial and outwards. (This outgoing wave condition is a consequence of the causal choice $t - t_0 > 0$.) Assuming $\partial N/\partial t = 0$ and integrating over a narrow cone with apex at the origin, we find from (4.32) that N decreases as $|\mathbf{r}|^{-2}$ in all directions. Variations in amplitude at fixed $|\mathbf{r}|$ are thus produced purely by the radiation pattern of the source at the origin.

Since the eikonal approximation breaks down near the singularity at $\mathbf{r} = 0$, the determination of the radiation pattern is beyond the scope of the present analysis and depends upon the geometry of the device exciting the sound waves. Nevertheless, let us consider what seems the most reasonable definition of an isotropic radiator. If the wave is represented by independently propagating infinitesimal spherical compression and rarefaction wavelets, the pressure disturbance is constant on a spherical wavefront and decreases as R^{-1}, where R is the magnitude of

$$\mathbf{R} \equiv \mathbf{r} - v_0 \mathbf{e}_1 (t - t_0). \tag{4.40}$$

Here $t - t_0$ is the positive root of $R = c_s(t - t_0)$, so from (4.37)

$$R = |\mathbf{r}| c_s \left[\left(c_s^2 - v_0^2 \sin^2 \theta \right)^{\frac{1}{2}} - v_0 \cos \theta \right] / \left(c_s^2 - v_0^2 \right) \tag{4.41}$$

in spherical polar coordinates. Consequently, the pressure perturbation is

$$p_1 = K \frac{\exp\left[i\left(k_0 R - \omega t \right) / \varepsilon \right]}{R} + \text{complex conjugate}, \tag{4.42}$$

where $K = \text{const}$ and $k_0 \equiv \omega / c_s$. Thus (4.33) yields

$$N = \frac{2K^2}{\rho {c_s}^2} \frac{1}{R^2 \omega'}, \tag{4.43}$$

and (4.34) yields

$$\hat{\mathbf{e}}_k = \left(\mathbf{c}_g - v_0 \hat{\mathbf{e}}_1 \right) / c_s = \mathbf{R}/R. \tag{4.44}$$

Since $\mathbf{k} \cdot \mathbf{c}_g = \omega$ from (4.20), we have from (4.39), (4.41) and (4.44) that

$$|\mathbf{k}| = \frac{\omega}{\left(\hat{\mathbf{e}}_k \cdot \mathbf{c}_g \right)} = \frac{\omega}{c_s} \frac{R^2}{\mathbf{r} \cdot \mathbf{R}}, \tag{4.45}$$

and hence

$$\omega' = |\mathbf{k}| c_s = \frac{\omega R^2}{\mathbf{r} \cdot \mathbf{R}} \tag{4.46}$$

and

$$N = \frac{2K^2 \omega}{\rho {c_s}^2} \frac{\mathbf{r} \cdot \mathbf{R}}{R^4}.$$

Thus the wave action density is lower in the upwind direction ($\theta = \pi$) where R is larger, than downwind ($\theta = 0$). Since $\mathbf{c}_g = c_s \mathbf{r}/R$, this is even more so for the action flux vector $N\mathbf{c}_g$. Indeed, for $\theta = \pi$ and $R \to \infty$ we have $N\mathbf{c}_g \to 0$ as $v_0 \to c_{s-}$, so the sonic limit corresponds to the point at which no wave information propagates upwind.

As illustrated in Fig. 4.1b, in supersonic flow ($v_0 > c_s$) the phase fronts intersect and occupy a cone in the downwind direction. The quadratic equation (4.37) then has two valid solutions, and $S(\mathbf{r}, t)$ becomes a "double valued" function. The surface that the phase fronts touch tangentially is the *Mach cone*, where the two solutions for $S(\mathbf{r}, t)$ match up—an example of a *caustic*, a region where the wave amplitude becomes exceptionally large. Indeed, here the Mach cone is a singular surface where the eikonal approximation breaks down in such a way that the amplitude goes to infinity, and recognisable as the cause of sonic booms behind jet aircraft for example.

A closely related phenomenon in supersonic fluid flow is the shock wave, where there are jumps in velocity, density, etc. (cf. Sect. 4.11). We can try to represent small-amplitude shocks using linear theory by making a Fourier superposition of monochromatic waves. However, this only works in the case of non-dispersive waves such as the sound waves studied in this section, because the location of the caustics is then independent of frequency. Dispersive waves are not so prone to shock formation, because wave dispersion can lead to spreading of an initial discontinuity into a smoother pattern at later times—although for large amplitude waves nonlinear effects can overcome dispersion, to produce wave steepening and ultimately shock formation. Wakes in a dispersive wave example (waves on ice sheets) are discussed in Sect. 4.6. Caustics and sometimes "supercaustics" are also an interesting feature of the flexural–gravity or capillary–gravity wave patterns briefly discussed there, provided the source on the surface moves sufficiently fast—cf. also [9].

4.5 Water Waves

The mathematical description of the familiar waves on an otherwise horizontal water surface under gravity, known as surface *water waves* or *gravity waves*, is an instructive introduction to transverse wave propagation [10]. Such waves are transverse because the vertical oscillation of the surface is perpendicular to the horizontal direction of the wave propagation. This terminology may prevail even for standing waves, which are due to appropriately interfering travelling waves, and for waves within a fluid that are usually distinguished as *internal* gravity waves. However, we now restrict our discussion to the surface waves commonly seen on the upper boundary of a body of water.

Water waves usually propagate either into a region at rest or through a uniform stream such that the essentially ideal subsonic flow remains irrotational (cf. Sect. 3.3), and therefore may be described via a velocity potential $\phi(\mathbf{r}, t)$ satisfying the Laplace equation (3.40). Thus the fluid velocity is again given by (3.29) and the pressure variation by (3.60), but the new feature is the perturbation $z = \zeta(x, y, t)$ of the original horizontal free surface $z = 0$ at time $t = 0$, if we adopt Cartesian coordinates.[1]

[1]Fluid particles near the water surface actually move forward as a wave peak passes, and then backward as a wave trough passes, so there are local circulations that flatten with depth into horizontal oscillatory motions within a surface layer. However, for small-amplitude waves it is reasonable to assume that the flow remains irrotational outside this narrow layer.

Let us proceed to formulate the relevant boundary conditions. The requirement that the fluid does not cross this free surface is

$$\hat{\mathbf{n}} \cdot \mathbf{v} = \hat{\mathbf{n}} \cdot \mathbf{v}_s \tag{4.47}$$

at any point moving with velocity $\mathbf{v}_s$ on the surface, where $\hat{\mathbf{n}}$ is the normal at the surface and $\mathbf{v}$ the local fluid velocity. Since it may be assumed that any point on this free surface moves in the z-direction such that $\mathbf{v}_s = \hat{\mathbf{k}}\,\partial\zeta/\partial t$, and the normal to the level surface $z - \zeta(x, y, t) = 0$ is $\hat{\mathbf{n}} = \nabla(z - \zeta) = \hat{\mathbf{k}} - \hat{\mathbf{i}}\,\partial\zeta/\partial x - \hat{\mathbf{j}}\,\partial\zeta/\partial y$, the kinematic condition (4.47) in Cartesian coordinates is

$$w(\mathbf{r}, t) - u\frac{\partial \zeta}{\partial x} - v\frac{\partial \zeta}{\partial y} = \frac{\partial \zeta}{\partial t} \tag{4.48}$$

where the fluid velocity field $\mathbf{v}$=(u, v, w)= $-\nabla\phi$. Linearising (4.48) and introducing the velocity potential ϕ therefore produces

$$-\frac{\partial \phi}{\partial z} = \frac{\partial \zeta}{\partial t}, \tag{4.49}$$

for small perturbations on water initially at rest. The corresponding linearised form of the Bernoulli equation (3.60) for incompressible irrotational flow provides the pressure variation at the surface $z = \zeta$

$$p = \rho\left(\frac{\partial \phi}{\partial t} - g\zeta\right), \tag{4.50}$$

on adopting the gravitational potential $V = gz$ for $\mathbf{g} = -\nabla V = -g\hat{\mathbf{k}}$ under (3.13). Then assuming the atmospheric pressure as constant, the pressure variation is set zero at the surface $z = \zeta$, whence

$$\frac{\partial \phi}{\partial t} - g\zeta = 0. \tag{4.51}$$

Moreover, in the linearised theory the free surface conditions (4.49) and (4.51) may be applied at $z = 0$ rather than $z = \zeta$, since only zeroth-order terms in Taylor expansions for the perturbation quantities about $z = 0$ need be retained, and combining the two conditions gives

$$\frac{\partial^2 \phi}{\partial t^2} + g\frac{\partial \phi}{\partial z} = 0 \quad \text{at } z = 0. \tag{4.52}$$

Finally, assuming the waves propagate on water above an impenetrable bed of constant depth $z = -H$ say, the corresponding boundary condition is

$$\frac{\partial \phi}{\partial z} = 0 \quad \text{at } z = -H. \tag{4.53}$$

The Fourier form for the solution of the Laplace equation (3.40) for the velocity potential is $\phi(x, y, z, t) = f(z) \exp[i(\mathbf{k} \cdot \mathbf{r} - \omega t)]$ where $\mathbf{k} = k_x \hat{\mathbf{i}} + k_y \hat{\mathbf{j}}$. Thus we have

$$\frac{d^2 f}{dz^2} - k^2 f = 0, \tag{4.54}$$

with solution $f(z) \sim \cosh k(z + H)$ to satisfy the boundary condition (4.53). The remaining condition (4.52) then yields the dispersion relation

$$\omega^2 = gk \tanh kH, \tag{4.55}$$

which has two branches $\omega = \pm\Omega(k)$ where $\Omega(k) = \sqrt{gk \tanh kH}$. We recall that the phase speed $c(k) = \omega/k$ defines the rate of propagation of the wave components, and the group speed $c_g(k) = d\omega/dk$ defines their associated energy transfer. In passing, let us also note that from (4.51) applied at $z = 0$ we have

$$\phi(x, y, z, t) = -\frac{ig}{\omega} a \frac{\cosh k(z + H)}{\cosh kH} \exp[i(\mathbf{k} \cdot \mathbf{r} - \omega t)], \tag{4.56}$$

corresponding to the elementary form for the displacement (4.2).

At shorter wavelengths on sufficiently deep water such that $kH \gg 1$ and hence $\tanh kH \simeq 1$, the dispersion relation (4.55) reduces to $\omega^2 \simeq gk$, when the waves are strongly dispersive corresponding to $c \simeq \sqrt{g/k} \simeq 2c_g$. This approximation applies to surface waves on a deep ocean for example, where it is well-known that ocean waves disperse and the longer wavelengths travel faster. Indeed, water waves are characteristically dispersive, but significantly less so at the longest wavelengths where $kH \ll 1$ such that $\tanh kH \simeq kH$, when Eq. (4.55) yields $\omega \simeq \pm\sqrt{gH}k$ such that $c \simeq c_g \simeq \pm\sqrt{gH}$.

The superposition to produce any general solution in linearised wave theory is typically an expression in Fourier series or Fourier integrals. For example, in the one-dimensional case of a surface initially at rest such that $\zeta(x, 0) = \zeta_0(x)$ say and $\partial_t \zeta = 0$ at $t = 0$ [consistent with $\phi(x, z, 0) = 0$], the water wave solution is

$$\zeta(x, t) = \frac{1}{2} \int_{-\infty}^{\infty} \tilde{\zeta}_0(k) \exp[i(kx - \Omega t)] dk + \frac{1}{2} \int_{-\infty}^{\infty} \tilde{\zeta}_0(k) \exp[i(kx + \Omega t)] dk \tag{4.57}$$

for $|x| < \infty$ and $t > 0$, involving

$$\tilde{\zeta}_0(k) = \frac{1}{2\pi}\int_{-\infty}^{\infty} \zeta_0(x)\exp(-ikx)dx.$$

Wave propagation to the right (in the positive x-direction) and to the left (in the negative x-direction) appear in both integrals in (4.57), since $\Omega(k)$ is an even function. Thus the initial surface displacement is split into equal components that propagate along characteristics defined by $x \pm c(k)t$ = constant where $c(k) = \Omega/k$, which has proven to be an especially useful concept in *nonlinear* wave theory [14]. Note that the dispersion changes the wave shape, as each component travels at a different phase speed.

Exercises

(Q1) Find data to justify the approximation $\omega^2 \simeq gk$ of dispersion relation (4.55) for most waves on a deep ocean, and then consider the case of tsunamis.

(Q2) A train of waves on a deep ocean (with wavelength much shorter than the ocean depth) is obliquely incident on a current channel. The current channel occupies the region $y > 0$ adjacent to a region $y < 0$ of still water, and the velocity field in the current is a shear flow $\mathbf{v} = Vy\,\hat{\mathbf{i}}$ with V constant (independent of depth), where (x, y) denote the relevant plane Cartesian coordinates on the ocean surface. Assuming the suitably modified water wave dispersion relation in the local rest frame of reference, use the ray equations to show that provided $k_x V > 0$ (there is downstream directed incidence) the wave number component k_x remains constant with respect to y but the other component k_y reverses sign at a finite value of y given by

$$y_c = \frac{\sqrt{g}[(k_0^2 + k_x^2)^{1/4} - |k_x|^{1/2}]}{k_x V},$$

where k_0 is the incident value of k_y. (This sign reversal implies that the waves are refracted out of the channel.)

(Q3) Assuming the general solution form

$$\zeta(x,t) = \int_{-\infty}^{\infty} a(k)\exp i[kx - \Omega(k)t]\,dk + \int_{-\infty}^{\infty} b(k)\exp i[kx + \Omega(k)t]\,dk$$

corresponding to superposition of Fourier components for the one-dimensional wave propagation, derive the result (4.57). Show that (4.57) may be re-expressed more concisely as

$$\zeta(x,t) = \int_{-\infty}^{\infty} \tilde{\zeta}_0(k)\exp(ikx)\cos(\Omega t)\,dk\ ;$$

and finally obtain the solution where the rightward and leftward are explicit, if the initial form $\zeta(x, 0) = \zeta_0(x)$ is an even function.

(Q4) A stone thrown into a pond produces small ripples that propagate in ever widening circles. Assuming the initial disturbance represented by

$$\zeta(x, 0) = \begin{cases} Q, & \text{if } |x| \leq R \\ 0, & \text{if } |x| > R \end{cases}$$

with $\partial_t \zeta = 0$ at $t = 0$ produces the propagation along any straight line through the origin, show that

$$\zeta(x, t) = \frac{Q}{2\pi} \int_{-\infty}^{\infty} \frac{\sin(kR)}{k} \exp[i(kx - \Omega t)]dk + \frac{Q}{2\pi} \int_{-\infty}^{\infty} \frac{\sin(kR)}{k} \exp[i(kx + \Omega t)]dk.$$

Noting that the exponential factor $\exp[ih(k)]$ where $h(k) = kx - \Omega(k)t$ varies rapidly when either x or t is sufficiently large, use the asymptotic ("stationary phase") result for a single stationary point $k_0 \in \mathcal{D}$

$$\int_{\mathcal{D}} f(k) \exp[ih(k)]\, dk \simeq \sqrt{\frac{2\pi}{h''(k_0)}}\, f(k_0) \exp[ih(k_0) + i\frac{\pi}{4} \operatorname{sgn} h''(k_0)]$$

where $h'(k_0) = 0$ but $h''(k_0) \neq 0$ to obtain the eventual wave form propagating away to the right on a deep pond as

$$\zeta_R(x, t) \simeq \frac{4Q}{t} \sqrt{\frac{x}{\pi g}} \sin\left(\frac{gt^2 R}{4x^2}\right) \cos\left(\frac{gt^2}{4x} - \frac{\pi}{4}\right).$$

4.6 Floating Flexible Plates and Surface Tension

As mentioned, water waves are often called gravity waves, which is because gravitation is the dominant restoring force towards the static equilibrium characterised by the pressure balance equation $\nabla p_0 = \rho \mathbf{g}$ implicitly assumed in the analysis of the previous section. In order to discuss a floating ice sheet for example [9], shown schematically in Fig. 4.2, the mathematical model may be extended to include a flexible plate at the surface that will introduce one or two additional restoring forces.

The governing equation for the deflexion $\zeta(\mathbf{r}, t)$ of a thin elastic plate is

$$D\nabla^4 \zeta + N\nabla^2 \zeta + \rho' h \frac{\partial^2 \zeta}{\partial t^2} = p - f, \tag{4.58}$$

which involves an elastic restoring force proportional to flexural rigidity D, in addition to an in-plane stress (tensile and restoring when $N < 0$, but compressive and just

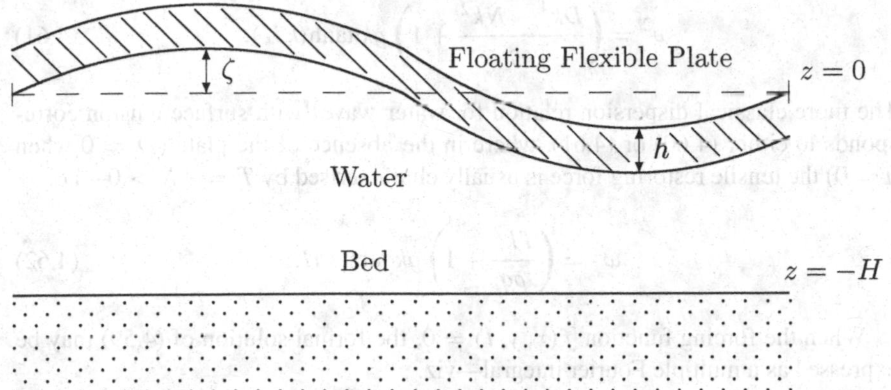

Fig. 4.2 Schematic of the deflected floating flexible plate. (Reproduced with permission of Cambridge University Press)

the opposite when the coefficient $N > 0$). The term $\rho' h \partial^2 \zeta / \partial t^2$ is the plate acceleration, where ρ' denotes the plate density and h its thickness; p denotes the underlying water pressure variation given by the linearised Bernoulli equation (4.50) as before, which together with the assumption that the water always remains in contact with the deflected plate (no cavitation) once again defines the gravitational restoring force; and the forcing function f represents some load located on the surface. Thus on invoking (4.50), the fundamental equation (4.58) for the deflexion at the surface becomes

$$D\nabla^4\zeta + N\nabla^2\zeta + \rho' h \frac{\partial^2 \zeta}{\partial t^2} = \rho \left.\frac{\partial \phi}{\partial t}\right|_{z=0} - \rho g \zeta - f, \tag{4.59}$$

with reference again to $z = 0$ rather than $z = \zeta$ in the linearised theory.

For free waves proportional to $\exp[i(\mathbf{k} \cdot \mathbf{r} - \omega t)]$, the consequent generalisation of dispersion relation (4.55) follows by first substituting the result (4.56) for the velocity potential ϕ into the kinematic non-cavitation condition (4.49) applied at $z = 0$. Adopting Cartesian coordinates as before, we therefore have

$$\frac{\partial \zeta}{\partial t} = k \tanh(kH)\phi(x, y, 0, t),$$

and substituting in (4.59) with the forcing function omitted ($f(x, y, t) \equiv 0$) yields

$$\omega^2 = \frac{Dk^5/\rho - Nk^3/\rho + gk}{kh' + \coth(kH)}, \quad \text{where } h' = \frac{\rho' h}{\rho}. \tag{4.60}$$

Provided the wavelength is large compared with plate thickness ($kh' \ll 1$), the plate acceleration term may be neglected, so the dispersion relation (4.60) reduces to

$$\omega^2 = \left(\frac{Dk^4 - Nk^2}{\rho g} + 1\right) gk \tanh(kH). \tag{4.61}$$

The more classical dispersion relation for water waves with surface tension corresponds to either (4.60) or (4.61), where in the absence of the plate ($D = 0$ when $h = 0$) the tensile restoring force is usually characterised by $T = -N > 0$—i.e.

$$\omega^2 = \left(\frac{Tk^2}{\rho g} + 1\right) gk \tanh kH. \tag{4.62}$$

When the forcing function $f(x, y, t) \neq 0$, the formal solution of (4.59) may be expressed as a multiple Fourier integral—viz.

$$\zeta(x, y, t) = -\int \frac{\tilde{f}(l, m, \omega) e^{i\,(lx+my-\omega t)} dl\, dm\, d\omega}{Dk^4 - Nk^2 + \rho g - \rho' h \omega^2 - (\rho\omega^2/k) \coth kH} \tag{4.63}$$

for $\sqrt{x^2 + y^2} < \infty$ and $t > 0$, where the wave vector $\mathbf{k} = l\,\hat{\mathbf{i}} + m\,\hat{\mathbf{j}}$ and

$$\tilde{f}(l, m, \omega) = \frac{1}{8\pi^3} \int f(x, y, t) e^{-i\,(lx+my-\omega t)} dl\, dm\, d\omega$$

is the associated multiple Fourier transform. Dispersion relation (4.60) corresponds to setting the denominator in the integrand of (4.63) to zero—but (4.63) yields wave patterns generated by a surface source, such as a steadily moving load representing a vehicle moving with uniform velocity $V\hat{\mathbf{i}}$ say, when the forcing function is of the form $f(x, y, t) = F(x - Vt, y)$. There are three typical length scales—a short scale characterised by the modified plate thickness h', a long scale characterised by the water depth H, and an intermediate scale corresponding to a reciprocal wave number $k_{\min}^{-1}$ where the phase speed has a minimum $c_{\min}$. We note also that the phase speed is generally reduced by compressive stress ($N > 0$)—until the stress has become so large that the plate buckles, despite the two restoring forces due to its flexibility and the underlying water foundation—cf. Exercise (Q1) below.

Neglecting the in-plane stress ($N = 0$), typical dispersion curves for sea ice at McMurdo Sound (Antarctica) are shown in Fig. 4.3, where the parameters are $D = 7.32422 \times 10^9\,\mathrm{Nm}^{-2}$, $\rho = 1000\,\mathrm{kg\,m}^{-3}$, $g = 9.8\,\mathrm{m\,s}^{-2}$ and $H = 350\,\mathrm{m}$.[2] The load speed $V = 30\,\mathrm{ms}^{-1}$ is representative for an aircraft landing on the ice sheet, and the unmarked points of intersection of the corresponding horizontal line with the phase speed curve define the wave numbers of the propagated waves. (The wave numbers k_A and k_B shown at the intersections with the group speed curve define points of stationary phase in time-dependent theory, which was pursued to confirm that the ice response to a uniformly moving load rapidly approaches a steady state as

[2] The relevant traditional analysis assumed a steadily moving line load corresponding to the forcing function $f(x, t) = \mathrm{const.}\,\delta(x - Vt)$, which essentially renders the response in the direction the load moves.

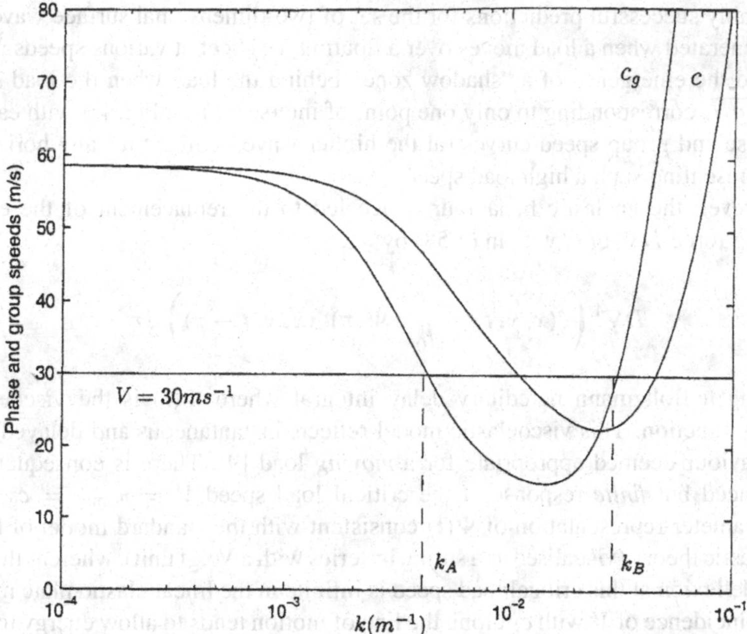

Fig. 4.3 The phase speed $c(k)$ and group speed $c_g(k)$ versus wave number k. (Reproduced with permission of Cambridge University Press)

mentioned again below.) Note that the group speed curve intersects the phase speed curve at its local minimum $c_{\min}$, and the two curves asymptotically coincide in the limit $k \to 0$ (when $c = c_g = \sqrt{gH}$, the long wavelength water wave speed identified in the previous section).

Flexural–gravity wave patterns are generated for all load speeds V greater than $c_{\min}$. For quite short waves (sufficiently large wave number k), the dispersion relation (4.60) reduces to

$$\omega^2 \simeq \frac{D}{\rho h'} k^4 = \frac{D}{\rho' h} k^4, \tag{4.64}$$

when the group speed is twice the phase speed. For longer waves and typical values of D and N in an ice sheet, the dispersion relation (4.60) or (4.61) renders the group speed less than the phase speed. Thus a characteristic flexural–gravity wave pattern has predominantly flexural waves of relatively short-wavelength ahead of a travelling pressure distribution, and rather longer predominantly gravity waves behind. In passing, we note this corresponds to the group speed characterising the energy propagation, relatively faster ahead of the load and slower behind. There is a pronounced localised response at the critical load speed $V = c_{\min} = c_g$, the speed at which energy tends to accumulate beneath the moving load. Asymptotic solutions at a distance from the load, which perhaps surprisingly may usually be treated as a concentrated point source (due to the relatively long wavelengths involved), produced remarkably successful predictions for the set of two-dimensional surface wave patterns generated when a load moves over a floating ice sheet at various speeds.[3] This included the emergence of a "shadow zone" behind the load when the load speed $V > \sqrt{gH}$, corresponding to only one point of intersection in Fig. 4.3 with each of the phase and group speed curves (at the higher wave number) for any horizontal line representing such a high load speed.

However, the anelastic behaviour of ice led to the replacement of the elastic restoring force $D\nabla^4\zeta(x, y, t)$ in (4.58) by

$$D\nabla^4 \left(\zeta(x, y, t) - \int_0^\infty \Psi(\tau)\zeta(x, y, t - \tau) \right) d\tau,$$

involving a Boltzmann hereditary delay integral where $\Psi(t)$ is the viscoelastic memory function. This viscoelastic model reflects instantaneous and delayed elastic behaviour deemed appropriate for a *moving* load [9]. There is consequently a pronounced but *finite* response at the critical load speed $V = c_{\text{min}} = c_g$ for a two-parameter representation of $\Psi(t)$ consistent with the standard model of linear viscoelastic theory (visualised as a spring in series with a Voigt unit), whereas the predicted deflexion at this critical load speed is infinite in the linear elastic plate model. (The coincidence of V with c_g along the line of motion tends to allow energy to continuously accumulate beneath the moving load.) The linear viscoelastic theory also predicts that the maximum deflexion generally occurs behind the load, and explains other observed phenomena—e.g. the asymmetric (rather than symmetric) quasi-static response when $V < c_{\text{min}}$, and the more severe attenuation of the predominantly flexural waves and why the two-dimensional flexural–gravity wave pattern may appear "swept back" to some extent (when $V > c_{\text{min}}$). Stratification of the underlying water has been considered, when there may be internal water waves generated. On the other hand, explicit consideration of finite plate thickness largely confirmed the validity of the thin plate assumption [9]. Later viscoelastic time-dependent theory[4] demonstrated the emergence of the pronounced but finite steady state response at the critical load speed $V = c_{\text{min}} = c_g$; and that the steady state is approached much more rapidly at all other load speeds, compared with earlier predictions from the linear elastic plate model. The upper part of Fig. 4.4 essentially illustrates the response in the line of motion of the load just above the critical speed (at $V = 22.5\,\text{m}\,\text{s}^{-1}$), for the parameters of Fig. 4.3 and the two-parameter representation of $\Psi(t)$. The lower part of Fig. 4.4 shows the wave patterns likewise obtained for various supercritical load speeds—viz. (a) $V = 22.5\,\text{m}\,\text{s}^{-1}$, (b) $V = 40.0\,\text{m}\,\text{s}^{-1}$, (c) $V = 58.6\,\text{m}\,\text{s}^{-1}$ and (d) $V = 70.0\,\text{m}\,\text{s}^{-1}$.

---

[3]The two-dimensional wave patterns were originally predicted by J.W. Davys, R.J. Hosking and A.D. Sneyd (*Journal of Fluid Mechanics* **158**, 269–287, 1985). The corresponding time-dependent asymptotic analysis for an impulsively started concentrated point source demonstrated that a steady state is approached for all load speeds other than $V = c_{\text{min}} = c_g$, including the load speed $V = \sqrt{gH}$ in the long wavelength limit $k \to 0$—cf. W.S. Nugroho, K. Wang, R.J. Hosking & F. Milinazzo (*Journal of Fluid Mechanics* **381**, 337–355, 1999).

[4]K. Wang, R.J. Hosking and F. Milinazzo (*Journal of Fluid Mechanics* **521**, 295–317, 2004).

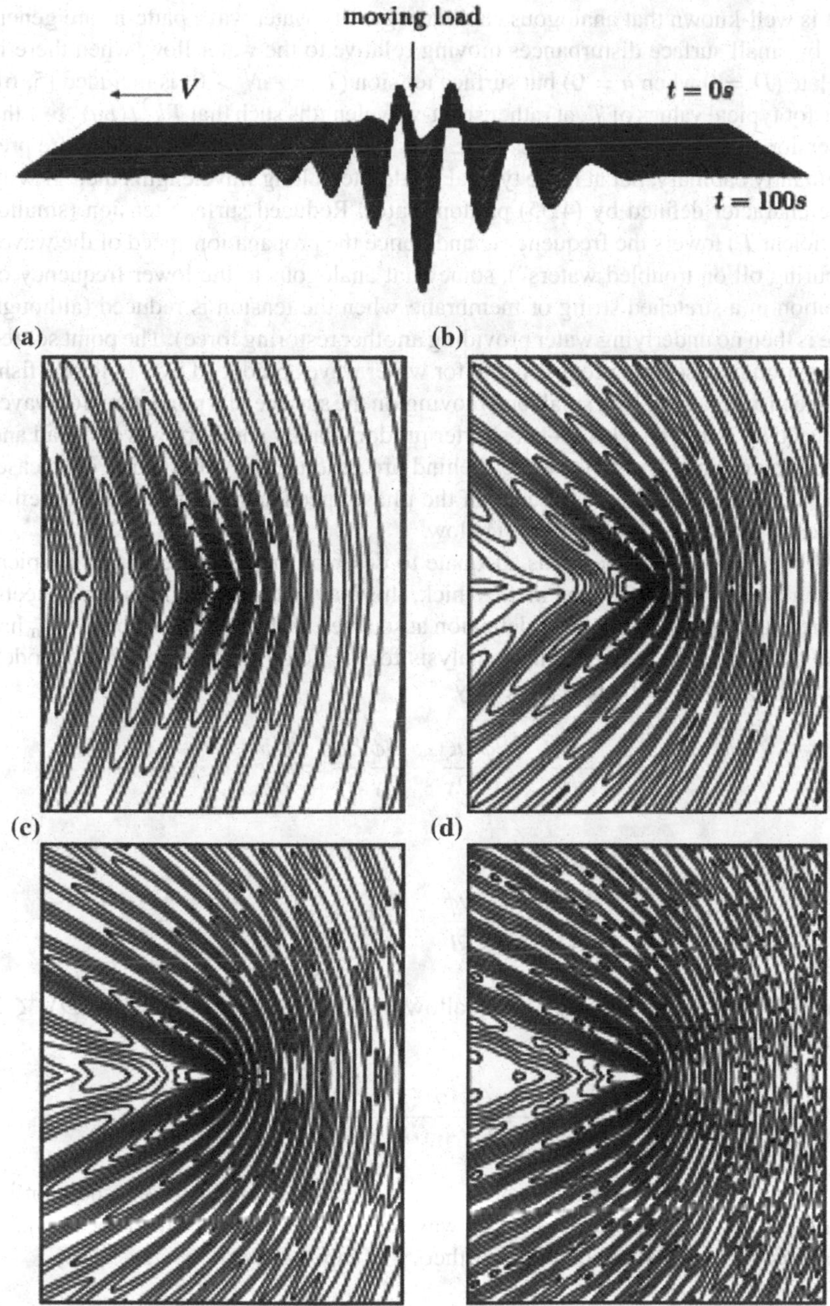

Fig. 4.4 Viscoelastic evolution at $c_{\min}$ (*above*) and $V > c_{\min}$ wave patterns (*below*). (Reproduced with permission of Cambridge University Press)

It is well-known that analogous capillary–gravity water wave patterns are generated by small surface disturbances moving relative to the water flow, when there is no plate ($D = 0$ when $h = 0$) but surface tension ($T = -N > 0$) is included [5, 6]. Thus for typical values of T, at rather short-wavelengths such that $Tk^2/(\rho g) \gg 1$ the dispersion relation (4.62) becomes $\omega^2 \simeq (Tk^3/\rho) \tanh kH$, when the waves are predominantly capillary, but at more typical moderate to long wavelengths their gravity wave character defined by (4.55) predominates. Reduced surface tension (smaller coefficient T) lowers the frequency ω and hence the propagation speed of the waves ("pouring oil on troubled waters"), somewhat analogous to the lower frequency of vibration in a stretched string or membrane when the tension is reduced (although there is then no underlying water providing another restoring force). The point source assumption may seem more justifiable for water waves produced by a twig or a fishing line in a stream, or by a small bug moving on the surface of a pond, than for waves produced by a ship on a lake—but shorter predominantly capillary waves ahead and longer predominantly gravity waves behind are nevertheless observed in each case. Once again, the relative magnitudes of the phase speed and the group speed define this feature—cf. also Exercise (Q2) below.

Although linear theory seems adequate to describe the response due to a typical moving load on sea ice several metres thick, this may not be so on thinner ice sheets. In particular, the relatively large deflexion associated with the critical speed $c_{\min}$ has led to the development of nonlinear analysis for the floating thin elastic plate model, where (4.49) and (4.59) are replaced by

$$-\frac{\partial \phi}{\partial t} - \frac{\partial \phi}{\partial x}\frac{\partial \zeta}{\partial x} - \frac{\partial \phi}{\partial y}\frac{\partial \zeta}{\partial y} = \frac{\partial \phi}{\partial z}$$

and

$$D\nabla^4\zeta + \rho' h \frac{\partial^2 \zeta}{\partial t^2} = \rho\left(\frac{\partial \phi}{\partial t} + \frac{1}{2}|\nabla\phi|^2\right)\bigg|_{z=0} - \rho g \zeta - f,$$

respectively[5]—and more recently, by allowing for plate curvature where $D\nabla^4\zeta$ is replaced by

$$D\frac{\partial}{\partial x^2}\left(\frac{\partial^2\zeta/\partial x^2}{1 + (\partial\zeta/\partial x)^2}\right).$$

The wave pattern at supercritical speed $V > c_{\min}$ is similar (cf. Fig. 4.5), but nonlinearity permits the evolution of solitary waves ("solitons") near critical speed analogous to earlier results in the nonlinear theory of capillary–gravity waves [12].

[5] E. Parau and F. Dias (*Journal of Fluid Mechanics* **460** 281–305, 2002); F. Bonnefoy, M.H. Meylan and P. Ferrant (*Journal of Fluid Mechanics* **621** 215–242, 2009); J.-M. Vanden-Broeck and E.I. Parau (*Philosophical Transactions of the Royal Society* A **369**, 2957–2972, 2011); and E.I. Parau and J.-M. Vanden-Broeck (*Philosophical Transactions of the Royal Society* A **369**, 2973–2988, 2011).

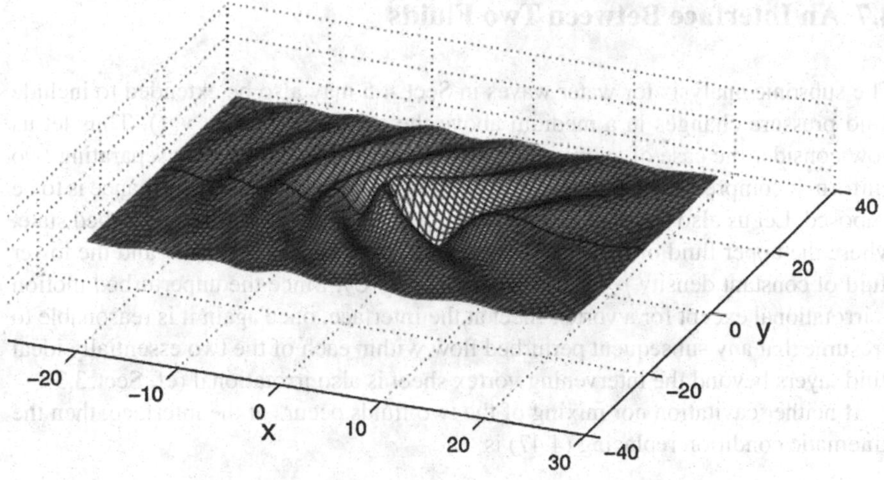

Fig. 4.5 Representative nonlinear deflexion for $V > c_{\min}$ (courtesy Emilian Parau)

Finally, we remark that practical applications of flexural–gravity waves include the deliberate breakup of ice sheets by moving sources (notably hovercraft travelling near critical speed and submarines operating below the ice), and in determining the hydroelastic behaviour of very large floating structures (VLFS) that can serve as aircraft runways.[6]

Exercises

(Q1) For intermediate wavelengths where $kh' \ll 1$ but $kH \gg 1$, show that the phase speed of flexural–gravity waves in a thin floating flexible plate is given by

$$c(k) = \pm\sqrt{\frac{Dk^3 - Nk}{\rho} + \frac{g}{k}}.$$

Then consider the positive branch, and show that the associated minimum phase speed is

$$c_{\min} = 2\left(\frac{Dg^3}{27\,\rho}\right)^{1/8} e^{-\epsilon/4}\sqrt{(3 - e^{2\epsilon})/2}, \text{ where } \epsilon = \sinh^{-1}\sqrt{\frac{N}{12\,\rho g D}}.$$

(Q2) From the dispersion relation (4.62) for capillary–gravity waves, deduce that $c < c_g$ for the shortest waves and that $c > c_g$ for the longer waves on water of finite depth, where c denotes the phase speed and c_g the group speed.

[6] V.A. Squire (*International Journal of Offshore and Polar Engineering*, **18**, 241–253, 2008).

4.7 An Interface Between Two Fluids

The subsonic analysis for water waves in Sect. 4.5 may also be extended to include fluid pressure changes in a medium above the surface $z = \zeta(x, y, t)$. Thus let us now consider the case of an initially plane horizontal surface $z = 0$ separating two uniform incompressible fluid layers, on which a small-amplitude disturbance is to be imposed. Let us also suppose that there is horizontal flow in their unperturbed state, where the upper fluid of constant density ρ_1 has uniform velocity $\mathbf{U}_1$ and the lower fluid of constant density ρ_2 has uniform velocity $\mathbf{U}_2$. Since the unperturbed motion is irrotational except for a vortex sheet at the interface, once again it is reasonable to presume that any subsequent perturbed flow within each of the two essentially ideal fluid layers beyond the intervening vortex sheet is also irrotational (cf. Sect. 3.3).

If neither cavitation nor mixing of the two fluids occurs at the interface, then the kinematic condition replacing (4.47) is

$$\hat{\mathbf{n}} \cdot \mathbf{v}_1 = \hat{\mathbf{n}} \cdot \mathbf{v}_2 = \hat{\mathbf{n}} \cdot \mathbf{v}_s . \tag{4.65}$$

Consequently, on recalling that the normal to the level surface $z - \zeta(x, y, t) = 0$ is $\hat{\mathbf{n}} = \nabla(z - \zeta) = \hat{\mathbf{k}} - \hat{\mathbf{i}}\,\partial\zeta/\partial x - \hat{\mathbf{j}}\,\partial\zeta/\partial y$, there are now two linearised free surface conditions generalising (4.49)—viz.

$$-\frac{\partial \phi_i}{\partial z} = \left(\frac{\partial}{\partial t} + \mathbf{U}_i \cdot \nabla\right)\zeta, \tag{4.66}$$

in terms of the perturbation velocity potentials ϕ_i ($i = 1, 2$) for each of the two fluids. However, we now have continuity of the pressure perturbation across the interface, assuming no shock wave is excited since the disturbance is presumed to be small (cf. Sect. 4.11). The Bernoulli equation (3.60) provides the pressure perturbations

$$p_i = \rho_i \left(\frac{\partial \phi_i}{\partial t} + \mathbf{U}_i \cdot \nabla\phi_i - g\zeta\right) \tag{4.67}$$

at the interface $z = \zeta$ as generalisations of (4.50), on again adopting the gravitational potential $V = gz$ for $\mathbf{g} = -\nabla V = -g\hat{\mathbf{k}}$.

Let us consider initial parallel mainstream flows in the x-direction (i.e. $\mathbf{U}_i = U_i\,\hat{\mathbf{i}}$), and one-dimensional perturbations of form $f(z)\exp[i(kx - \omega t)]$. The elementary solutions of Fourier form are then the surface displacement $\zeta = a\exp[i(kx - \omega t)]$ and the respective perturbation velocity potentials $\phi_1 = C_1\exp[i(kx - \omega t) - kz]$ and $\phi_2 = C_2\exp[i(kx - \omega t) + kz]$, where a, C_1 and C_2 are constants—i.e. with ϕ_1 and ϕ_2 satisfying the Laplace equation and hence $f(z)$ given by (4.54), and each suitably decaying away from the interface provided the two fluids are many wavelengths deep (effectively both semi-infinitely deep such that $\phi_1 \to 0$ as $z \to \infty$ and $\phi_2 \to 0$ as $z \to -\infty$, assuming $k > 0$). Substituting these elementary solutions into the three conditions resulting from (4.66) and (4.67) with pressure continuity ($p_1 = p_2$), once again applied at $z = 0$ in the linearised theory for small perturbations, we have respectively

$$-i(\omega - kU_1)a = kC_1, \quad i(\omega - kU_2)a = kC_2, \tag{4.68}$$

$$\rho_1[i(\omega - kU_1)C_1 + ga] = \rho_2[i(\omega - kU_2)C_2 + ga]. \tag{4.69}$$

Eliminating C_1 and C_2 from (4.69) using (4.68) gives

$$\rho_1[(\omega - kU_1)^2 + gk] = \rho_2[-(\omega - kU_2)^2 + gk]\,; \tag{4.70}$$

and expanding this result yields the dispersion relation

$$(\rho_2+\rho_1)\omega^2 - 2k\omega(\rho_2 U_2+\rho_1 U_1)+k^2(\rho_2 U_2^2+\rho_1 U_1^2) - gk(\rho_2-\rho_1) = 0, \tag{4.71}$$

a quadratic with two roots such that

$$\frac{\omega}{k} = \frac{\rho_2 U_2 + \rho_1 U_1}{\rho_2 + \rho_1} \pm \sqrt{\frac{g}{k}\frac{\rho_2 - \rho_1}{\rho_2 + \rho_1} - \frac{\rho_2\rho_1}{(\rho_2+\rho_1)^2}(U_2 - U_1)^2}. \tag{4.72}$$

Thus once again there are two branches of the dispersion relation, but there is a new feature that the roots are complex conjugates if

$$(U_2 - U_1)^2 > \frac{g}{k}\frac{\rho_2^2 - \rho_1^2}{\rho_2\rho_1}. \tag{4.73}$$

When there is no flow ($U_1 = U_2 = 0$), the dispersion relation (4.71) reduces to

$$\omega^2 = \frac{\rho_2 - \rho_1}{\rho_2 + \rho_1} gk, \tag{4.74}$$

and the roots are either real when the lower fluid is denser than the upper fluid ($\rho_2 > \rho_1$), or pure imaginary when it is less dense ($\rho_1 > \rho_2$). In the limit $\rho_1 \to 0$, dispersion relation (4.74) reduces to the surface gravity wave result $\omega^2 = gk$, which is independent of the lower fluid density and consistent with dispersion relation (4.55) in the short-wavelength approximation $kH \gg 1$—such as for waves on a deep ocean previously considered in Sect. 4.5, where the density of the upper fluid (air) is neglected. Similarly, if there is a layer of lighter water above a layer of denser water due to different salinities or temperatures ($\rho_1 < \rho_2$), then (4.74) shows that internal gravity waves can propagate along the interface at somewhat lower speeds.[7] On the other hand, if the upper fluid is heavier (4.74) yields

[7] Atmospheric internal gravity waves, where the buoyancy of the air due to a vertical potential temperature gradient rather than gravity alone is the restoring force, are of course quite different—and the alternative term buoyancy waves is actually more appropriate. Buoyancy waves have the remarkable property that their group velocity is perpendicular to the direction of phase propagation,

$$\omega = \pm i\,\frac{\rho_1 - \rho_2}{\rho_1 + \rho_2}\sqrt{gk}, \tag{4.75}$$

so the disturbance is now proportional to $\exp(\pm\,|\omega|\,t)$ and the positive exponent corresponds to an exponentially growing instability. This departure from an initial static equilibrium state is often called Rayleigh–Taylor instability, and the ratio $(\rho_1 - \rho_2)/(\rho_1 + \rho_2)$ when $\rho_1 > \rho_2$ is sometimes called the Atwood number. Thus a denser upper fluid tends to sink towards the bottom, as it penetrates the lower fluid in a phenomenon known as "fingering". According to (4.75), the largest growth rate at any particular wavelength can approach $\sqrt{gk}$ (when $\rho_1 \gg \rho_2$). It also appears that short-wavelengths grow fastest since the growth rate $|\omega| \to \infty$ as $k \to \infty$, but there is an upper bound to $|\omega|$ since surface tension is important at shorter wavelengths (cf. also Sect. 4.6 and Chandrasekhar [1]).

When the two fluids are in relative motion, the inequality (4.73) implies that the interface can also be unstable even if the lower fluid is denser than the upper fluid ($\rho_2 > \rho_1$) provided $|U_1 - U_2|$ is sufficiently large. Such a shear velocity driven disruption of an equilibrium flow is often called Kelvin–Helmholtz instability [1, 3]. If $\rho_2 U_2 + \rho_1 U_1 \neq 0$, the real part on the right-hand side of (4.72) implies the surface oscillates as the disturbances grow—when the system is often called "overstable" (cf. also Chap. 6). Condition (4.73) can always be satisfied by taking k large enough, predicting that the flow is always unstable to short waves, but once again we note that surface tension has been neglected (and also finite interface thickness).

Exercise

(Q1) The velocity field magnitude for the two unperturbed parallel mainstream flows in the x-direction may be expressed as $U(z) = U_2 + (U_1 - U_2)H(z)$, where $H(z)$ is the Heaviside step function. As discussed in Sect. 1.10, the derivative dH/dz is the impulse or Dirac delta function $\delta(z)$. Obtain the vorticity ω in the vortex sheet at $z = 0$, and then verify (3.24) over a rectangular region $R : |x| \leq l/2,\ |z| \leq b/2$ at the interface.

4.8 Local Averaging

There is a general basic formalism for wave motion if no interaction with emitters or boundaries is considered, which avoids complications specific to particular kinds of waves. Let us consider a slowly varying background flow with a superimposed short-wavelength wave, or a spectrum of short-wavelength waves. For mathematical

---

(Footnote 7 continued)
so they can propagate energy high into the atmosphere. The rotation of the Earth provides another restoring force, leading to further important atmospheric wave types (e.g. "inertio-gravity" and Rossby waves) not discussed here, and the interested reader is referred to the book by Holton cited in the previous chapter.

convenience, let us also also suppose the flow extends to infinity (i.e. its domain is $\mathbb{R}^3$.)

The first consideration is how to distinguish the background component from the fluctuation, given that both are varying in space and time. The key assumption is that there be an asymptotic ordering $\lambda/L = \mathrm{O}(\varepsilon)$, where λ is a typical wavelength and L a background scale length. One then seeks to prove asymptotic equality between certain expressions as discussed below, where asymptotic equality to all orders in ε between two functions f and g (say) is denoted by

$$f \sim g \tag{4.76}$$

and implies that $f - g = \mathrm{O}(\varepsilon^N)\ \ \forall N$. For example, when $f - g = c_1 \exp(-c_2/\varepsilon)$ for $\varepsilon \to 0^+$ and $c_2 > 0$, f is not exactly equal to g but the error approaches zero exceedingly rapidly. Then introducing the idea of local averaging, we can decompose an arbitrary function f into a slowly varying "average part" $\langle f \rangle$ and a "fluctuating part" $\tilde{f}$ with local average zero to all orders in ε. An operational method to separate off the slowly varying part would be to apply a low wavelength bandpass filter to f—i.e. to convolute f with a smoothing function of width intermediate between the λ and L scales. Rather than specify an arbitrary smoothing function, we may adopt a more general approach.

Suppressing the t-dependence for simplicity, let us define an acceptable local average of an arbitrary function $f(\mathbf{r}, \mathbf{r}/\varepsilon, \varepsilon)$ to be the slowly varying function $\langle f \rangle(\mathbf{r}, \varepsilon)$, such that

$$\frac{1}{\langle f \rangle} \frac{\partial^N \langle f \rangle}{\partial x_i \cdots \partial x_k} = \mathrm{O}(1)\ \ \forall \text{ arbitrary integer } N \text{ and } i, \ldots, k \in \{1, 2, 3\} \tag{4.77}$$

and

$$\int_{\mathbb{R}^3} (f - \langle f \rangle)\phi(\mathbf{r})\, d^3\mathbf{r} \sim 0\ \ \forall \phi(\mathbf{r}) \in \mathcal{T}_3, \tag{4.78}$$

where any member of the class of test functions $\mathcal{T}_3$ chosen here satisfies the requirement $\int_{\mathbb{R}^3} \left|\partial^N \phi / \partial x_i \partial x_k\right| d^3\mathbf{r} < \infty\ \forall N,\ i, \ldots, k \in \{1, 2, 3\}$. Clearly (4.77) is easy enough to satisfy; and to show that (4.78) can be satisfied to all orders for the functions of interest, let us first define the fluctuating part

$$\tilde{f} = f - \langle f \rangle, \tag{4.79}$$

so that (4.78) becomes the "orthogonality" condition

$$\int_{\mathbb{R}^3} \tilde{f}\phi\, d^3\mathbf{r} \sim 0\ \ \forall \phi \in \mathcal{T}_3. \tag{4.80}$$

Now if

$$\tilde{f} = \sum_j a_j \exp\left[\frac{iS_j}{\varepsilon}\right] + \text{c.c.}, \tag{4.81}$$

where a_j and S_j are slowly varying and "cc" denotes the complex conjugate, then (4.80) follows from a slight generalisation of the following result:

$$\int_{-\infty}^{\infty} \exp\left(\frac{ix}{\varepsilon}\right) \phi(x)\, dx \sim 0 \ \ \forall \phi(x) \in \mathcal{T}_1, \tag{4.82}$$

where any member of the class $\mathcal{T}_1$ here is such that $\int_{-\infty}^{\infty} |\phi^{(n)}|\, dx < \infty \ \forall n$.

To prove (4.82), which is similar to the well-known Riemann–Lebesgue lemma but with the integral extending over an infinite interval, we may use integration by parts:

$$\begin{aligned}\int_{-\infty}^{\infty} \exp\left(\frac{ix}{\varepsilon}\right) \phi(x)\, dx &= -i\varepsilon \int_{-\infty}^{\infty} \phi(x) \frac{d}{dx} \exp\left(\frac{ix}{\varepsilon}\right) dx \\ &= i\varepsilon \int_{-\infty}^{\infty} \phi'(x) \exp\left(\frac{ix}{\varepsilon}\right) dx,\end{aligned}$$

since the finiteness of $\int_{-\infty}^{\infty} |\phi(x)|\, dx$ implies that $|\phi(x)|$ must decrease faster than $|x|^{-1}$ as $|x| \to \infty$, so the endpoint contributions vanish. Repeated integration by parts yields

$$\int_{-\infty}^{\infty} \exp\left(\frac{ix}{\varepsilon}\right) \phi(x)\, dx = (i\varepsilon)^N \int_{-\infty}^{\infty} \exp\left(\frac{ix}{\varepsilon}\right) \phi^{(N)}(x)\, dx \tag{4.83}$$

for arbitrary integer N. Thus

$$\left| \int_{-\infty}^{\infty} \exp\left(\frac{ix}{\varepsilon}\right) \phi(x)\, dx \right| < \varepsilon^N \int_{-\infty}^{\infty} \left| \phi^{(N)}(x) \right| dx,$$

or

$$\int_{-\infty}^{\infty} \exp\left(\frac{ix}{\varepsilon}\right) \phi(x)\, dx = \mathrm{O}(\varepsilon^N) \ \ \forall N,$$

which by (4.76) is the meaning of (4.82). [This does not mean that the left-hand side is actually zero, but merely that it is "exponentially small" as $\varepsilon \to 0$, for instance $\exp(-1/\varepsilon)$]. The extension to prove (4.80) is straightforward. Thus the space of rapidly varying functions for which the local average exists is not null; and it is relevant if the rapidly varying parts are wave-like or oscillatory, such as for subsonic flows without vortex sheets or density discontinuities.

To complete the theory, another result is needed—analogous to the fundamental lemma of variational calculus: thus if $\bar{f}$ is a slowly varying function satisfying (4.77), and

$$\int_{\mathbb{R}^3} \bar{f}\,\phi\, d^3\mathbf{r} \sim 0 \;\; \forall \phi \in \mathcal{T}_3, \tag{4.84}$$

then

$$\bar{f} \sim 0. \tag{4.85}$$

This can be proven by choosing a sequence $\{\phi_n(\mathbf{r})\}$ of test functions, each of which approximates the complex conjugate f^* over some region Ω but is zero outside, so that

$$0 \sim \int_{\mathbb{R}^3} f\phi_n\, d^3\mathbf{r} \to \int_{\Omega} |\bar{f}|^2\, d^3\mathbf{r} \geq 0,$$

which can only be consistent if $f \sim 0$. The result can now be used to prove several basic properties of local averaging.
Linearity:

$$\langle af + bg\rangle \sim a\langle f\rangle + b\langle g\rangle, \quad \text{where } a \text{ and } b \text{ are constants.} \tag{4.86}$$

This property follows by writing

$$\langle af + bg\rangle - (a\langle f\rangle + b\langle g\rangle) \equiv \langle af + bg\rangle - (af + bg) + af - a\langle f\rangle + bg - b\langle g\rangle$$

and then applying (4.78) to the various local averages on the right-hand side, such that the left-hand side satisfies (4.84) and therefore vanishes to all orders in ε. Any reasonable prescription for performing local averaging would indeed be linear rather than merely asymptotic, but this demonstrates consistency with that requirement.
Projective Property:

$$\langle\langle f\rangle\rangle \sim \langle f\rangle. \tag{4.87}$$

This property follows immediately by applying (4.83) to $\langle f\rangle$, and a corollary from the definition (4.79) of $\tilde{f}$ and linearity is

$$\langle \tilde{f}\rangle \sim 0. \tag{4.88}$$

(If this property were not satisfied, the approach would not fit our intuitive concept of local averaging.)

Commutativity of Averaging and Differentiation:

$$\frac{\partial\langle f\rangle}{\partial t} \sim \langle\frac{\partial f}{\partial t}\rangle, \tag{4.89}$$

$$\nabla\langle f\rangle \sim \langle\nabla f\rangle. \tag{4.90}$$

These properties follow by subtracting and adding $\partial_t f$ and ∇f, respectively:

$$\frac{\partial\langle f\rangle}{\partial t} - \langle\frac{\partial f}{\partial t}\rangle \equiv \frac{\partial}{\partial t}(\langle f\rangle - f) + \left(\frac{\partial f}{\partial t} - \langle\frac{\partial f}{\partial t}\rangle\right),$$

$$\nabla\langle f\rangle - \langle\nabla f\rangle \equiv \nabla(\langle f\rangle - f) + (\nabla f - \langle\nabla f\rangle).$$

Multiplying by ϕ and integrating over $\mathbb{R}^3$, commuting the first ∂_t on the right-hand side outside the integral, integrating the first ∇ on the right-hand side by parts, and using (4.78) (noting that $\hat{\mathbf{e}}_i \cdot\nabla\,\phi \in \mathcal{T}_3$), shows that the left-hand sides of these two expressions satisfy the required condition and thus vanish to all orders.

4.9 Wave Reaction on a Background Mean Flow

The commutation results (4.89) and (4.90) may be used to develop conservation equations for the system of background flow and waves. Let ρ_*, p_* and v_* denote the "excited state" variables—i.e. the background flow with waves imposed, and let us first define the background density to be the Eulerian mean density

$$\bar{\rho} \equiv \langle\rho_*\rangle. \tag{4.91}$$

Then the background velocity may be defined to be the mass-weighted local Eulerian average velocity

$$\bar{\mathbf{v}} \equiv \frac{\langle\rho_*\mathbf{v}_*\rangle}{\langle\rho_*\rangle}. \tag{4.92}$$

This is completely analogous to the definition of the macroscopic fluid velocity from the mass-weighted microscopic velocities in Sect. 2.2. In the next section, it is shown to closely approximate the Lagrangian mean or oscillation-centre velocity—and to be the natural velocity to assign to the background fluid, because (4.89) and (4.90) applied to the continuity equation (2.3) immediately yield

$$\frac{\partial\bar{\rho}}{\partial t} + \nabla\cdot(\bar{\rho}\bar{\mathbf{v}}) = 0. \tag{4.93}$$

Thus just as a fluid may be regarded as a collection of fluid elements, the background fluid may be regarded as consisting of fictitious fluid elements moving along trajectories (oscillation-centre trajectories) from which the wave oscillations have been averaged out. This concept is developed further in the next section.

Since ε has been used as the eikonal expansion parameter, let us use the symbol α to denote the expansion parameter expressing the smallness of the wave amplitude. Because the natural length for measuring the nonlinearity of oscillations is λ rather than L, we assume $\lambda = \mathrm{O}(1)$, $L = \mathrm{O}(\varepsilon^{-1})$ rather than the ordering adopted in Sect. 4.8. The results of Sect. 4.8 are unaffected however, since they are valid to all orders. Expanding in powers of α,

$$\rho_* = \rho_0 + \alpha\rho_1 + \alpha^2\rho_2 + \dots \tag{4.94}$$

$$p_* = p_0 + \alpha p_1 + \alpha^2 p_2 + \dots \tag{4.95}$$

$$\mathbf{v}_* = \mathbf{v}_0 + \alpha\mathbf{v}_1 + \alpha^2\mathbf{v}_2 + \dots. \tag{4.96}$$

The linear $\mathrm{O}(\alpha)$ terms average to zero, since they are of the form (4.81), so (4.88) applies. The $\mathrm{O}(\alpha^2)$ terms do not in general average to zero, but modify ("renormalise") the unperturbed quantities

$$\bar\rho \equiv \langle\rho_*\rangle \sim \rho_0 + \alpha^2\langle\rho_2\rangle + \mathrm{O}\left(\alpha^4\right), \tag{4.97}$$

$$\bar\rho\bar{\mathbf{v}} \equiv \langle\rho_*\mathbf{v}_*\rangle \sim \rho_0\mathbf{v}_0 + \alpha^2\left(\rho_0\langle\mathbf{v}_2\rangle + \langle\rho_1\mathbf{v}_1\rangle + \langle\rho_2\rangle\mathbf{v}_0\right) + \mathrm{O}\left(\alpha^4\right); \tag{4.98}$$

then dividing (4.98) by (4.97) yields

$$\bar{\mathbf{v}} \sim \mathbf{v}_0 + \alpha^2\left(\langle\mathbf{v}_2\rangle + \frac{\langle\rho_1\mathbf{v}_1\rangle}{\rho_0}\right) + \mathrm{O}\left(\alpha^4\right), \tag{4.99}$$

using the $\sim$ symbol for asymptotic equality to all orders as in Sect. 2.2. Note that $\bar{\mathbf{v}}$ differs from the Eulerian mean velocity $\langle\mathbf{v}_*\rangle$ by the term $\alpha^2\langle\rho_1\mathbf{v}_1\rangle/\rho_0$.

We can also calculate the momentum flux dyadic

$$\begin{aligned}\langle\rho_*\mathbf{v}_*\mathbf{v}_*\rangle &\sim \rho_0\mathbf{v}_0\mathbf{v}_0 + \alpha^2\left(\rho_0\langle\mathbf{v}_1\mathbf{v}_1\rangle + \langle\rho_1\mathbf{v}_1\rangle\mathbf{v}_0 + \mathbf{v}_0\langle\rho_1\mathbf{v}_1\rangle\right.\\ &\quad\left. + \langle\rho_2\rangle\mathbf{v}_0\mathbf{v}_0 + \rho_0\langle\mathbf{v}_2\rangle\mathbf{v}_0 + \rho_0\mathbf{v}_0\langle\mathbf{v}_2\rangle\right) + 0\left(\alpha^4\right),\\ &= \bar\rho\bar{\mathbf{v}}\bar{\mathbf{v}} + \alpha^2\rho_0\langle\mathbf{v}_1\mathbf{v}_1\rangle + \mathrm{O}\left(\alpha^4\right),\end{aligned}$$

using (4.83) and (4.84)—i.e.

$$\langle\rho_*\mathbf{v}_*\mathbf{v}_*\rangle = \bar\rho\bar{\mathbf{v}}\bar{\mathbf{v}} + \alpha^2\bar\rho\langle\mathbf{v}_1\mathbf{v}_1\rangle + \mathrm{O}\left(\alpha^4\right). \tag{4.100}$$

Assuming for simplicity that an adiabatic equation of state $p = K\rho^\gamma$ applies, and defining the background pressure $\bar p$ by

$$\bar p \equiv K\bar\rho^\gamma, \tag{4.101}$$

from (4.98) the renormalised pressure to $\mathrm{O}(\alpha^2)$ is

$$\bar{p} = K\rho_0^\gamma\left(1+\alpha^2\gamma\frac{\langle\rho_2\rangle}{\rho_0}\right). \tag{4.102}$$

This is to be compared with

$$\langle p_*\rangle = K\rho_0^\gamma\left\langle 1+\alpha\gamma\frac{\rho_1}{\rho_0}+\alpha^2\left[\gamma\frac{(\gamma-1)}{2}\frac{\rho_1^2}{\rho_0^2}+\gamma\frac{\rho_2}{\rho_0}\right]\right\rangle,$$

or

$$\langle p_*\rangle = \bar{p}\left(1+\alpha^2\gamma\frac{(\gamma-1)}{2}\frac{\langle\rho_1^2\rangle}{\bar{\rho}^2}\right)+\mathrm{O}\left(\alpha^4\right). \tag{4.103}$$

These results can be used to find an equation of motion for the background flow, by averaging the ideal equation of motion (3.5) written in the conservation form (2.17). From the commutation results (4.89) and (4.90), together with (4.100) and (4.103), we have

$$\frac{\partial}{\partial t}(\bar{\rho}\bar{\mathbf{v}})+\nabla\cdot(\bar{\rho}\bar{\mathbf{v}}\bar{\mathbf{v}}+\bar{p}\mathbf{I}+\mathbf{P}_W)=\bar{\rho}\mathbf{g}, \tag{4.104}$$

where the *wave* or *radiation* stress dyadic $\mathbf{P}_W$ is given by

$$\begin{aligned}\mathbf{P}_W &= \alpha^2\left[\bar{\rho}\langle\mathbf{v}_1\mathbf{v}_1\rangle+\frac{\gamma(\gamma-1)}{2}\bar{p}\frac{\langle\rho_1^2\rangle}{\bar{\rho}^2}\mathbf{I}\right]+0\left(\alpha^4\right),\\ &= \alpha^2\bar{\rho}\left[\langle\mathbf{v}_1\mathbf{v}_1\rangle+\frac{\gamma-1}{2}\langle|\mathbf{v}_1|^2\rangle\mathbf{I}\right],\end{aligned}$$

since $\rho_1=\bar{\rho}|\mathbf{v}_1|/c_\mathrm{s}$ and $c_\mathrm{s}^2=\gamma\bar{p}/\bar{\rho}$.

For a monochromatic wave, $\langle\mathbf{v}_1\mathbf{v}_1\rangle = 2|\hat{\mathbf{v}}_1^{(0)}|\hat{\mathbf{e}}_k\hat{\mathbf{e}}_k$ where $\hat{\mathbf{e}}_k\equiv\mathbf{k}/|\mathbf{k}|$. Comparing with (4.33), we have

$$\mathbf{P}_W=\alpha^2N\omega'\left[\hat{\mathbf{e}}_k\hat{\mathbf{e}}_k+\frac{(\gamma-1)}{2}\mathbf{I}\right]+\mathrm{O}\left(\alpha^4\right). \tag{4.105}$$

If the momentum conservation equation (4.102) is regarded as an equation of motion for the background flow, the term $-\nabla\cdot\mathbf{P}_W$ is the radiation pressure or "ponderomotive force" density acting, arising from the nonlinear reaction of the waves on the background. Although nonlinear $\mathrm{O}(\alpha^2)$, the only surviving terms are products of linear terms, as the perturbation expansions were expressed in terms of the "renormalised" background quantities ($\bar{\rho}$, $\bar{p}$ and $\bar{\mathbf{v}}$), so such calculations are often called quasilinear.

4.10 Oscillation-Centre (OC) Description

The procedure adopted in the previous section is straightforward and applicable to other than ideal fluids, but it is a little cumbersome. The more elegant Oscillation-Centre (OC) description discussed in this section is especially suited (but not restricted) to ideal fluids, and leads to variational formulations in stability analysis to be considered later.

The OC description is related to the Lagrangian description in Sect. 2.4. For convenience, let $\mathbf{r}_*$ denote the true fluid element position and $\mathbf{r}$ the averaged position or oscillation-centre. The mapping

$$\mathbf{r}_* = \mathbf{r} + \boldsymbol{\xi}(\mathbf{r}, t), \tag{4.106}$$

from $\mathbf{r}$ to $\mathbf{r}_*$ in terms of the fluid displacement $\boldsymbol{\xi}$, changes the variables $\mathbf{r}_*, t$ in the Eulerian description to $\mathbf{r}, t$ in the OC description. Let us use $\rho(\mathbf{r}, t)$, $p(\mathbf{r}, t)$ and $\mathbf{v}(\mathbf{r}, t)$ to denote the OC position, density and velocity. The point of view now is that a fictitious OC fluid, which is a subsystem of the total system, interacts weakly with the wave subsystem. Then

$$\mathbf{v}_*(\mathbf{r}_*, t) \equiv \left(\frac{\partial \mathbf{r}_*}{\partial t}\right)_{\mathbf{r}_0} = \mathbf{v}(\mathbf{r}, t) + \dot{\boldsymbol{\xi}}(\mathbf{r}, t), \tag{4.107}$$

$$\rho_*(\mathbf{r}_*, t) = \rho(\mathbf{r}, t)/J(\mathbf{r}), \tag{4.108}$$

$$p_*(\mathbf{r}_*, t) = p(\mathbf{r}, t)/J^{\gamma}(\mathbf{r}), \tag{4.109}$$

where

$$\dot{\boldsymbol{\xi}}(\mathbf{r}, t) \equiv \left[\frac{\partial}{\partial t} + \mathbf{v}(\mathbf{r}, t)\cdot\nabla\right] \boldsymbol{\xi}(\mathbf{r}, t) \tag{4.110}$$

and

$$J(\mathbf{r}, t) \equiv \det\left(\frac{\partial \mathbf{r}_*}{\partial \mathbf{r}}\right) = \det\left(\mathbf{I} + \nabla\boldsymbol{\xi}\right) \tag{4.111}$$

is the Jacobian of the transformation (4.106)—i.e. the ratio of the volume elements $\Delta\tau_*$ and $\Delta\tau$. The relation (4.108) simply expresses mass conservation: $\rho_*\Delta\tau_* = \rho\Delta\tau$, while (4.109) ensures that $p_*/\rho_*^{\gamma} = p/\rho^{\gamma}$.

The essential simplification achieved by this approach is that the two scalar variables ρ_* and p_*, together with the vector $\mathbf{v}_*$, have been reduced to functionals of the single vector variable $\boldsymbol{\xi}$. In the case of ρ_* and p_*, it has been recognised that the Eqs. (3.1) and (3.3) can be integrated to give holonomic constraints on ρ_* and p_* (constraints that depend explicitly on the position coordinates). Calculations with these representations are greatly facilitated by two results itemised as follows.

- If $D_1(\lambda) \equiv \det(\mathbf{I} + \lambda\mathbf{A})$ and $D_2(\lambda) \equiv \exp\left[\mathrm{Tr}\ln(\mathbf{I} + \lambda\mathbf{A})\right]$, where λ is an arbitrary scalar and $\mathbf{A}$ an arbitrary dyadic, then

$$D_1(\lambda) = D_2(\lambda).$$

Proof $D_1'(\lambda) = \sum_{ij} A_{ij}C_{ij}$, where C_{ij} are the cofactors of $(\delta_{ij} + \lambda A_{ij})$. This follows from varying each element of $(\delta_{ij} + \lambda A_{ij})$ in turn, using the appropriate expansion of the determinant in terms of cofactors. Thus

$$\begin{aligned} &D_1'(\lambda)/D_1(\lambda) \\ &\quad = \sum_{ij} A_{ij}(\mathbf{I} + \lambda\mathbf{A})^{-1}_{ji} \\ &\quad \text{from the standard expression for the matrix inverse} \\ &\quad \equiv \mathrm{Tr}\left[\mathbf{A}\cdot(\mathbf{I} + \lambda\mathbf{A})^{-1}\right]; \\ \text{but}\quad &D_2'(\lambda)/D_2(\lambda) = \mathrm{Tr}\left[\mathbf{A}\cdot(\mathbf{I} + \lambda\mathbf{A})^{-1}\right] \end{aligned}$$

and $D_1(0) = D_2(0) = 1$, so both $D_1(\lambda)$ and $D_2(\lambda)$ obey the same first-order ordinary differential equation with the same initial conditions. Provided that $D(\lambda') \neq 0$ anywhere in the interval $0 < \lambda' \leq \lambda$, the uniqueness theorem for an ordinary differential equation initial value problem implies $D_2 \equiv D_1$. □

This result is useful for evaluating the Jacobian (4.111), since

$$\mathrm{Tr}\ln(\mathbf{I} + \nabla\boldsymbol{\xi}) = \nabla\cdot\boldsymbol{\xi} - \frac{1}{2}\nabla\boldsymbol{\xi} : \nabla\boldsymbol{\xi} + \mathrm{O}\left(\alpha^3\right),$$

on recalling that the double dot product $\mathbf{A} : \mathbf{B} \equiv \mathrm{Tr}(\mathbf{A}\cdot\mathbf{B})$ from Chap. 1: thus

$$\begin{aligned} J &= \exp\left(\nabla\cdot\boldsymbol{\xi} - \frac{1}{2}\nabla\boldsymbol{\xi} : \nabla\boldsymbol{\xi}\right) + \mathrm{O}\left(\alpha^3\right) \\ &= 1 + \nabla\cdot\boldsymbol{\xi} + \frac{1}{2}\left[(\nabla\cdot\boldsymbol{\xi})^2 - \nabla\boldsymbol{\xi} : \nabla\boldsymbol{\xi}\right] + \mathrm{O}\left(\alpha^3\right). \end{aligned} \tag{4.112}$$

- Let $f_*(\mathbf{r}, t)$ be an arbitrary function, varying on both the fast and slow scales of Sect. 4.8. Then on suppressing the t-dependence,

$$\begin{aligned} \langle f_*(\mathbf{r})\rangle \sim \langle J(\mathbf{r}) f_*(\mathbf{r}_*)\rangle - \nabla\cdot\langle\boldsymbol{\xi} J(\mathbf{r}) f_*(\mathbf{r}_*)\rangle + \frac{1}{2!}\nabla\nabla : \langle\boldsymbol{\xi}\boldsymbol{\xi} J(\mathbf{r}) f_*(\mathbf{r}_*)\rangle \\ - \frac{1}{3!}\sum_{ijk}\frac{\partial}{\partial x_i}\frac{\partial}{\partial x_j}\frac{\partial}{\partial x_k}\langle\xi_k\xi_j\xi_i J(\mathbf{r}) f_*(\mathbf{r}_*)\rangle + \ldots \end{aligned} \tag{4.113}$$

Proof

$$
\begin{aligned}
&\int_{\mathbb{R}^3} \phi(\mathbf{r}) f_*(\mathbf{r}) d^3\mathbf{r} \\
&= \int_{\mathbb{R}^3} f_*(\mathbf{r}_*)\phi(\mathbf{r}_*) d^3\mathbf{r}_* && \text{[changing dummy]} \\
&= \int_{\mathbb{R}^3} J(\mathbf{r}) f_*(\mathbf{r}_*)\phi(\mathbf{r}+\boldsymbol{\xi}) d^3\mathbf{r} && \text{[using (4.106)]} \\
&= \int_{\mathbb{R}^3} J(\mathbf{r}) f_*(\mathbf{r}_*) \left[\phi(\mathbf{r}) + \boldsymbol{\xi}\cdot\nabla\,\phi \frac{1}{2!}\boldsymbol{\xi}\boldsymbol{\xi} : \nabla\nabla\phi + \ldots\right] d^3\mathbf{r} \\
&= \int_{\mathbb{R}^3} \phi(\mathbf{r}) \Bigg\{ J(\mathbf{r}) f_*(\mathbf{r}_*) - \nabla\cdot[\boldsymbol{\xi} J(\mathbf{r}) f_*(\mathbf{r}_*)] \\
&\quad + \nabla\nabla : \left[\frac{1}{2}\boldsymbol{\xi}\boldsymbol{\xi} J(\mathbf{r}) f_*(\mathbf{r}_*)\right] + \ldots \Bigg\} d^3\mathbf{r} && \text{[parts integration]}.
\end{aligned}
$$

From (4.78) the coefficients of $\phi(\mathbf{r})$ in the first and last expressions above may be replaced by their local averages, and then from (4.85) and (4.90) the desired result follows. □

This asymptotic equality (4.113) is useful to average quantities involving J.

Various Eulerian averages can be evaluated—e.g. from (4.108) and (4.113),

$$
\langle \rho_*(\mathbf{r}, t)\rangle \sim \rho(\mathbf{r}, t) - \nabla\cdot\langle\boldsymbol{\xi}\rho\rangle + \frac{1}{2}\nabla\nabla : \langle\boldsymbol{\xi}\boldsymbol{\xi}\rho\rangle + \mathrm{O}\left(\alpha^2\varepsilon^2\right).
$$

To make the OC unique to all orders, we may impose the condition

$$
\langle\boldsymbol{\xi}(\mathbf{r}, t)\rangle \sim 0. \tag{4.114}
$$

Thus $\nabla\cdot\langle\xi\rho\rangle \sim 0$ by (4.78) since $\rho\phi \in \mathcal{T}_3$, so that

$$
\bar{\rho}(\mathbf{r}, t) \equiv \langle\rho_*(\mathbf{r}, t)\rangle \sim \rho(\mathbf{r}, t) + \mathrm{O}\left(\alpha^2\varepsilon^2\right), \tag{4.115}
$$

on using the bar notation for background quantities as defined by Eulerian averaging in (4.80) and (4.81). Thus the Eulerian average definition and the OC definition of background density are equivalent, to good approximation. Similarly,

$$
\bar{\rho}\bar{\mathbf{v}} \equiv \langle\rho_* v_*\rangle \sim \rho\mathbf{v} + \frac{1}{2}\nabla\nabla : \langle\boldsymbol{\xi}\boldsymbol{\xi}\rho\mathbf{v}\rangle - \nabla\cdot\langle\boldsymbol{\xi}\rho\dot{\boldsymbol{\xi}}\rangle + \ldots,
$$

such that

$$
\bar{\mathbf{v}} = \mathbf{v} + \mathrm{O}\left(\alpha^2\varepsilon\right). \tag{4.116}
$$

Thus the mass-weighted Eulerian average velocity is, to a good approximation, equal to the OC velocity. (For linearly polarised waves such as sound waves, the error is actually $\mathrm{O}(\alpha^2\varepsilon)$ because the leading order contribution to $\langle \boldsymbol{\xi}\rho\dot{\boldsymbol{\xi}}\rangle$ vanishes.)

From these expected identifications of $\bar{\rho}$ and $\bar{\mathbf{v}}$, we can now proceed to calculate the averages required for the averaged momentum equation. Using (4.107) and (4.108) in (4.113),

$$\begin{aligned}
\langle \rho_*\mathbf{v}_*\mathbf{v}_*\rangle &\sim \left\langle \rho\left(\mathbf{v}+\dot{\boldsymbol{\xi}}\right)\left(\mathbf{v}+\dot{\boldsymbol{\xi}}\right)\right\rangle - \nabla\cdot\left\langle \rho\boldsymbol{\xi}\left(\mathbf{v}+\dot{\boldsymbol{\xi}}\right)\left(\mathbf{v}+\dot{\boldsymbol{\xi}}\right)\right\rangle \\
&\quad + \frac{1}{2}\nabla\nabla : \left\langle \rho\boldsymbol{\xi}\boldsymbol{\xi}\left(\mathbf{v}+\dot{\boldsymbol{\xi}}\right)\left(\mathbf{v}+\dot{\boldsymbol{\xi}}\right)\right\rangle + \dots \\
&= \rho\left(\mathbf{v}\mathbf{v} + \langle\dot{\boldsymbol{\xi}}\dot{\boldsymbol{\xi}}\rangle\right) - \nabla\cdot\left[\rho\left\langle\boldsymbol{\xi}\left(\mathbf{v}\dot{\boldsymbol{\xi}}+\dot{\boldsymbol{\xi}}\mathbf{v}\right)\right\rangle\right] + \frac{1}{2}\nabla\nabla : \left[\langle\boldsymbol{\xi}\boldsymbol{\xi}\rangle\rho\mathbf{v}\mathbf{v}\right] + \dots \\
&= \rho\left(\mathbf{v}\mathbf{v} + \langle\dot{\boldsymbol{\xi}}\dot{\boldsymbol{\xi}}\rangle\right) + \mathrm{O}\left(\alpha^2\varepsilon\right),
\end{aligned} \tag{4.117}$$

expanding (4.117) up to $\mathrm{O}(\alpha^2)$. From a comparison of (4.96) and (4.107), we have that $\dot{\boldsymbol{\xi}} = \alpha\mathbf{v}_1$ to $\mathrm{O}(\alpha)$. Thus (4.117) is in agreement with (4.100), to leading order in ε.

Finally, the local average pressure follows by using (4.109) in (4.113), yielding

$$\langle p_*\rangle - \langle pJ^{1-\gamma}\rangle - \nabla\cdot\langle\boldsymbol{\xi}pJ^{1-\gamma}\rangle + \frac{1}{2}\nabla\cdot\langle\boldsymbol{\xi}\boldsymbol{\xi}pJ^{1-\gamma}\rangle + \dots; \tag{4.118}$$

and from (4.112),

$$\begin{aligned}
J^{1-\gamma} &= \exp\left[-(\gamma-1)\left(\nabla\cdot\boldsymbol{\xi} - \frac{1}{2}\nabla\boldsymbol{\xi}:\nabla\boldsymbol{\xi}\right)\right] + \dots \\
&= 1 - (\gamma-1)\left(\nabla\cdot\boldsymbol{\xi} - \frac{1}{2}\nabla\boldsymbol{\xi}:\nabla\boldsymbol{\xi}\right) + \frac{1}{2}(\gamma-1)^2(\nabla\cdot\boldsymbol{\xi})^2 + \dots \\
&= 1 - (\gamma-1)\nabla\cdot\boldsymbol{\xi} - \frac{1}{2}\gamma(\gamma-1)(\nabla\cdot\boldsymbol{\xi})^2 \\
&\quad + \frac{1}{2}(\gamma-1)\nabla\cdot(\boldsymbol{\xi}\cdot\nabla\boldsymbol{\xi} - \boldsymbol{\xi}\nabla\cdot\boldsymbol{\xi})
\end{aligned}$$

to $\mathrm{O}(\alpha^3)$, on invoking the identity

$$\nabla\boldsymbol{\xi}:\nabla\boldsymbol{\xi} \equiv (\nabla\cdot\boldsymbol{\xi})^2 + \nabla\cdot(\boldsymbol{\xi}\cdot\nabla\boldsymbol{\xi} - \boldsymbol{\xi}\nabla\cdot\boldsymbol{\xi}). \tag{4.119}$$

Proof

$$\begin{aligned}
\nabla\boldsymbol{\xi}:\nabla\boldsymbol{\xi} &= \nabla\cdot(\boldsymbol{\xi}\cdot\nabla\boldsymbol{\xi}) - \boldsymbol{\xi}\cdot\nabla\nabla\cdot\boldsymbol{\xi} \\
&= \nabla\cdot(\boldsymbol{\xi}\cdot\nabla\boldsymbol{\xi}) - \nabla\cdot(\boldsymbol{\xi}\nabla\cdot\boldsymbol{\xi}) + (\nabla\cdot\boldsymbol{\xi})^2
\end{aligned}$$

Substituting (4.119) in (4.118) yields

$$\begin{aligned}\langle p_* \rangle &\sim p + \frac{1}{2}\gamma(\gamma-1)p\langle(\nabla\cdot\boldsymbol{\xi})^2\rangle + \frac{1}{2}(\gamma-1)p\nabla\cdot\langle\boldsymbol{\xi}\cdot\nabla\boldsymbol{\xi} - \boldsymbol{\xi}\cdot\nabla\boldsymbol{\xi}\rangle \\ &\quad + (\gamma-1)\nabla\cdot[p\langle\boldsymbol{\xi}\nabla\cdot\boldsymbol{\xi}\rangle] + \ldots \\ &= p\left[1 + \frac{1}{2}\gamma(\gamma-1)\langle(\nabla\cdot\boldsymbol{\xi})^2\rangle\right] + \mathrm{O}\left(\varepsilon\alpha^2\right) \end{aligned} \tag{4.120}$$

Using (4.112) in (4.108) and comparing with (4.94), we have $\alpha\rho_1 = \rho\nabla\cdot\boldsymbol{\xi}$ such that (4.120) agrees with (4.103).

4.11 Shock Waves

In this chapter, we first discussed linear wave propagation that corresponds to small perturbations from some equilibrium state—notably longitudinal sound waves as disturbances in a background flow of an adiabatic (compressible) fluid, and transverse gravity waves as sub-sonic (incompressible) disturbances of an hydrostatic equilibrium. We noted that gravity waves are usually dispersive, but sound waves are not. However, the sound speed in any flow depends on the fluid density and pressure.

Moreover, as indicated in Sect. 4.4, supersonic flow leads to the formation of interfaces within the fluid, where field variations become so large that the linear theory must break down. Indeed, because there is no natural length scale in the inviscid equation of motion, one might expect that discontinuities or even "infinities" can occur. One such discontinuity has already been encountered—viz. the tangential discontinuity at a vortex sheet, where the tangential velocity at the interface is discontinuous (cf. Sect. 4.7). As mentioned there, a *shock* occurs when the *normal* velocity is discontinuous.

Since field variable derivatives do not exist at discontinuities, it might seem that the corresponding differential forms of the conservation equations break down. However, as indicated in Sect. 1.10, they do apply in what is known as the weak sense. The relevant test functions now consist of all sufficiently differentiable functions $\phi(\mathbf{r}, t)$ with compact support, nonzero over finite regions of space–time so that there is again no need to consider boundary conditions at infinity. For simplicity, it is sufficient to consider the case of one space dimension, although the discussion could be extended to higher dimensions. Thus a *weak solution* of a conservation equation

$$\frac{\partial u}{\partial t} + \frac{\partial f(u)}{\partial x} = 0 \tag{4.121}$$

is a solution $u(x, t)$ of the integral equation

$$\int_{-\infty}^{\infty}\int_{-\infty}^{\infty}\left(\frac{\partial u}{\partial t} + \frac{\partial f(u)}{\partial x}\right)\phi(x,t)\,dxdt = 0 \quad \forall\phi \in \mathcal{T}, \tag{4.122}$$

where $\mathcal{T}$ is the set of test functions and the integral is interpreted as

$$\int_{-\infty}^{\infty}\int_{-\infty}^{\infty}\left(\frac{\partial u}{\partial t}+\frac{\partial f(u)}{\partial x}\right)\phi(x,t)\,dxdt \equiv -\int_{-\infty}^{\infty}\int_{-\infty}^{\infty}\left(u\frac{\partial \phi}{\partial t}+f(u)\frac{\partial \phi}{\partial x}\right)dxdt. \tag{4.123}$$

Equation (4.123) is a natural definition, because it is a result obtained by integrating the left-hand side by parts when u is sufficiently smooth such that the integral in (4.122) is well defined. Since the support of ϕ in x, t-space may be arbitrarily localised about any point, (4.122) is equivalent to the usual localised (pointwise) interpretation of (4.121) when u is sufficiently smooth—i.e. the test function acts as a probe, which tests the value of the expression at any point. The integral in (4.122) and the left-hand side of (4.123) is not classically defined when u is discontinuous, because its derivative is then not defined pointwise everywhere. On the other hand, the form on the right-hand side of (4.123) is uniquely and classically defined, because the derivatives now act on the well-behaved test functions.

We now proceed to derive *jump conditions* for a plane (one-dimensional) shock. Thus let us suppose that at $x = 0$ the otherwise classically differentiable momentum flux ρv has a jump—i.e. across $x = 0$,

$$[\![\rho v]\!] = (\rho v)|_{x=0+} - (\rho v)|_{x=0-} \neq 0 \quad \text{such that} \quad \rho v = (\rho v)|_{x=0-} + [\![\rho v]\!]\, H(x)$$

where $H(x)$ is the Heaviside step function, hence

$$\frac{\partial(\rho v)}{\partial x} = [\![\rho v]\!]\,\delta(x) + \text{a continuous function} \tag{4.124}$$

where $\delta(x)$ is the Dirac delta function (cf. Sect. 1.10). The fluid equations are Galilean invariant, so we may conveniently adopt a reference frame moving with the discontinuity such that the one-dimensional mass conservation equation (2.3) reduces to

$$\frac{\partial(\rho v)}{\partial x} = 0. \tag{4.125}$$

Equation (4.124) is only compatible with (4.125) if the coefficient of the delta function vanishes, yielding

$$[\![\rho v]\!] = 0 \tag{4.126}$$

as the first jump condition. Thus ρv is unchanged across the shock, so the mass flux entering equals the mass flux leaving. This is precisely the condition obtained by integrating (4.125) across the shock—i.e. from the integral form of the mass conservation equation. Indeed, a weak solution of a conservation equation in differential form (2.1) generally corresponds to the physical interpretation in its integral form (2.2).

Let us now consider the ideal (inviscid) equation (3.5), written in momentum conservation form—cf. (2.19). Thus with one space dimension and ignoring gravity, we have

$$\frac{\partial}{\partial t}(\rho v) + \frac{\partial}{\partial x}(\rho v^2 + p) = 0, \tag{4.127}$$

whence the second jump condition in the reference frame moving with the shock:

$$\left[\!\left[\rho v^2 + p\right]\!\right] = 0. \tag{4.128}$$

The conservation of energy equation (3.9) in one space dimension likewise yields a third jump condition—i.e.

$$\frac{\partial}{\partial t}\left(\frac{1}{2}\rho v^2 + \frac{p}{\gamma - 1}\right) + \frac{\partial}{\partial x}\left(\frac{1}{2}\rho v^3 + \frac{\gamma}{\gamma - 1}pv\right) = 0 \tag{4.129}$$

produces

$$\left[\!\left[\frac{1}{2}\rho v^3 + \frac{\gamma}{\gamma - 1}pv\right]\!\right] = 0, \tag{4.130}$$

or using (4.126)

$$\left[\!\left[\frac{1}{2}v^2 + \frac{\gamma}{\gamma - 1}\frac{p}{\rho}\right]\!\right] = 0. \tag{4.131}$$

The set of three jump conditions (4.126), (4.128) and (4.131) involving ρ, v and p are known as the *Rankine–Hugoniot equations.*

It is notable that continuity of any one of the variables v, ρ and p implies continuity of the other two—so all three must change across a shock. Note also that it is important to use the conservation form of the relevant macroscopic equation, expressing the vanishing of a space–time divergence, in deriving the corresponding jump condition—one should not divide (4.125) by ρ for example, since the weak interpretation of the resulting equation does not necessarily give the same result. The second law of thermodynamics (the entropy of an isolated system can never decrease) implies that

$$\left[\!\left[p\rho^{-\gamma}\right]\!\right] \geq 0. \tag{4.132}$$

An important consequence is that steady flow through a shock proceeds from a low pressure (often supersonic) region to a high pressure subsonic region.

Exercise

(Q1) (a) Deduce (4.132) from the second law of thermodynamics, and note why this result can be reconciled with the equation of state (3.3).
(b) Assuming an entropy increase across the shock such that the flow is supersonic behind (upstream) and subsonic ahead (downstream), show that the temperature behind the shock exceeds the temperature ahead of it.

4.12 Shock Structure

In the previous section, weak solutions of ideal fundamental conservation equations of form

$$\frac{\partial \rho}{\partial t}+\frac{\partial q}{\partial x}=0, \tag{4.133}$$

led to jump conditions across shock discontinuities. However, it must be expected that fluid viscosity and heat conduction are important within a more realistic narrow transition region across any such shock, where fundamental field quantities vary rapidly. It follows that corresponding dissipative terms must be included in the fluid model to determine the *shock structure* in this transition region.
For a brief insight here, let us summarise some discussion from Whitham [14], where more detail may be found. Thus suppressing the velocity field for simplicity such that $q = Q(\rho) - \nu \partial\rho/\partial x$ where ν is a constant, the conservation form (4.133) becomes

$$\frac{\partial \rho}{\partial t}+c(\rho)\frac{\partial \rho}{\partial x}=\nu\frac{\partial^2 \rho}{\partial x^2}, \quad \text{where } c(\rho)=Q'(\rho). \tag{4.134}$$

Mathematical aspects of shock structure can thus be explored to some extent in the context of a quadratic $Q(\rho)$ such that $Q''(\rho) = 0$, when multiplying (4.134) through by $c'(\rho)$ yields

$$\frac{\partial c}{\partial t}+c\frac{\partial c}{\partial x}=\nu\frac{\partial^2 c}{\partial x^2}. \tag{4.135}$$

This is often called Burgers' equation, a model equation combining nonlinearity on the left-hand side with diffusion on the right-hand side. An explicit solution of (4.135) follows from the Cole–Hopf transformation, which can be represented in two steps, such that on introducing $c = \partial\psi/\partial x$ a first integral of equation (4.133) is

$$\frac{\partial \psi}{\partial t}+\frac{1}{2}\left(\frac{\partial \psi}{\partial x}\right)^2=\nu\frac{\partial^2 \psi}{\partial x^2}, \tag{4.136}$$

and then introducing $\psi = -2\nu \ln \phi$ produces the well-known linear diffusion equation

$$\frac{\partial \phi}{\partial t} = \nu \frac{\partial^2 \phi}{\partial x^2}. \tag{4.137}$$

Assuming the initial condition $c = F(x)$ at $t = 0$, which corresponds to

$$\phi(x, 0) = \Phi(x) \equiv \exp\left(-\frac{1}{2\nu}\int_0^x F(\eta)d\eta\right),$$

under the composite Cole–Hopf transformation

$$c = -2\nu \frac{\partial \phi/\partial x}{\phi}$$

the solution of the diffusion equation (4.137) is

$$\phi(x, t) = \frac{1}{\sqrt{4\pi\nu t}}\int_{-\infty}^{\infty} \Phi(\eta) \exp\left(-\frac{(x-\eta)^2}{4\nu t}\right) d\eta, \tag{4.138}$$

and hence the solution of (4.135) is

$$c(x, t) = \frac{\int_{-\infty}^{\infty} (x-\eta)\, e^{-G/2\nu}/t \, d\eta}{\int_{-\infty}^{\infty} e^{-G/2\nu} d\eta} \tag{4.139}$$

where

$$G(\eta; x, t) = \int_0^{\eta} F(\eta')d\eta' + \frac{(x-\eta)^2}{2t}. \tag{4.140}$$

In the limit $\nu \to 0$ with x, t and $F(x)$ held fixed, the dominant contributions to the integrals in (4.138) come from the neighbourhood of the stationary points of G—i.e. points where

$$\frac{\partial G}{\partial \eta} = F(\eta) - \frac{x-\eta}{t} = 0. \tag{4.141}$$

If $\eta = \xi(x, t)$ is such a stationary point, then the classical Laplace asymptotic formula is

$$\int_{-\infty}^{\infty} g(\eta) e^{-G(\eta)/2\nu} d\eta \simeq g(\xi)\sqrt{\frac{4\pi\nu}{|G''(\xi)|}} \exp(-G(\xi)/2\nu) \quad \text{as } \nu \to 0.$$

For example, if there is only one stationary point $\xi(x, t)$ we have

$$\int_{-\infty}^{\infty} \frac{x-\eta}{t} \mathrm{e}^{-G/2\nu} d\eta \simeq \frac{x-\xi}{t} \sqrt{\frac{4\pi\nu}{|G''(\xi)|}}$$

and

$$\int_{-\infty}^{\infty} \mathrm{e}^{-G/2\nu} d\eta \simeq \sqrt{\frac{4\pi\nu}{|G''(\xi)|}},$$

so that (4.139) yields

$$c \simeq \frac{x-\xi}{t}, \tag{4.142}$$

where $\eta = \xi(x, t)$ satisfies (4.141). This asymptotic solution may be rewritten

$$c = F(\xi), \qquad x = \xi + F(\xi)t, \tag{4.143}$$

which is the parametric solution of the nonlinear dissipationless ($\nu = 0$) first-order wave equation

$$\frac{\partial c}{\partial t} + c\frac{\partial c}{\partial x} = 0, \tag{4.144}$$

where the stationary point $\xi(x, t)$ is identified as its characteristic variable. Equation (4.143) may have a multi-valued solution after a sufficient time has elapsed, when weak solutions and shock discontinuities are introduced, although the solution (4.139) is single-valued and continuous $\forall\ t$. The explanation is that a multi-valued solution of (4.139) corresponds to there being more than one stationary point satisfying (4.141).

Bibliography

1. S. Chandrasekhar, *Hydrodynamic and Hydromagnetic Stability* (Dover, New York, 1981). (The classic treatise mentioned in the Preface and referenced in Chapter 2)
2. R. Courant, K.O. Friedrichs, *Supersonic Flow and Shock Waves* (Springer, New Jersey, 1977). (Reprint of the classic treatise originally published in 1948, but still remarkably suitable advanced textbook treating the dynamics of compressible fluids)
3. P.G. Drazin, W.H. Reid, *Hydrodynamic Stability*, 2nd edn. (Cambridge University Press, Cambridge, 2004). (First referenced in Chapter 3)
4. J.P. Goedbloed, S. Poedts, *Principles of Magnetohydrodynamics* (Cambridge University Press, Cambridge, 2004). (First referenced in Chapter 2)
5. H. Lamb, *Hydrodynamics*, 6th edn. (Cambridge University Press, Cambridge, 1993). (First referenced in Chapter 2)

6. J. Lighthill, *Waves in Fluids* (Cambridge University Press, Cambridge, 2001). (Valuable source on sound waves, shock waves and water waves)
7. P.D. Miller, *Applied Asymptotic Analysis* (American Mathematical Society, Providence, 2006). (First referenced in Chapter 3)
8. J.W.S. Rayleigh, *The Theory of Sound: Volume One, Unabridged* 2nd Rev. edn. (Dover, New York, 1976). (Classic text by the distinguished British scientist)
9. V.A. Squire, R.J. Hosking, A.D. Kerr, P.J. Langhorne, *Moving Loads on Ice Plates* (Kluwer, Dordrecht, 1996). (Mathematical and experimental field work undertaken to identify key phenomena for vehicles and aircraft to safely travel over floating ice sheets)
10. J.J. Stoker, *Water Waves: The Mathematical Theory with Applications* (Interscience, New York, 1992). (Well known presentation on linear gravity waves in liquids with a free surface)
11. J.S. Turner, *Buoyancy Effects in Fluids* (Cambridge University Press, Cambridge, 1979). (On motion due to buoyancy forces in a non-rotating stratified fluid under gravity)
12. J.-M. Vanden-Broeck, *Gravity-Capillary Free-Surface Flows* (Cambridge University Press, Cambridge, 2010). (Extensive discussion on surface tension effects in flows and waves)
13. R. von Mises, *Mathematical Theory of Compressible Fluid Flow* (Dover, New York, 2004). (A textbook where three chapters by von Mises are supplemented by two more chapters and notes written by others after he died, also directed to supersonic flow and shocks)
14. G.B. Whitham, *Linear and Nonlinear Waves* (Wiley, New York, 2011). (Highly recommended discussion of hyperbolic and dispersive waves first published in 1974, with extensive applications previously investigated by the author—including flood waves in rivers, waves in glaciers, traffic flow, sonic booms, blast waves, and ocean waves from storms)

Chapter 5
Magnetohydrodynamics (MHD)

We mentioned in the Preface that Fluid Mechanics and MHD often draw upon much the same mathematics and yield many closely related results. The mathematical kinship of the fundamental mathematical models was quite evident in Chap. 2, where novelties for MHD nevertheless emerged—viz. additional terms arising in the macroscopic equations (notably the Lorentz force in the equation of motion), the distinctive anisotropic plasma pressure tensor due to a magnetic field and the necessity to invoke suitable electromagnetic equations. This chapter explores the origin of the ideal and non-ideal MHD models briefly mentioned there, and then various important topics in MHD that are often prerequisite for our subsequent discussion of MHD stability theory. The additional bibliography for this chapter provides further background reading.

5.1 Introduction

In Chap. 3, we discussed the ideal model for fluid motion and the important consequences when fluid shear viscosity is introduced, before proceeding to consider wave propagation in fluids in Chap. 4. Now we build upon the discussion in Chap. 2 to describe the motion not only of an electrically conducting fluid but also of a plasma in the "fluid approximation"—viz. via *magnetohydrodynamics*, or *MHD* for short.[1] Once again, there is an ideal model to consider, and then non-ideal modifications.

While some early authors restricted the application of MHD theory to liquids (cf. Sect. 5.5), MHD models have been used much more extensively in describing laboratory and astrophysical plasmas—cf. Sect. 5.6 and thereafter, and also Refs. [2, 7] for example. Indeed, Alfvén invented MHD to explain energy transport in the Sun, and the field of plasma astrophysics quickly evolved [1]. Moreover, geophysical

[1]MHD is sometimes called hydromagnetics, less frequently but rather more comprehensively magneto-fluid-dynamics, and occasionally magnetogasdynamics in the context of dense partially ionised gases.

R.J. Hosking and R.L. Dewar, *Fundamental Fluid Mechanics and Magnetohydrodynamics*, DOI 10.1007/978-981-287-600-3_5

questions such as the maintenance of the Earth's magnetic field and the interaction of the solar wind with the magnetosphere continue to be addressed [3, 4]. The relatively late emergence of MHD and plasma physics during the twentieth century may also seem an historical irony, given the ubiquity of the plasma state beyond our Earth and the primarily classical nature of the theory.

It is generally accepted that the energy radiated by the Sun and other stars (for much of their lives) is produced by thermonuclear fusion of the largely hydrogen and helium plasma they contain, which is confined by their self-gravitational fields. The goal of designing and building a controlled thermonuclear fusion reactor to produce substantial clean sustainable energy on Earth has been a major stimulus for interest in MHD over more than fifty years, in the approach where extremely hot plasma is to be isolated from the material walls of the reactor by appropriately designed magnetic fields under various *magnetic confinement* schemes. Indeed, magnetic confinement research has led to many contributions to MHD from an extraordinarily large number of research scientists worldwide. We therefore often have in mind high-temperature plasmas of interest in fusion research, in addition to some astrophysical applications.

Let us now recall that the equation of motion (2.56) in a conducting fluid or a plasma introduces the electromagnetic body force $\mathbf{j} \times \mathbf{B}$, where $\mathbf{j}$ denotes the current density and $\mathbf{B}$ the magnetic field. As mentioned in Sect. 2.12, the "pre-Maxwell" electromagnetic equation (2.91) may be invoked to relate those two vector fields, and this aspect associated with Galilean invariance is now considered more closely in Sects. 5.2–5.4. Liquid metals such as mercury or liquid sodium are briefly considered in Sect. 5.5, followed by some remarks in Sect. 5.6 on the plasma context subsequently emphasised in this book. We consider the ideal MHD model in Sect. 5.7, including the interpretation of important terms and a summary of associated momentum and energy conservation—and then discuss the associated "frozen-in" magnetic field concept in Sect. 5.8, leading to our description of Alfvén and magnetosonic wave propagation in Sect. 5.9. A discussion of magnetohydrostatics is presented in Sect. 5.10, and magnetic coordinates are described in Sect. 5.11. Ideal MHD-advected discontinuities are discussed in Sect. 5.12, and plasma magnetic confinement experiments in Sect. 5.13. The resistive and Hall MHD models foreshadowed in Sect. 2.12 are then explored in Sects. 5.14 and 5.15, where the respective modifications to the equation of magnetic induction assume a prominent role, and corresponding non-ideal advected discontinuities are discussed in the optional (starred) Sect. 5.16.

5.2 Maxwell Equations

Rewritten in SI units (Systéme International d'Unités) and modern vector notation, the electromagnetic equations assembled by Maxwell are

$$\nabla \cdot \mathbf{B} = 0 \qquad \text{("no magnetic monopoles")}, \tag{5.1}$$

$$\left(\epsilon^0\right)$$

$$\nabla \times \mathbf{B} = \mu_0 \mathbf{j} + \frac{1}{c^2}\frac{\partial \mathbf{E}}{\partial t} \qquad \text{("modified Oersted/Ampère law")}, \tag{5.2}$$

$$\left(\epsilon^0\right) \quad \left(\epsilon^0\right) \quad \left(\epsilon^2\right)$$

$$\nabla \cdot \mathbf{E} = q/\epsilon_0 \qquad \text{("Poisson/Coulomb law")}, \tag{5.3}$$

$$\left(\epsilon^0\right) \quad \left(\epsilon^0\right)$$

$$\nabla \times \mathbf{E} = -\frac{\partial \mathbf{B}}{\partial t} \qquad \text{("Faraday law of induction")},$$

$$\left(\epsilon^0\right) \quad \left(\epsilon^0\right) \tag{5.4}$$

where the additional field variables introduced here are the electric field $\mathbf{E}$ and the total electric charge density q. The constant $c = \sqrt{\mu_0 \varepsilon_0}$ denotes the speed of light, where μ_0 and ε_0 are the magnetic permeability and electric permittivity in a vacuum, respectively. It is presumed that the medium is not significantly polarisable or magnetisable, which is appropriate for highly conducting media where currents are predominantly due to free electrons. The superscripted ϵ in brackets beneath each equation are explained below, and should not be confused with the electric permittivity ε_0.

5.3 Pre-Maxwell Equations

Typical low frequency MHD behaviour, on a time scale T much longer than the transit time for light waves across a system of length scale L, corresponds to the ordering

$$L/cT = \mathrm{O}(\epsilon), \quad \epsilon \ll 1. \tag{5.5}$$

Thus for characteristic length and time scales $L = \mathrm{O}(1)$ and $T = \mathrm{O}(1)$, we have $c = \mathrm{O}(\epsilon^{-1})$—i.e. the speed of light is much larger than any characteristic velocity. Also $\nabla \times \mathbf{E} \sim \mathbf{B}/T$ from (5.4), where $\sim$ may be interpreted as "scales like" rather than strict asymptotic equality. Moreover, MHD phenomena of interest are not primarily electrostatic, so that $\nabla \times \mathbf{E} \sim \mathbf{E}/L$ (except for static fields). Consequently, in Maxwell's equations (5.1)–(5.4) above, the orderings of the terms with respect to ϵ are as indicated in the aforementioned brackets beneath each of them. All terms are of the same zeroth-order, except for the displacement current term $c^{-2}\partial \mathbf{E}/\partial t$ that is only $\mathrm{O}(\epsilon^2)$. Substituting the following formal expansions in ϵ for the electromagnetic variables

$$\mathbf{E} = \mathbf{E}^{(0)} + \epsilon \mathbf{E}^{(1)} + \cdots, \tag{5.6}$$

$$\mathbf{B} = \mathbf{B}^{(0)} + \epsilon \mathbf{B}^{(1)} + \cdots, \tag{5.7}$$

$$\mu_0 \mathbf{j} = (\mu_0 \mathbf{j})^{(0)} + \epsilon\,(\mu_0 \mathbf{j})^{(1)} + \cdots, \tag{5.8}$$

$$q/\epsilon_0 = (q/\epsilon_0)^{(0)} + \epsilon\,(q/\epsilon_0)^{(1)} + \cdots, \tag{5.9}$$

into Maxwell's equations (5.1)–(5.4) therefore yields to lowest order

$$\nabla \cdot \mathbf{B}^{(0)} = 0, \tag{5.10}$$

$$\nabla \times \mathbf{B}^{(0)} = (\mu_0 \mathbf{j})^{(0)}, \quad (\text{“Ampère's law”}) \tag{5.11}$$

$$\nabla \cdot \mathbf{E}^{(0)} = (q/\epsilon_0)^{(0)}, \tag{5.12}$$

$$\nabla \times \mathbf{E}^{(0)} = -\frac{\partial \mathbf{B}^{(0)}}{\partial t}, \tag{5.13}$$

and also to first order $O(\epsilon)$ with superscripts $^{(1)}$ replacing the superscripts $^{(0)}$. We call the set (5.10)–(5.13) "pre-Maxwell equations" because one of Maxwell's famous contributions to electromagnetism has been undone by dropping the displacement current from (5.2), a characteristic feature of classical MHD.

An immediate consequence of the divergence of (5.11) and its first order-counterpart is that

$$\nabla \cdot (\mu_0 \mathbf{j})^{(0)} = \nabla \cdot (\mu_0 \mathbf{j})^{(1)} = 0, \tag{5.14}$$

hence the charge conservation equation obtained from (5.2) and (5.3) becomes

$$\frac{\partial}{\partial t}(q/\varepsilon_0)^{(0)} + c^2 \nabla \cdot (\mu_0 \mathbf{j})^{(2)} = 0 \tag{5.15}$$

to lowest order. Consequently, except in a vacuum where $(\mu_0 \mathbf{j})^{(2)} \equiv 0$, the zero order charge density $q^{(0)}$ cannot be determined directly by integrating (5.15) because it involves the unknown second order current density. However, it turns out that (5.12) is not needed to determine $\mathbf{E}^{(0)}$ except in a vacuum, so this equation can be used to give $q^{(0)}$ if desired. On the other hand, there is usually no need to know the charge density, so (5.12) is often omitted from the set of essential pre-Maxwell equations.

5.4 Galilean Invariance

Let us, henceforth, omit the implicit ordering superscript from the essential pre-Maxwell equations (5.10), (5.11) and (5.13). Thus we adopt

$$\nabla \cdot \mathbf{B} = 0, \tag{5.16}$$

$$\nabla \times \mathbf{B} = \mu_0 \mathbf{j}, \tag{5.17}$$

$$\nabla \times \mathbf{E} = -\frac{\partial \mathbf{B}}{\partial t}, \tag{5.18}$$

a set of model equations that may be expected to have invariance and conservation properties, although not the same as for the fundamental Maxwell equations. Indeed, since the ordering where $c = O(\epsilon^{-1}) \to \infty$ is basically non-relativistic, we might anticipate that the pre-Maxwell equations share with Newton's laws the property of *Galilean invariance* under transformation to another uniformly moving coordinate system (inertial reference frame) as demonstrated below.

The Galilean transformation from a reference frame with Cartesian coordinates (x', y', z') moving with constant velocity $\mathbf{V} = V\hat{\mathbf{i}}$ relative to another with coordinates (x, y, z) is

$$x' = x - Vt,\ y' = y,\ z' = z,\ t' = t. \tag{5.19}$$

Applying the chain rule to functions of the primed variables identifies

$$\begin{aligned}\nabla &= (\nabla x')\frac{\partial}{\partial x'} + (\nabla y')\frac{\partial}{\partial y'} + (\nabla z')\frac{\partial}{\partial z'} + (\nabla t')\frac{\partial}{\partial t'}\\ &= \hat{\mathbf{i}}\frac{\partial}{\partial x'} + \hat{\mathbf{j}}\frac{\partial}{\partial y'} + \hat{\mathbf{k}}\frac{\partial}{\partial z'}\\ &\equiv \nabla'\end{aligned} \tag{5.20}$$

and
$$\begin{aligned}\frac{\partial}{\partial t} &= \left(\frac{\partial x'}{\partial t}\right)\frac{\partial}{\partial x'} + \left(\frac{\partial y'}{\partial t}\right)\frac{\partial}{\partial y'} + \left(\frac{\partial z'}{\partial t}\right)\frac{\partial}{\partial z'} + \left(\frac{\partial t'}{\partial t}\right)\frac{\partial}{\partial t'}\\ &= \frac{\partial}{\partial t'} - \mathbf{V}\cdot\nabla'\\ &\equiv \frac{\partial}{\partial t'} - \mathbf{V}\cdot\nabla\end{aligned} \tag{5.21}$$

from (5.19), such that

$$\frac{\partial}{\partial t'} = \frac{\partial}{\partial t} + \mathbf{V}\cdot\nabla. \tag{5.22}$$

This is similar to the advective derivative d/dt introduced in Sect. 2.4, although here $\mathbf{V}$ is not necessarily a fluid velocity. Consequently, equations (5.16)–(5.18) transform to

$$\nabla'\cdot\mathbf{B} = 0, \tag{5.23}$$

$$\nabla'\times\mathbf{B} = \mu_0\mathbf{j}, \tag{5.24}$$

$$\left(\frac{\partial}{\partial t'} - \mathbf{V}\cdot\nabla'\right)\mathbf{B} + \nabla'\times\mathbf{E} = 0 \quad \text{or} \quad \frac{\partial\mathbf{B}}{\partial t'} + \nabla'\times(\mathbf{E} + \mathbf{V}\times\mathbf{B}) = 0, \tag{5.25}$$

on using the identity

$$\begin{aligned}\nabla'\times(\mathbf{V}\times\mathbf{B}) &= \mathbf{V}\,\nabla'\cdot\mathbf{B} - \mathbf{V}\cdot\nabla'\mathbf{B}\\ &= -\mathbf{V}\cdot\nabla'\mathbf{B}\end{aligned}$$

from (5.23). Equations (5.23)–(5.25) have the same form as (5.16)–(5.18), provided the electric and magnetic fields transform as

$$\mathbf{B}' = \mathbf{B}, \tag{5.26}$$

$$\mathbf{E}' = \mathbf{E} + \mathbf{V} \times \mathbf{B}. \tag{5.27}$$

The pre-Maxwell equations (5.16)–(5.18) are thus invariant under Galilean transformation, where the electric and magnetic fields in the moving frame are, respectively, defined by (5.26) and (5.27).[2] Equations (5.16)–(5.18) are therefore compatible with the macroscopic equations discussed in Chap. 2, and so may be combined with an appropriate subset to constitute a suitable MHD model as foreshadowed in Sect. 2.12. A common characterisation of MHD is that electric currents modify the flow of a conducting fluid or plasma while at the same time the flow induces electric currents in the medium. Thus the principal field variables describing this interaction are the velocity $\mathbf{v}$ and the magnetic field $\mathbf{B}$, and to a lesser extent the density ρ and pressure p. Nevertheless, sometimes the electric field $\mathbf{E}$ or the current density $\mathbf{j}$ are also relevant in the discussion. While $\mathbf{j}$ is given directly by the pre-Maxwell equation (5.17), Eq. (5.18) is insufficient to give $\mathbf{E}$ even if the relevant boundary conditions are known, as we do not know $\nabla \cdot \mathbf{E}$. Instead. $\mathbf{E}$ must be found from a constitutive equation—viz. Ohm's law (2.92) or the generalised Ohm's law (2.98).

5.5 Conducting Liquids

As indicated in Sect. 5.1, MHD refers to the dynamics of electrically conducting fluids and plasmas in a magnetic field, but let us first consider the simpler case of electrically conducting liquids such as molten metals (e.g. iron in the core of the Earth, considered to produce the terrestrial magnetic field by dynamo action) or electrolytes (e.g. flowing seawater producing a measurable voltage due to its interaction with the magnetic field of the Earth). Their flow is usually described by an extended Navier–Stokes model, involving the additional Lorentz electromagnetic body force term $\mathbf{j} \times \mathbf{B}$ on the right-hand side of the equation of motion. Thus the governing equations include the incompressible approximation for a liquid or subsonic gas

$$\nabla \cdot \mathbf{v} = 0, \tag{5.28}$$

and the modified Navier–Stokes equation of motion

$$\rho \frac{d\mathbf{v}}{dt} + \nabla p = \rho \mathbf{g} + \mu_0^{-1} (\nabla \times \mathbf{B}) \times \mathbf{B} + \mu \nabla^2 \mathbf{v} \tag{5.29}$$

---

[2] Since $\mathbf{E}$ and $\mathbf{B}$ appear only as derivatives in (5.16)–(5.18), it might seem that we could add constants to $\mathbf{E}'$ and $\mathbf{B}'$, but this possibility is ruled out because the Lorentz force $\mathbf{E} + \mathbf{v} \times \mathbf{B}$ on a particle must be invariant.

obtained from (2.56) for constant shear viscosity coefficient μ, on adopting the classical fluid pressure tensor (2.16) for a neutral fluid and invoking (5.17).

The term ∇p can be annihilated from (5.29) by taking its curl when the fluid density ρ is constant throughout the flow, but of course another equation to account for $\mathbf{B}$ is required to close the system. As indicated in Sect. 2.12, the simple constitutive relation $\mathbf{j} = \sigma \mathbf{E}'$ referred to as Ohm's law is often assumed to hold between the current density $\mathbf{j}$ and the electric field $\mathbf{E}'$ in the local rest frame of the fluid, where σ denotes the electrical conductivity of the fluid. Thus from the Galilean transformation (5.27), on setting $\mathbf{V} = \mathbf{v}$, we find an expression for the electric field $\mathbf{E} = \mathbf{j}/\sigma - \mathbf{v} \times \mathbf{B}$, to provide (2.92) and hence the resistive equation of magnetic induction (2.94) reproduced here for convenience:

$$\frac{\partial \mathbf{B}}{\partial t} = \nabla \times (\mathbf{v} \times \mathbf{B}) - \nabla \times (\eta \nabla \times \mathbf{B}). \tag{5.30}$$

The flow is typically over solid boundaries, which in ducts are often (although not always) considered to be perfect insulators, such that any magnetically induced electric currents in the liquid remain inside it. The earliest laboratory experiments used mercury, corresponding to a relatively small magnetic Reynolds number $\mathrm{Rm} \equiv UL/\eta$ (resembling the Reynolds number Re of Chap. 3), so the magnetic field diffusion governed by the resistive term $\nabla \times (\eta \nabla \times \mathbf{B})$ in (5.30) is relatively large compared to the inductive term $\nabla \times (\mathbf{v} \times \mathbf{B})$. This was frustrating, since MHD theory at the time emphasised ideal models that *inter alia* neglect resistivity ($\mathrm{Rm} \to \infty$), and the resistive term is significant in mercury (cf. Sect. 5.14). A Hall term similar to that previously identified for plasmas in Sect. 2.12 may also be significant, as it produces topologically important contributions.[3]

Exercises

(Q1) A finitely conducting liquid swirls with velocity $\mathbf{v} = v(r)\hat{\mathbf{e}}_\theta$ in the presence of a magnetic field $\mathbf{B} = \nabla \times (\psi \hat{\mathbf{e}}_z)$ specified by the stream function analogue $\psi(r, \theta, t)$, in terms of cylindrical coordinates. Show that

$$\frac{\partial \psi}{\partial t} + \frac{v}{r}\frac{\partial \psi}{\partial \theta} = \eta \nabla^2 \psi$$

where $\eta = (\mu_0 \sigma)^{-1}$ is the constant resistivity, provided there is no prevailing electric field. When

$$v(r) = \begin{cases} k/r\,, & \text{if } r < a \\ 0\,, & \text{if } r > a \end{cases}$$

[3] The conducting liquid is often but not always treated as uniform, such that all essential coefficients (of viscosity, resistivity or the Hall term) are taken to be known and constant. A non-uniform liquid with position-dependent coefficients may be considered, to discuss MHD motion in the Earth's core for example—given of course suitable modifications and additional definitive equations such as $d\rho/dt = 0$ to account for the variable density and the variable coefficients.

where k and a are constants, and $\mathbf{B} \to B_0\hat{\mathbf{i}}$ (where B_0 is constant) as $r \to \infty$ in analogy with uniform ideal flow past a cylinder, obtain the following steady solution in the region $r < a$:

$$\psi(r,\theta) = \frac{2B_0 a}{\sqrt{(1+\alpha)^2+\beta^2}}\left(\frac{r}{a}\right)^{\alpha} \sin[\beta \ln\left(\frac{r}{a}\right) + \theta - \epsilon],$$

where $\alpha + i\beta = \sqrt{1+ik/\eta}$ and $\tan\epsilon = \beta/(1+\alpha)$, with $\alpha > 0$.

Hint: First write $\psi = f(r)\exp(i\theta)$, and proceed to consider its imaginary part.

(Q2) (Hartmann flow) A viscous finitely conducting uniform liquid occupies a rigid straight duct of rectangular cross-section, with vertical dimension $|z| < a$ much smaller than its horizontal dimension. If the liquid is subjected to a constant axial pressure gradient $-G\hat{\mathbf{i}}$ and a constant vertical magnetic field $B_0\hat{\mathbf{k}}$, so the resulting component of the current density is $j_y = \sigma(E_0 - B_0 v)$ where E_0 denotes the y-component of an electric field, show that the velocity field $\mathbf{v} = v(z)\hat{\mathbf{i}}$ in the bulk of the steady flow satisfies the differential equation

$$\frac{d^2v}{dz^2} - \frac{\mathrm{M}^2}{a^2}v = -\frac{1}{\mu}(G + \sigma B_0 E_0),$$

where $\mathrm{M} = B_0 a\sqrt{\sigma/\mu}$ is the Hartmann number. Hence obtain the velocity profile

$$v(z) = \frac{a^2}{\mathrm{M}^2\mu}(G + \sigma E_0 B_0)\left(1 - \frac{\cosh(\mathrm{M}z/a)}{\cosh \mathrm{M}}\right).$$

If the distant side walls of the duct are insulated from each other so that the net electric current flow in the y-direction is zero, show that

$$E_0 = \frac{G}{\sigma B_0}\left(\frac{\mathrm{M}}{\tanh \mathrm{M}} - 1\right).$$

When $\mathrm{M} \gg 1$, demonstrate that $v(0) \simeq U_0/\mathrm{M}$ for an insulated duct but that $v(0) \simeq U_0/\mathrm{M}^2$ if the duct is perfectly conducting such that $E_0 = 0$, where $U_0 = Ga^2/\mu$.

5.6 Plasma MHD

By the middle of the last century, it was widely appreciated that hydrogen is the predominant element in the Universe, and that in many cases it can be treated as a simple ion-electron plasma because the kinetic temperature is high enough to ensure almost complete ionisation.[4] As discussed in detail in Chap. 2, a plasma may be described in terms of more fundamental kinetic theory that explicitly involves a distribution function in phase space for each of its microscopic constituents, but an MHD model can often be adopted [3, 4].

We mentioned in Sect. 5.1 that MHD was associated at first with the question of solar energy transport—and theories for the sunspot cycle and sunspot equilibria, the solar wind and its interaction with the Earth's magnetosphere, magnetic fields in interstellar plasma and other cosmic questions were addressed [1]. The kinetic temperature of laboratory plasma can also be high enough that it is essentially fully ionised almost everywhere, as in various magnetic confinement devices being explored in controlled thermonuclear fusion research programmes. Wave propagation and stability, topics central to the earliest astrophysical applications, are also of major interest in the laboratory context. As mentioned in the previous section, the resistivity (finite conductivity) of mercury was found to significantly influence wave propagation in early experiments and it soon emerged that resistivity and other non-ideal effects have important consequences in plasmas. However, it is appropriate to begin with the ideal MHD model and related matters such as ideal MHD wave propagation, before introducing non-ideal effects into the discussion.

5.7 Ideal MHD Model

As first mentioned in Sect. 2.12, both the inviscid (zero viscosity) and the perfect conductivity (zero resistivity) assumptions are implicit in the ideal MHD model. One may also continue to neglect heat flow (especially perpendicular to the local magnetic field), treating the plasma as a compressible medium with p/ρ^γ constant for any given fluid element, as in the earlier fluid mechanics discussion (cf. Sect. 3.2). Thus the ideal compressible MHD equations comprise the continuity equation (2.55), the ideal (inviscid) equation of motion (3.5) with the $\mathbf{j} \times \mathbf{B}$ term added to the right-hand side as first envisaged in (2.56), the adiabatic equation of state (3.3), Ohm's law (2.92) with its right-hand side set to zero, and the pre-Maxwell equations of Sect. 5.4.

From (5.17), the MHD forcing term $\mathbf{j} \times \mathbf{B}$ entering the equation of motion can be written as

$$\begin{aligned}\mu_0^{-1}(\nabla \times \mathbf{B}) \times \mathbf{B} &= \mu_0^{-1}\mathbf{B} \cdot \nabla\mathbf{B} - \mu_0^{-1}\nabla(B^2/2) \\ &= \frac{B^2}{\mu_0}\hat{\mathbf{b}} \cdot \nabla\hat{\mathbf{b}} + \hat{\mathbf{b}}\hat{\mathbf{b}} \cdot \nabla\left(\frac{B^2}{2\mu_0}\right) - \nabla\left(\frac{B^2}{2\mu_0}\right) \\ &= \frac{B^2}{\mu_0}\boldsymbol{\kappa} - \nabla_\perp\frac{B^2}{2\mu_0}. \qquad (5.31)\end{aligned}$$

[4]However, not *all* plasmas are simple and fully ionised—e.g. in the MHD power generation context, the MHD fluid is a complex mix of partially ionised plasma and burning coal dust!

Here $\boldsymbol{\kappa}$ denotes the magnetic field line curvature, the derivative of the parallel unit vector $\hat{\mathbf{b}} = \mathbf{B}/B$ with respect to length along the magnetic field line as previously defined in (2.81), and

$$\nabla_{\perp} = \nabla - \hat{\mathbf{b}}\hat{\mathbf{b}} \cdot \nabla \tag{5.32}$$

is the perpendicular gradient operator. Thus the essential term that encapsulates the electric current modification of the plasma dynamics consists of a magnetic tension $(B^2/\mu_0)\,\boldsymbol{\kappa}$ acting towards the centre of magnetic field curvature, and a magnetic pressure gradient $\nabla_{\perp} B^2/(2\mu_0)$. These two contributions on the right-hand side of (5.31) also produce definitive waves in Hall MHD, briefly discussed later—cf. Sect. 5.15.

In summary therefore, on invoking (5.17) the fundamental ideal compressible equations in the fields ρ, $\mathbf{v}$, p, $\mathbf{E}$ and $\mathbf{B}$ are

$$\frac{d\rho}{dt} + \rho\nabla\cdot\mathbf{v} = 0, \tag{5.33}$$

$$\rho\frac{d\mathbf{v}}{dt} + \nabla p = \rho\mathbf{g} + \mu_0^{-1}(\nabla\times\mathbf{B})\times\mathbf{B}, \tag{5.34}$$

$$\frac{dp}{dt} + \gamma p\nabla\cdot\mathbf{v} = 0, \tag{5.35}$$

$$\mathbf{E} + \mathbf{v}\times\mathbf{B} = 0, \tag{5.36}$$

$$\nabla\times\mathbf{E} = -\frac{\partial\mathbf{B}}{\partial t}. \tag{5.37}$$

The gravitational term $\rho\mathbf{g}$ in the equation of motion (5.34) is often quite important in astrophysical applications—and it has also been used as a pseudo-centrifugal external force to model the actual magnetic field curvature term $(B^2/\mu_0)\,\boldsymbol{\kappa}$ of (5.31) in MHD stability analyses undertaken in Cartesian geometry (cf. Chap. 6), although this notion proved to be rather less satisfactory than had been hoped.

This set of ideal compressible equations (5.33)–(5.37) is complete. In particular, we observe that the divergence of (5.37) implies $\partial(\nabla\cdot\mathbf{B})/\partial t = 0$. Thus the pre-Maxwell equation (5.16) is not a constraint, but merely an initial condition that is propagated to later times by an equation of evolution for $\mathbf{B}$—viz. the relevant ideal equation of magnetic induction

$$\frac{\partial\mathbf{B}}{\partial t} = \nabla\times(\mathbf{v}\times\mathbf{B}), \tag{5.38}$$

obtained by eliminating $\mathbf{E}$ between (5.36) and (5.37) as noted previously in Sect. 2.12. The notable similarity of (5.38) and (3.25) suggests there is a strong analogy between the vorticity and magnetic fields (cf. Sect. 5.8).

The reduced system of ideal MHD equations (5.33)–(5.35) and (5.38) obtained on eliminating $\mathbf{E}$ is of course a complete set in the field variables ρ, $\mathbf{v}$, p and $\mathbf{B}$ that were specified at the end of Sect. 5.4. Moreover, recalling that compressible fluids are effectively incompressible to leading order in the slow subsonic expansion introduced in Sect. 3.4, we might anticipate that compressible plasma may often also behave in an approximately incompressible manner. Rather than assuming incompressibility ad hoc, this may be shown to follow from the compressible MHD equations under an analogous consistent ordering scheme. As previously indicated, there are no plasma viscosity terms retained in the equation of motion (5.34), and the neglect of resistivity in the equation of magnetic induction (5.38) is sometimes said to correspond to infinite magnetic Reynolds number Rm (cf. Sect. 5.5).

We can re-express the equation of motion (5.34) as

$$\rho\frac{d\mathbf{v}}{dt} = -\nabla\cdot(p\,\mathsf{I} + \mathcal{T}) + \rho\mathbf{g}, \tag{5.39}$$

which involves the *magnetic stress tensor*

$$\mathcal{T} = \frac{B^2}{2\mu_0}\mathsf{I} - \frac{\mathbf{BB}}{\mu_0}, \tag{5.40}$$

upon invoking the two identities $(\nabla\times\mathbf{B})\times\mathbf{B} = \mathbf{B}\cdot\nabla\mathbf{B} - \nabla(B^2/2)$ and $\nabla\cdot(\mathbf{BB}) = \mathbf{B}\cdot\nabla\mathbf{B}$ since $\nabla\cdot\mathbf{B} = 0$. Alternatively, we may write (5.39) in momentum conservation form as

$$\frac{\partial(\rho\mathbf{v})}{\partial t} + \nabla\cdot\rho\mathbf{vv} = -\nabla\cdot(p\,\mathsf{I} + \mathcal{T}) + \rho\mathbf{g} \tag{5.41}$$

on recalling the material derivative $d/dt = \partial_t + \mathbf{v}\cdot\nabla$ and invoking (5.33). Thus the MHD forcing term $\mathbf{j}\times\mathbf{B}$ produces an isotropic magnetic pressure $B^2/2\mu_0$, which in combination with the hydrostatic pressure p constitutes a total isotropic pressure; and in addition there is an anisotropic second term $\mathbf{BB}/\mu_0$, which corresponds to a magnetic tension, twice the strength of the isotropic magnetic pressure, acting along the field lines. We also note that the magnetic stress tensor may be rewritten

$$\mathcal{T} = \frac{B^2}{2\mu_0}\mathsf{I}_\perp - \frac{B^2}{2\mu_0}\hat{\mathbf{b}}\hat{\mathbf{b}}, \tag{5.42}$$

where $\mathsf{I}_\perp = \mathsf{I} - \hat{\mathbf{b}}\hat{\mathbf{b}}$ as in (2.74), giving a net parallel magnetic tension equal to the perpendicular magnetic pressure. Further, since $\nabla\cdot\mathbf{B} = 0$ the divergence of either (5.40) or (5.42) reproduces (5.31), with the previously associated alternative interpretation.

We have retained the gravitational body force $\rho\mathbf{g}$ in (5.39) and (5.41), although it is usually omitted in considering laboratory applications—i.e. unless this term is used to simulate magnetic field curvature, as previously mentioned. However, gravity is of course instrumental in many astrophysical applications, and the term $\rho\mathbf{g}$ is retained in representing the gravitational field of a neighbouring massive body for example. On the other hand, for a self-gravitating system one may introduce the gravitational field stress tensor (2.18), such that the right-hand sides of (5.39) and (5.41) may be rewritten as

$$\nabla\cdot(p\,\mathsf{I} + \mathcal{T} + \mathsf{G})$$

if desired—cf. also (2.19) and Ref. [7]. Both self-gravitation and the case of a central gravitational force are considered in our discussion of magnetorotational instability later—cf. Sect. 6.8.

Dotting (5.34) with $\mathbf{v}$ yields the mechanical subsystem energy equation generalising (3.7)—i.e.

$$\frac{\partial}{\partial t}\left(\frac{1}{2}\rho v^2\right) + \nabla\cdot\left(\frac{1}{2}\rho v^2\mathbf{v}\right) - p\nabla\cdot\mathbf{v} = \rho\,\mathbf{g}\cdot\mathbf{v} - \nabla\cdot(p\mathbf{v}) + \mathbf{j}\cdot\mathbf{E}, \tag{5.43}$$

where (5.17) and (5.36) have been used to rewrite the $\mu_0^{-1}(\nabla\times\mathbf{B})\times\mathbf{B}\cdot\mathbf{v}$ term as the familiar electric power term $\mathbf{j}\cdot\mathbf{E}$ (supplementing the rate of work done by the gravity and hydrostatic pressure on the right-hand side). The thermodynamic energy Eq. (3.8) remains appropriate here, since none of the $\mathbf{j}\cdot\mathbf{E}$ power goes into the thermodynamic subsystem (there is no Ohmic dissipation term ηE^2 in the ideal MHD model where the resistivity $\eta = 0$).

A conservation equation for the magnetic subsystem is also necessary for closure, in order to obtain the equation of total energy conservation. Dotting (5.37) with $\mathbf{B}$ readily yields

$$\frac{\partial}{\partial t}\left(\frac{1}{2}B^2\right) + \nabla\cdot(\mathbf{E}\times\mathbf{B}) + \mathbf{E}\cdot\nabla\times\mathbf{B} = 0,$$

so invoking (5.17) we have the magnetic energy equation

$$\frac{\partial}{\partial t}\left(\frac{B^2}{2\mu_0}\right) + \nabla\cdot\left(\frac{\mathbf{E}\times\mathbf{B}}{\mu_0}\right) = -\mathbf{j}\cdot\mathbf{E}, \tag{5.44}$$

where the magnetic energy flux term $\mu_0^{-1}\mathbf{E}\times\mathbf{B}$ is recognised as the Poynting vector of electromagnetic theory, and the magnetic energy density $B^2/2\mu_0$ is the surviving component of the electromagnetic energy (formerly $B^2/2\mu_0 + \varepsilon_0 E^2/2$) in the pre-Maxwell MHD ordering. Invoking (3.8) and adding Eqs. (5.43) and (5.44) then provides the equation for the total energy density (kinetic plus adiabatic plus magnetic energy densities)

$$\frac{\partial}{\partial t}\left(\frac{1}{2}\rho v^2 + \frac{p}{\gamma-1} + \frac{B^2}{2\mu_0}\right) + \nabla\cdot\left(\frac{1}{2}\rho v^2\mathbf{v} + \frac{\gamma}{\gamma-1}p\mathbf{v} + \frac{\mathbf{E}\times\mathbf{B}}{\mu_0}\right) = \rho\mathbf{g}\cdot\mathbf{v}$$

generalising (3.9); and eliminating $\mathbf{E}$ using (5.36) produces the alternative form

$$\begin{aligned}\frac{\partial}{\partial t}\left(\frac{1}{2}\rho v^2 + \frac{p}{\gamma-1} + \frac{B^2}{2\mu_0}\right)& \\ + \nabla\cdot\left[\left(\frac{1}{2}\rho v^2 + \frac{\gamma}{\gamma-1}p + \frac{B^2}{2\mu_0}\right)\mathbf{v} + \mathcal{T}\cdot\mathbf{v}\right] &= \rho\mathbf{g}\cdot\mathbf{v}.\end{aligned} \tag{5.45}$$

Finally, it is notable that the ideal equation of magnetic induction (5.38) can be rewritten as

$$\frac{d\mathbf{B}}{dt} = \mathbf{B}\cdot(\nabla\mathbf{v} - \mathbf{I}\nabla\cdot\mathbf{v}) = \mathbf{B}\cdot\nabla\,\mathbf{v} - \mathbf{B}\nabla\cdot\mathbf{v}. \tag{5.46}$$

This evolution equation for the vector field **B** is similar in form to Eqs. (5.33) and (5.35) for the scalar fields ρ and p, and can be integrated via a generalisation of the procedure outlined in Sect. 4.10. Furthermore, (5.46) is isomorphic to the vorticity equation (3.26), as foreshadowed with reference to (5.38) and (3.25) before.

Exercise

(Q1) Consider steady unidirectional ideal MHD flow where the magnetic field is perpendicular to the plane of the flow. Show that $(d/dt)(B/\rho) = 0$; and also that

$$\frac{1}{2}|\nabla\phi|^2 + h + V + \frac{B^2}{2\mu_0} = \text{constant}$$

throughout, the modified form of the Bernoulli equation (3.30) for steady irrotational flow.

5.8 "Frozen-In" Magnetic Field

A useful concept associated with ideal MHD is that the magnetic field is "frozen-in", and advects with the ideal MHD fluid. The "frozen-in" flux condition was first introduced by Alfvén in his pioneering paper on MHD published in 1942 (cf. also Ref. [1]), and the analogy between vortex flux tubes and magnetic flux tubes implicit in the isomorphic forms (3.26) and (5.46) eventually became more widely appreciated.

Let us consider a surface S in the MHD fluid bounded by the closed material curve C, which is advected from the contour $C_t = \partial S_t$ in space at time t to a new contour ∂S_{t+dt} at time $t + dt$ as the material surface S is continuously advected from the surface S_t at time t to the corresponding new surface $C_{t+dt} = S_{t+dt}$ at time $t + dt$ (cf. Fig. 5.1). From (5.16) and the Divergence Theorem (1.60), the magnetic flux leaving the volume swept out is equal to the sum of the flux entering through S at time t and the surface strip swept out by C in the time interval dt. The time rate of change of the magnetic flux $\Phi(t)$ threading the material surface is therefore

$$\begin{aligned}
\frac{d\Phi}{dt} &= \left[\int_{S_{t+dt}} \mathbf{B}\cdot d\mathbf{S} - \int_{S_t} \mathbf{B}\cdot d\mathbf{S}\right] \\
&= \int_S \frac{\partial\mathbf{B}}{\partial t}\cdot d\mathbf{S} + \int_C \mathbf{B}\cdot\mathbf{v}\times d\mathbf{r} \\
&= -\int_S \nabla\times\mathbf{E}\cdot d\mathbf{S} - \int_C \mathbf{v}\times\mathbf{B}\cdot d\mathbf{r} \\
&= -\int_S (\mathbf{E} + \mathbf{v}\times\mathbf{B})\cdot d\mathbf{r}
\end{aligned} \tag{5.47}$$

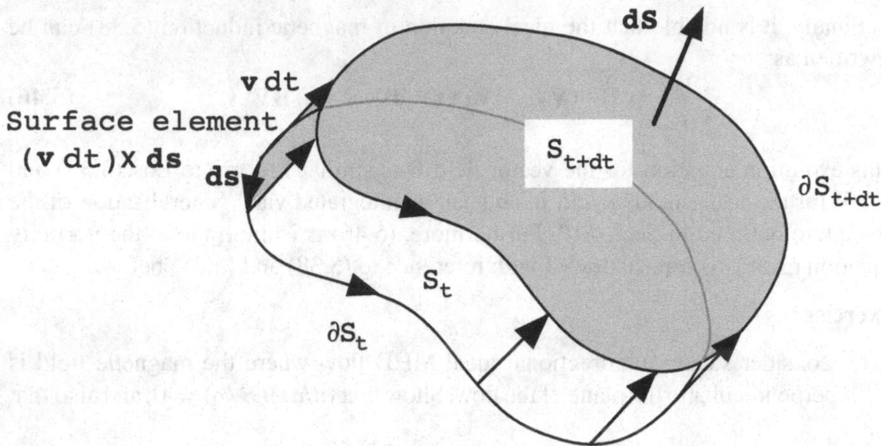

Fig. 5.1 Advection of a small surface element S_t of MHD fluid, of area dS and unit normal $\hat{\mathbf{n}}$, to S_{t+dt} during the infinitesimal time interval $(t, t + dt)$. The vector surface element $\mathbf{dS} = \hat{\mathbf{n}}dS$ is indicated, as well as the side vector surface element $\mathbf{v}dt \times \mathbf{ds}$ as the boundary $\partial\mathbf{S}_t$ advects to $\partial\mathbf{S}_{t+dt}$

by (5.37) and Stokes Theorem, whence from (5.36) the rate of change of the material integral is zero—i.e.

$$\frac{d\Phi}{dt} = \frac{d}{dt}\int_S \mathbf{B}\cdot d\mathbf{S} = 0. \tag{5.48}$$

At the second line of (5.47), note that our discussion in Sect. 2.5 has now been extended to the rate of change of a material surface integral.

Equation (5.48) is the "frozen-in flux condition" such that the magnetic flux linking a closed curve carried along by an ideal MHD fluid is conserved, provided the topological integrity of the closed curve is preserved by the flow. Indeed, the magnetic flux Φ is invariant under any topologically equivalent deformation of the surface S that leaves its boundary curve C unchanged, since $\nabla\cdot\mathbf{B} = 0$, so we may choose to consider a surface S that entirely advects with the fluid (i.e. corresponds to a material surface). Further, by considering an arbitrarily small tube of magnetic field lines, (5.48) is seen to be consistent with the idea that the field lines themselves are carried along with the flow—i.e. unless the flow takes the lines through a singular non-ideal plasma region where the topological integrity of a magnetic flux tube is destroyed, when the magnetic field lines break and so may reconnect as the plasma construct is said to *tear*. Moreover, since the circulation in Keivin's Theorem (3.27) is $K \equiv \oint_C \mathbf{v}\cdot d\mathbf{r} = \int_S \boldsymbol{\omega}\cdot d\mathbf{S}$, the result (5.48) is another manifestation of the analogy between the vorticity $\boldsymbol{\omega}$ and the magnetic field $\mathbf{B}$.

As discussed in Sect. 5.13 and depicted in Fig. 5.2, in the laboratory it is of interest[5] to consider a ribbon-like surface Σ_k in a vacuum, bounded on one side by an ideal plasma and on the other side by perfectly conducting surfaces, with a gap across

[5] R.L. Dewar (*Nuclear Fusion* **18**, 1541–1553, 1978).

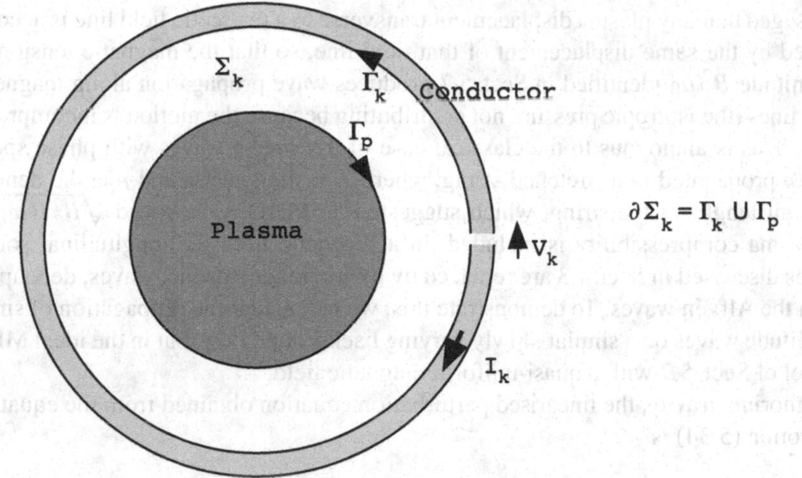

Fig. 5.2 Schematic of the plasma-vacuum-conductor model for a confined high-temperature plasma, to be discussed in Sect. 5.13. Although the conductors (including the vacuum vessel wall) on which the loops Γ_k lie are often taken to be perfectly conducting, in describing slow large-scale plasma evolution they may be regarded as having breaks that allow magnetic flux to enter or leave as the plasma interacts with the external circuitry

which there is a voltage V_k. Thus with its boundary $C_k = \partial\Sigma_k$ consisting of a contour Γ_k on the conductor and a contour Γ_p on the plasma, the analysis leading to (5.47) holds on Γ_k (where of course $\mathbf{v} = 0$), so that

$$\frac{d\Phi}{dt} = -\int_{\Gamma_k} \mathbf{E} \cdot d\mathbf{r}. \tag{5.49}$$

The application of a suitable boundary condition introduced later (cf. Sect. 5.12) then yields

$$\frac{d\Phi}{dt} = V_k, \tag{5.50}$$

which may be interpreted as saying that the magnetic flux is trapped between the plasma and the conductor, except for flux injected through the gap at the rate V_k.

5.9 Ideal MHD Waves

Since thermal conduction could not account for the rapid energy transfer observed through the interior of the Sun, and it could not be due to electromagnetic waves because they are strongly absorbed in highly ionised media, Alfvén initiated MHD in proposing the famous new transverse wave type subsequently named after him [1]. In the "frozen-in" scenario of the ideal MHD model discussed in Sect. 5.8, he envisaged that any plasma displacement transverse to a magnetic field line is accompanied by the same displacement of that field line, so that the magnetic tension of magnitude B^2/μ_0 identified in Sect. 5.7 produces wave propagation along magnetic field lines (the isotropic pressure not contributing because the motion is incompressible). This is analogous to the classical case of transverse waves with phase speed $\sqrt{T/\rho}$ propagated in a stretched string, where T is the tension and ρ is the density per unit length of the string, which suggested an MHD wave speed $\sqrt{B^2/(\mu_0\rho)}$. If plasma compressibility is included, in a magnetic field the longitudinal sound waves discussed in Sect. 4.3 are replaced by hybrid magnetosonic waves, decoupled from the Alfvén waves. To demonstrate this, we reconsider the propagation of small amplitude waves on a similar slowly varying background flow, but in the ideal MHD model of Sect. 5.7 with a quasi-uniform magnetic field.

Ignoring gravity, the linearised perturbation equation obtained from the equation of motion (5.34) is

$$\rho\omega'\mathbf{v}_1 - \mathbf{k}p_1 = \mu_0^{-1}\mathbf{B}\times(\mathbf{k}\times\mathbf{B}_1) \tag{5.51}$$

where $\omega' = \omega - \mathbf{k}\cdot\mathbf{v}$ is the frequency in the local rest frame of the plasma, analogous to the case of the fluid in Sect. 4.3. From the adiabatic Eq. (5.35), we again have

$$p_1 = \frac{\gamma p}{\omega'}\,\mathbf{k}\cdot\mathbf{v}_1, \tag{5.52}$$

but now, in addition, the linearised perturbation equation from the ideal equation of magnetic induction (5.38)—viz.

$$\omega'\mathbf{B}_1 = \mathbf{k}\times(\mathbf{B}\times\mathbf{v}_1) = (\mathbf{B}\,\mathbf{k} - \mathbf{k}\cdot\mathbf{B}\,\mathbf{I})\cdot\mathbf{v}_1, \tag{5.53}$$

on omitting the term $\mathbf{B}_1\cdot\nabla\mathbf{v}$ since the background flow is slowly varying.

Writing $\mathbf{B}\times(\mathbf{k}\times\mathbf{B}_1)$ as $(\mathbf{k}\,\mathbf{B} - \mathbf{k}\cdot\mathbf{B}\,\mathbf{I})\cdot\mathbf{B}_1$ and multiplying by ω', on using (5.52) and (5.53) to eliminate p_1 and $\mathbf{B}_1$ from (5.51) we have

$$\mathsf{D}\cdot\mathbf{v}_1 = 0, \tag{5.54}$$

where

$$\mathsf{D} \equiv \rho\omega'^2\,\mathsf{I} - \gamma p\,\mathbf{k}\mathbf{k} - \mu_0^{-1}(\mathbf{k}\,\mathbf{B} - \mathbf{k}\cdot\mathbf{B}\,\mathbf{I})\cdot(\mathbf{B}\mathbf{k} - \mathbf{k}\cdot\mathbf{B}\,\mathbf{I}). \tag{5.55}$$

When $\mathbf{k}$ is not parallel to $\mathbf{B}$, the vectors

$$\mathbf{e}_1 = \mathbf{k},\quad \mathbf{e}_2 = \mathbf{B},\quad \mathbf{e}_3 = \mathbf{k}\times\mathbf{B} \tag{5.56}$$

form a natural (but non-orthogonal) basis for converting (5.54) into a matrix eigenvalue problem. From (1.6), the reciprocal basis is

$$\mathbf{e}^1 = \frac{\mathbf{B} \times (\mathbf{k} \times \mathbf{B})}{|\mathbf{k} \times \mathbf{B}|^2}, \quad \mathbf{e}^2 = \frac{(\mathbf{k} \times \mathbf{B}) \times \mathbf{k}}{|\mathbf{k} \times \mathbf{B}|^2}, \quad \mathbf{e}^3 = \frac{\mathbf{k} \times \mathbf{B}}{|\mathbf{k} \times \mathbf{B}|^2}.$$

Thus on writing

$$\mathbf{v}_1 = v_1^1 \mathbf{e}_1 + v_1^2 \mathbf{e}_2 + v_1^3 \mathbf{e}_3$$

and resolving $\mathbf{D} \cdot \mathbf{v}_1$ onto the reciprocal basis vectors $\mathbf{e}^i$, from (5.54) we have the system of linear equations ($i = 1,\ 2,\ 3$)

$$\sum_{j=1}^{3} D^i_j v_1^j = 0, \tag{5.57}$$

where the coefficient matrix with entries $D^i_j \equiv \mathbf{e}^i \cdot \mathbf{D} \cdot \mathbf{e}_j$ is

$$\left[D^i_j\right] = \begin{bmatrix} \rho\omega'^2 - \left(\gamma p + \mu_0^{-1} B^2\right) k^2 & -\gamma p\, \mathbf{k} \cdot \mathbf{B} & 0 \\ \mu_0^{-1} k^2\, \mathbf{k} \cdot \mathbf{B} & \rho\omega'^2 & 0 \\ 0 & 0 & \rho\omega'^2 - \mu_0^{-1} (\mathbf{k} \cdot \mathbf{B})^2 \end{bmatrix}. \tag{5.58}$$

Consequently

$$\begin{aligned} \det[D^i_j] &= \left(\rho\omega'^2 - \mu_0^{-1} (\mathbf{k} \cdot \mathbf{B})^2\right) \det\begin{pmatrix} \rho\omega'^2 - \left(\gamma p + \mu_0^{-1} B^2\right) k^2 & -\gamma p\, \mathbf{k} \cdot \mathbf{B} \\ \mu_0^{-1} k^2\, \mathbf{k} \cdot \mathbf{B} & \rho\omega'^2 \end{pmatrix} \\ &= \rho^3 \left(\omega'^2 - k_\parallel^2 c_A^2\right) \left[\omega'^4 - k^2 \left(c_s^2 + c_A^2\right) \omega'^2 + k^2 k_\parallel^2 c_s^2\, c_A^2\right], \end{aligned} \tag{5.59}$$

where

$$c_s \equiv \sqrt{\frac{\gamma p}{\rho}}$$

is the sound speed as before,

$$c_A \equiv \sqrt{\frac{B^2}{\mu_0 \rho}} \tag{5.60}$$

defines the anticipated Alfvén speed analogous to the phase speed $\sqrt{T/\rho}$ in the stretched string, and the parallel wave number $k_\parallel$ is defined by $k_\parallel \equiv \mathbf{k} \cdot \hat{\mathbf{b}}$ where $\hat{\mathbf{b}} \equiv \mathbf{B}/B$ as before. The MHD dispersion relation $\det[D^i_j] = 0$, corresponding to a nontrivial solution of (5.57), factorises into three separate relations for the MHD waves—respectively

$$\text{Alfvén waves:} \qquad \omega'^2 = k_\parallel^2 c_A^2, \tag{5.61a}$$

$$\text{slow MS waves:} \qquad \omega'^2 = \frac{1}{2} k^2 \left(c_s^2 + c_A^2\right)(1 - \sqrt{1-\alpha^2}), \tag{5.61b}$$

$$\text{fast MS waves:} \qquad \omega'^2 = \frac{1}{2} k^2 \left(c_s^2 + c_A^2\right)(1 + \sqrt{1-\alpha^2}), \tag{5.61c}$$

where "MS" stands for "magnetosonic" and

$$\alpha^2 = 4\,\frac{k_\parallel^2}{k^2}\,\frac{c_s^2 c_A^2}{(c_s^2 + c_A^2)^2}. \tag{5.62}$$

In passing, let us note that $\mathbf{D} \cdot \mathbf{v}_1$ could have been resolved onto the basis vectors $\mathbf{e}_i$ such that the dispersion relation would have come from the determinant of the symmetric matrix D_{ij}, which only differs from $\det[D^i_j]$ by a positive factor and therefore has the same zeros.

In laboratory plasmas, the pressure is typically much less than the magnetic pressure, which is usually expressed as $\beta \ll 1$ where $\beta \equiv 2\mu_0 \bar{p}/\bar{B}^2$, with $\bar{p}$ and $\bar{B}$ denoting representative values of p and B. Consequently, $c_s \ll c_A$ and $(1-\alpha^2)^{1/2} \simeq 1 - \alpha^2/2$, such that the slow magnetosonic wave tends to propagate at the sound speed ($\omega'^2 \simeq k_\parallel^2 c_s^2$) and the fast magnetosonic wave speed approaches the Alfvén speed ($\omega'^2 \simeq k^2 c_A^2$). Indeed, $c_s < c_A$ is usually assumed in making polar plots of the rest frame phase speeds $c' = \omega'/k$ (i.e. plots of each phase speed versus the direction of propagation). On the other hand, in the incompressible limit $\gamma \to \infty$ the slow magnetosonic wave degenerates to the Alfvén wave ($\omega'^2 \to k_\parallel^2 c_A^2$), and the fast magnetosonic wave speed becomes infinite.[6]

Exercise

(Q1) As discussed in the following section, there is no background flow in magnetohydrostatic configurations, which are often considered in MHD stability investigations (cf. Chap. 6). Indeed, in linearised theory, a Lagrangian displacement $\boldsymbol{\xi}$ such that $\mathbf{v}_1(\mathbf{r}, t) = \partial\boldsymbol{\xi}/\partial t$ is often introduced (cf. Sect. 6.5). When the parallel wave number component $k_\parallel = \mathbf{k} \cdot \hat{\mathbf{b}}$ is much less than the magnitude of the perpendicular wave number component $k_\perp = |\mathbf{k} \times \hat{\mathbf{b}}|$ (where $\hat{\mathbf{b}} = \mathbf{B}/B$), deduce the following results for the Alfvén, slow and fast magnetosonic branches ($\omega = \omega_A, \omega_S$ and ω_F) of the dispersion relation and their respective polarisation vectors (arbitrary normalisation) to be invoked in Sect. 6.7:

[6] This is discussed further in B.F. McMillan, R.L. Dewar and R.G. Storer (*Plasma Physics and Controlled Fusion*, **46**, 1027–1038, 2004).

$$\rho\omega_{\rm A}^2 = \mu_0^{-1}(\mathbf{k}\cdot\mathbf{B})^2, \qquad \boldsymbol{\xi}_{\rm A} = \mathbf{k}\times\mathbf{B}/(kB),$$

$$\rho\omega_{\rm S}^2 = \frac{\mu_0^{-1}\gamma p(\mathbf{k}\cdot\mathbf{B})^2}{\mu_0^{-1}B^2+\gamma p}, \boldsymbol{\xi}_{\rm S} = (1+\mu_0\gamma p/B^2)\mathbf{B}/B - \mu_0\gamma p\mathbf{k}\cdot\mathbf{B}\mathbf{k}_\perp/(k_\perp^2 B^3),$$

$$\rho\omega_{\rm F}^2 = (\mu_0^{-1}B^2+\gamma p)k^2, \boldsymbol{\xi}_{\rm F} = \mu_0\gamma p\mathbf{k}\cdot\mathbf{B}\mathbf{B}/(k_\perp B^4) + (1+\mu_0\gamma p/B^2)\mathbf{k}_\perp/k_\perp.$$

5.10 MHD Equilibria (Magnetohydrostatics)

An equilibrium state is an unperturbed steady configuration—i.e. there exists a rest frame in which all fields are time-independent. Thus on setting $\partial\mathbf{v}/\partial t = 0$ in (5.34), we obtain

$$\rho\mathbf{v}\cdot\nabla\mathbf{v} + \nabla p = \rho\mathbf{g} + \mu_0^{-1}(\nabla\times\mathbf{B})\times\mathbf{B}, \tag{5.63}$$

expressing the balance of the inertia, pressure gradient and the gravitational and Lorentz forces. In fluid mechanics, an equilibrium where there is no flow ($\mathbf{v} = 0$) in the rest frame is called static. Here we again omit the gravitational term, and restrict our attention to *magnetohydrostatic* equilibria—i.e. we also neglect the term $\rho\mathbf{v}\cdot\nabla\mathbf{v}$, although recently there has been increasing interest in equilibria with strong flows [5]. Thus we consider the equilibrium equation (5.63) reduced to

$$\nabla p = \mu_0^{-1}(\nabla\times\mathbf{B})\times\mathbf{B} \equiv \mathbf{j}\times\mathbf{B}, \tag{5.64}$$

which is the form usually considered in the laboratory plasma literature, and its general solution is far from trivial. Indeed, just how to pose the corresponding three-dimensional magnetohydrostatic problem such that solutions can be shown to exist remains a topic of current research. Explicit solutions have been identified for some special axisymmetric cases, and computational procedures may generally be used to find accurate numerical solutions in other axisymmetric cases.

To better understand the implications of (5.64), let us resolve it in the magnetic field and current vector directions. Thus on dotting both sides with $\mathbf{B}$ and $\mathbf{j}$, we obtain the pair of scalar necessary conditions $\mathbf{B}\cdot\nabla p = 0$ and $\mathbf{j}\cdot\nabla p = 0$, respectively. There are two standard alternatives for satisfying these conditions throughout a finite plasma volume $\mathcal{P}$—viz. (a) $p = \text{const}$ in $\mathcal{P}$, so $\nabla p \equiv 0$; or (b) p is a continuously varying differentiable function in $\mathcal{P}$, so $\nabla p \neq 0$ almost everywhere. The simpler alternative (a) will be discussed in Sect. 5.10.2, while alternative (b) is discussed below. A third hybrid case is even more interesting for magnetic confinement than (a), and less problematic for treating non-axisymmetric equilibria than (b)—viz. (c) p is a piecewise constant function, constant in subregions of $\mathcal{P}$ but changing discontinuously across interfaces separating these subregions. However, except to remark on a relevant force-free aspect below, we shall not discuss case (c) further in this book,[7] although we do give the relevant jump conditions at discontinuities in Sect. 5.12.

5.10.1 Magnetically Confined Plasma Equilibria

The aim of many plasma experiments is to confine a plasma with a high temperature at its centre. For instance, in a successful fusion reactor the temperature T must be at least $10^8\,^\circ$K near the centre of the plasma, so although the particle density n may only be about a millionth of that in atmospheric air the corresponding pressure (given by $p = nkT$) is of the order of an atmosphere. However, near the walls of the confining vessel, the temperature can only be of the order of $10^3\,^\circ$K (for otherwise the walls would vaporise); so, the pressure must vary by at least 5 orders of magnitude across the plasma!

Suppose $p(\mathbf{r})$ is a continuous function corresponding to such a magnetically confined plasma, monotonically decreasing from the hottest region such that $\nabla p \neq 0$, so there must be level surfaces $p =$ const *nesting* the hottest region. The force balance conditions $\mathbf{B} \cdot \nabla p = 0$ and $\mathbf{j} \cdot \nabla p = 0$ imply the magnetic field $\mathbf{B}$ and its curl (in the direction of the current $\mathbf{j}$) are everywhere tangential to these level surfaces—i.e. these pressure isosurfaces are traced out by the magnetic and current field lines, and so are called *magnetic flux surfaces* (or magnetic surfaces or flux surfaces, for short).

Let $\hat{\mathbf{n}} \equiv \nabla p/|\nabla p|$ be the unit normal at each point on these magnetic surfaces, and consider the implications of resolving the pressure balance condition (5.64) in this remaining independent direction—i.e. $|\nabla p| = \hat{\mathbf{n}} \cdot \mathbf{j} \times \mathbf{B}$. The magnetic confinement condition $\nabla p \neq 0$ is then seen to imply that $\mathbf{j}$ and $\mathbf{B}$ can neither be parallel to each other nor *vanish* anywhere. The non-vanishing of these fields tangential to the magnetic surfaces has profound topological implications for the design of any magnetic plasma confinement device (at least, any that can be described by MHD), due to an important result in algebraic topology—viz. *any tangent field on a sphere must have at least one zero.*[8] Thus we cannot confine a topologically spherical plasma magnetically, as the required *non*-vanishing of $\mathbf{j}$ and $\mathbf{B}$ is inconsistent with this topological result. Fortunately, this does not apply to tori and henceforth we shall assume that *the magnetic surfaces are toroidal.*

The *integrability* condition that vector fields such as $\mathbf{j}$ and $\mathbf{B}$ are everywhere tangential to smoothly nested tori is rather special, except in the case where the system

[7] A detailed discussion of case (c) is given in S. R. Hudson, R. L. Dewar, G. Dennis, M. J. Hole, M. McGann, G. von Nessi and S. Lazerson (*Physics of Plasmas* **19**, 112502, 2012) and references therein; and we also note that the "stepped pressure model" discussed there was envisaged much earlier by David Potter (in *Methods in Computational Physics* **16** edited by John Killeen, pp. 43–83, Academic Press, 1976).

[8] This result is amusingly called the "Hairy Ball Theorem," for it may be visualised as the impossibility of combing finite-length hairs on such a ball flat to its surface everywhere—it must have a bald spot or cowlick.

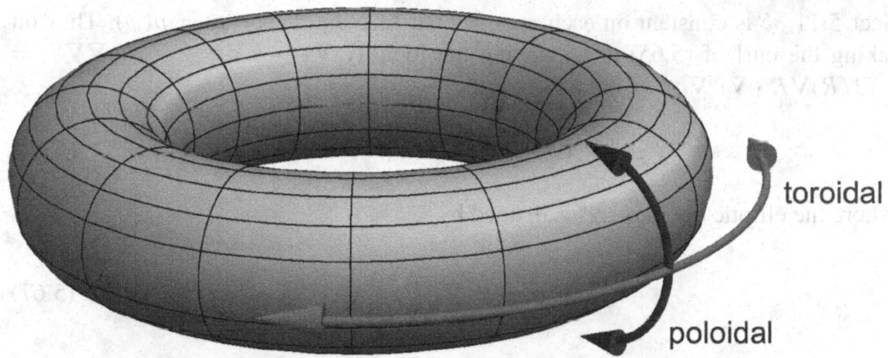

Fig. 5.3 An axisymmetric torus, depicting the two topologically distinct directions—viz. toroidal (the "long" way), and poloidal (the "short" way). See Glossary 3 and Sect. 5.11 for further discussion of this terminology

has a continuous symmetry. We therefore restrict our attention to systems that are invariant under rotation about a line—i.e. to *axisymmetric* configurations. (Without such a symmetry assumption we would be led into chaos theory, which is beyond the scope of this book.) Axisymmetric equilibrium models are of interest as a first approximation in both the laboratory (e.g. the tokamak configuration discussed—cf. Sect. 5.13) and in astrophysics (e.g. planetary magnetospheres). The magnetic field lines wind helically on these tori so that, while the vector **B** at each point on a magnetic surface has no normal component, it has both a *toroidal* component (i.e. in the direction the long way around the torus) and a *poloidal* component (i.e. in the direction the short way around)—cf. Fig. 5.3.

Axisymmetric configurations in fusion research are typically described using cylindrical coordinates denoted by (R, ϕ, Z), with R the distance from the axis of symmetry and the *toroidal angle* ϕ the angle of rotation about this axis. That is, with respect to a Cartesian coordinate system, x, y, z, $x = R\cos\phi$, $y = R\sin\phi$, $z = Z$. Then the representation

$$\mathbf{B} = \nabla\phi \times \nabla\psi + F(\psi)\nabla\phi \tag{5.65}$$

is used to enforce $\nabla\cdot\mathbf{B} = 0$. Thus the divergence of the first term (the poloidal magnetic field) vanishes quite generally since $\nabla\times\nabla f \equiv 0$ (with f representing ϕ and ψ); and the divergence of the second term vanishes because $F'(\psi)\nabla\psi\cdot\nabla\phi = 0$ and $\nabla^2\phi = 0$, for the assumed axisymmetric toroidal magnetic field such that $\psi = \psi(R, Z)$.

The so-called *poloidal flux function* ψ is again analogous to the velocity stream function in fluid mechanics—cf. Sects. 3.9, 3.10 and 3.13. As discussed further in

Sect. 5.11, ψ is constant on each magnetic surface and hence $p = p(\psi)$. Thus on taking the curl of (5.65) and invoking the identity $\nabla\psi \cdot \nabla\nabla\phi - \nabla\phi \cdot \nabla\nabla\psi = -(2/R)\nabla R \cdot \nabla\psi\nabla\phi$, we obtain

$$\mu_0 \mathbf{j} = \Delta^*\psi \, \nabla\phi + F'(\psi)\nabla\psi \times \nabla\phi, \tag{5.66}$$

where the elliptic operator Δ^* is defined by

$$\Delta^*\psi \equiv R^2\nabla \cdot \left(\frac{\nabla\psi}{R^2}\right). \tag{5.67}$$

Consequently, the equilibrium condition $\nabla p = \mathbf{j} \times \mathbf{B}$ is satisfied if and only if the *Grad-Shafranov* equation

$$\Delta^*\psi + F(\psi)F'(\psi) + R^2\mu_0 p'(\psi) = 0 \tag{5.68}$$

for static axisymmetric MHD equilibria is satisfied. In general, this nonlinear partial differential equation can only be solved numerically, given the two arbitrary "profile" functions $p(\psi)$ and $F(\psi)$ and appropriate boundary conditions—but as indicated earlier, there are analytic solutions for some special choices of $p(\psi)$ and $F(\psi)$. We also recall the fluid mechanics analogy on writing the Grad-Shafranov equation as (3.76), mentioned in the discussion leading to Hill's spherical vortex in Sect. 3.9.

5.10.2 Force-Free Equilibria

Force-free magnetic fields [8] are defined by the property

$$\nabla \times \mathbf{B} = \alpha(\mathbf{r})\mathbf{B}, \tag{5.69}$$

implying that $\mathbf{j}$ is parallel to $\mathbf{B}$ such that $\mathbf{j} \times \mathbf{B} = \nabla p \equiv 0$ and hence p is constant (e.g. zero). Taking the divergence of both sides of (5.69), since $\nabla \cdot (\nabla \times \mathbf{B}) = 0$ and $\nabla \cdot \mathbf{B} = 0$ we have $\mathbf{B} \cdot \nabla\alpha = 0$ such that α is constant along each field line.

A *linear* force-free field is such that $\alpha = \text{const}$ throughout a finite volume $\mathcal{V}$, and arises in particular when field lines wander chaotically throughout $\mathcal{V}$. A *nonlinear* force-free field is such that α is not constant throughout the entire volume $\mathcal{V}$, and, when the field lines are integrable(lie on nested flux surfaces), can be described by the Grad-Shafranov equation (5.68) with p' set to zero. Force-free magnetic fields were originally discussed in astrophysics, but also have application in modelling fusion plasma equilbria even if p is piecewise constant rather than continuous.

5.10.3 Variational Principles for MHD Equilibrium

We recall Eq. (5.45) for the conservation of energy, from which we can identify the total plasma energy density as the sum of the energy density for an ideal gas (3.10) and the energy density $B^2/2\mu_0$ of the magnetic field entrained in the plasma. Integrating over the plasma volume $\mathcal{P}$, under suitable boundary conditions, we find the total energy of the plasma system $W_{\rm tot}$ is the sum of the kinetic energy $\int_{\mathcal{P}} \frac{1}{2}\rho v^2 d\tau$ and a *potential energy*

$$W = \int_{\mathcal{P}} \left(\frac{p}{\gamma - 1} + \frac{B^2}{2\mu_0} \right) d\tau, \tag{5.70}$$

the sum of the thermodynamic (adiabatic) and magnetic energies.

We can approach the equilibrium energy principle in an intuitive way through a "thought experiment".[9] Suppose the plasma is initially stationary (so its initial kinetic energy is zero) but not in equilibrium so there are unbalanced forces that cause it to move, converting potential energy into kinetic energy—i.e. *the potential energy is reduced*, since the total energy $W_{\rm tot}$ is conserved. Now we imagine a friction-like interaction with some background medium is switched on, which removes momentum and absorbs the friction-generated heat such that each fluid element of the inviscid fluid continues to obey the isentropic equation of state (2.59). Our hypothetical damping mechanism consequently removes kinetic energy but not the potential energy W directly, which is only reduced through its conversion to kinetic energy. Thus assuming a stable equilibrium exists, it is reached when all possible motion is eventually damped out, corresponding to W having reached a *minimum* where no more potential energy can be converted to kinetic energy. The equilibrium variational energy principle may then be stated as:

A stable static ideal MHD equilibrium is found by minimising the potential energy W subject to the ideal constraints of conservation of mass, entropy and magnetic flux.

If extra constraints are imposed, an equilibrium can still be constructed from this principle—but in particular, it may not be stable to symmetry-breaking perturbations if axisymmetry is imposed. The investigation of stability is then often a further step, using a variational principle for linearised perturbations about a symmetric equilibrium (cf. Sect. 6.5).

A variant of this energy principle is often used for force-free fields, based on the observation that the above thought experiment can also be applied to cases where not all of the constraints of ideal MHD are strictly conserved, such that W relaxes to a lower energy. In particular, a single *magnetic helicity* functional

$$K_0 = \frac{1}{2\mu_0} \int_{\mathcal{P}} \mathbf{A} \cdot \mathbf{B} \, d\tau \tag{5.71}$$

[9]This is essentially the "imagined experiment" described by M.D. Kruskal and R.M. Kulsrud, (*Physics of Fluids* **1**, 265–274, 1958).

with **A** a magnetic potential for **B** (i.e. $\mathbf{B} = \nabla \times \mathbf{A}$) has been proposed as the robust constraint, in allowing for a much wider class of perturbations where field lines can break and reconnect.[10] Then on introducing a Lagrange multiplier λ, the corresponding equilibrium variational principle for a linear force-free-field equilibrium is that the free energy

$$W - \lambda K_0 \tag{5.72}$$

is a minimum under variations in **A**, where p is taken to be constant and not varied.

Exercises

(Q1) ("Straight cylinder") For a cylindrical plasma of arbitrary cross-section (*not* necessarily circular), verify that $\mathbf{B} = \nabla z \times \nabla \psi + F(\psi)\nabla z$ is a general representation for the equilibrium magnetic field, where $\psi = \psi(x, y)$ is the flux function and $\{x, y, z\}$ are Cartesian coordinates with z directed along the axis of symmetry. Show that the corresponding analogue of the Grad-Shafranov equation is

$$\nabla^2\psi + F(\psi)F'(\psi) + \mu_0 p'(\psi) = 0, \tag{5.73}$$

where ∇^2 is the two-dimensional Laplacian $\partial^2/\partial x^2 + \partial^2/\partial y^2$.

(Q2) In a force-free region where $\nabla \times \mathbf{B} = \alpha(\mathbf{r})\mathbf{B}$, show that the electric current lines lie on surfaces of constant α.

(Q3) Show that (5.64) is the Euler–Lagrange equation corresponding to the ideal energy principle involving (5.70); and (5.69) with $\alpha = \lambda$ is the Euler equation corresponding to the force-free energy principle involving (5.72), where p is not varied.

5.11 Magnetic Coordinates

We indicated in the previous section that the calculation of an axisymmetric MHD equilibrium is often just a first step towards a stability investigation, to be taken up in the next chapter. As foreshadowed in Sect. 1.2, because of the extreme anisotropy imposed on plasma dynamics by a strong magnetic field, a stability analysis is preferably carried out in a curvilinear coordinate system (usually non-orthogonal) that conforms to the field line geometry as much as possible. Moreover, since the magnetic field lines lie in flux surfaces, it is natural to include a coordinate such as the radius r that uniquely labels magnetic surfaces in cylindrical geometry. Accordingly, in toroidal coordinates we adopt as one coordinate an analogous parameter s (say) that labels the continuously nested flux surfaces, with the value zero at the "magnetic axis" in the middle of the plasma (i.e. where the innermost flux surface degenerates

---

[10] This is often called "Taylor relaxation". For a review see J.B. Taylor (*Reviews of Modern Physics*, **58**, 741–763, 1986).

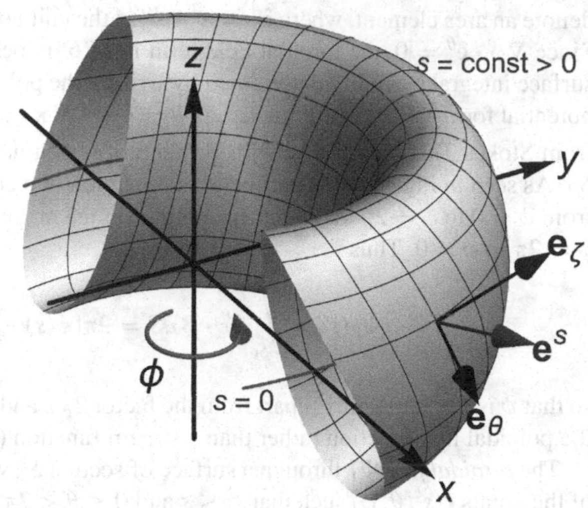

Fig. 5.4 A schematic showing half a toroidal flux surface $s =$ const and the circular magnetic axis $s = 0$, in relation to Cartesian coordinates x, y, z. Also shown are level curves of the generalised angle coordinates θ and ζ on the magnetic surface, tangential basis vectors $\mathbf{e}_\theta$ and $\mathbf{e}_\zeta$ (not to scale) from Eq. (1.5) for the contravariant representation and the normal basis vector $\mathbf{e}^s = \nabla s$ to the surface $s =$ constant (cf. Sect. 1.5)

to a closed circular field line) and increasing values outwards to the edge of the plasma. It is sometimes convenient to use ψ as this parameter, since $\mathbf{B} \cdot \nabla\psi = 0$ from (5.65) implies that the level surfaces of $\psi(R, Z)$ introduced in the previous section are magnetic surfaces—but for a general discussion it is preferable to leave $\psi(s)$ as a function to be determined by both geometry and physics.[11]

The "radial" coordinate s can be complemented by two generalised angles $\theta(R, Z)$ and $\zeta(R, \phi, Z)$, which define a two-dimensional coordinate system on each torus $s =$ const—cf. Fig. 5.4. The poloidal angle θ shown is a polar angle the "short way" around, directed so that $\{\nabla s, \nabla\theta, \nabla\phi\}$ form a right-handed basis; and the toroidal angle $\zeta = \phi + f(R, Z)$ shown is a polar angle the "long way" around each torus, with the function f as yet arbitrary.[12] In the reciprocal basis $\{\mathbf{e}^s, \mathbf{e}^\theta, \mathbf{e}^\zeta\} = \{\nabla s, \nabla\theta, \nabla\zeta\}$, from (1.36) the vector differential operator is then

$$\nabla = (\nabla s)\frac{\partial}{\partial s} + (\nabla\theta)\frac{\partial}{\partial\theta} + (\nabla\zeta)\frac{\partial}{\partial\zeta}, \tag{5.74}$$

and from (1.9) and (1.41) the Jacobian is $J = (\nabla s \cdot \nabla\theta \times \nabla\zeta)^{-1} > 0$.

The *poloidal flux* Ψ_P contained within a magnetic surface $s =$ const is the magnetic flux through any ribbon-like surface S_P where $\theta =$ const, extending from the magnetic axis $s = 0$ to $s =$ const at the given magnetic surface. Let $d\mathbf{S} = \hat{\mathbf{e}}^\theta dS$

[11] The parameter s has been adopted in some important computer codes, and of course should not be confused with our previous usage as a "line length".

[12] We have followed the convention illustrated in Fig. 1 of R. C. Grimm, R. L. Dewar and J. Manickam (*Journal of Computational Physics* **49**, 94–117, 1983). The choice of direction of $\nabla\theta$ is not universal in the literature—cf. O. Sauter and S. Yu. Medvedev (*Plasma Physics Communications* **184**, 293–302, 2013). However, it *is* usual for $\{\nabla s, \nabla\theta, \nabla\zeta\}$ to be a right-handed set, so if the angle θ is reversed then so too is ζ—i.e. to increase in the opposite sense to ϕ.

denote an area element, where $\hat{\mathbf{e}}^\theta = \mathbf{e}^\theta/|\mathbf{e}^\theta|$ is the unit normal in the direction of $\nabla\theta$. Since $\nabla\phi \cdot \hat{\mathbf{e}}^\theta = 0$, the toroidal field term in (5.65) does not contribute to Ψ_P. The surface integral is readily determined by writing the poloidal field term in the vector potential form $-\nabla\times(\psi\nabla\phi)$ such that $\Psi_\mathrm{P} = -\int \nabla\times(\psi\nabla\phi)\cdot d\mathbf{S} = -\oint_\phi \psi\nabla\phi\cdot d\mathbf{r}$ from Stokes Theorem (1.57), with $\oint_\phi$ denoting integration around the perimeter of S_P. As seen in Fig. 5.4, this perimeter is the union of a contour on the magnetic axis from $\phi = 0$ to $\phi = 2\pi$ and a return contour on the magnetic surface $s = \text{const}$ from $\phi = 2\pi$ to $\phi = 0$. Thus

$$\Psi_\mathrm{P}(s) = \int_{S_\mathrm{P}} \hat{\mathbf{e}}^\theta \cdot \mathbf{B}\, dS = 2\pi[\psi(s) - \psi(0)], \tag{5.75}$$

so that ψ is a poloidal flux (apart from the factor 2π), and therefore commonly called the poloidal flux function rather than a stream function (as previously mentioned).

The *toroidal flux* Ψ_T through a surface of section S_T where $\phi = \text{const}$, composed of the points $\mathbf{r}(s', \theta, \phi)$ such that $s' \le s$ and $0 \le \theta < 2\pi$, is similarly (taking $\zeta = \phi$)

$$\Psi_\mathrm{T}(s) = \int_{S_\mathrm{T}} \hat{\mathbf{e}}^\phi \cdot \mathbf{B}\, dS = \int_0^s \oint_\theta \frac{ds'|d\mathbf{r}_\theta|}{|\nabla s|(s', \theta)} \frac{F(s')}{R(s', \theta)}, \tag{5.76}$$

where $\hat{\mathbf{e}}^\phi = \nabla\phi/|\nabla\phi|$ and $d\mathbf{r}_\theta = d\theta\, \partial\mathbf{r}/\partial\theta = d\theta\, \mathbf{e}_\theta(s', \theta, \phi)$ is a line element on a poloidal loop around the surface $s' = \text{const}$, on which θ runs from 0 to 2π. Here the distance between the two surfaces $s' = \text{const}$ and $s' = \text{const} + ds'$ is $ds'/|\nabla s|(s', \theta)$, so that $dS = ds'|d\mathbf{r}_\theta|/|\nabla s|(s', \theta)$ is the surface element.

It is often convenient to specialise the (θ, ζ) coordinate system on each flux surface such that the slope

$$q(s) = \frac{d\zeta}{d\theta} \tag{5.77}$$

of the magnetic field lines of the torus in the (θ, ζ)-plane (sometimes called the *covering space*) is *constant*—cf. Fig. 5.5. To relate this requirement to the geometry of the magnetic field lines, suppose the angle differentials in (5.77) correspond to a spatial displacement $d\mathbf{r} = dl\,\mathbf{B}/B$ a distance dl along a field line, so that $d\theta = d\mathbf{r}\cdot\nabla\theta$ and $d\zeta = d\mathbf{r}\cdot\nabla\zeta = q(s)\, d\theta$. Then (5.77) is seen as the condition that $\mathbf{B}\cdot\nabla\zeta/\mathbf{B}\cdot\nabla\theta$ is the constant $q(s)$ at all points on the flux surface $s = \text{const}$. Since ∇s is normal to a flux surface, we also require $\mathbf{B}\cdot\nabla s = 0$ at all points on the surface. These two conditions are evidently satisfied by the general representation

$$\mathbf{B} = \nabla\zeta \times \nabla\psi + q(s)\nabla\psi \times \nabla\theta, \tag{5.78}$$

such that from (5.74) we have

$$\mathbf{B}\cdot\nabla = \frac{\psi'(s)}{J}\left(\frac{\partial}{\partial\theta} + q(s)\frac{\partial}{\partial\zeta}\right). \tag{5.79}$$

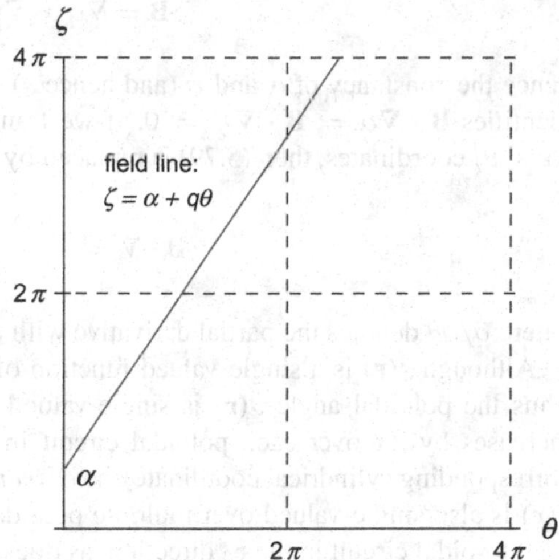

Fig. 5.5 Graph of a magnetic field line plotted in straight-field-line coordinates θ and ζ on the covering space of a toroidal flux surface, as discussed in the text

In passing, we note that (5.78) satisfies the propagated condition $\nabla \cdot \mathbf{B} = 0$.

Such coordinates are appropriately called *straight-field-line coordinates*,[13] with the slope $q(\psi)$ called the *safety factor*. This historical terminology reflects the stability requirement $q > 1$ at the plasma edge, which is a rough criterion (the *Kruskal-Shafranov condition*) that limits the amount of toroidal current a tokamak plasma can safely carry without disrupting.[14] In the theory of devices (stellarators) that do not rely on toroidal current to generate poloidal magnetic field, its inverse $\iota\!\!\!-\equiv 1/q = d\theta/d\zeta$ called the *rotational transform*[15] is often used instead of q to characterise the *pitch* of the magnetic field lines. In either type of device, *rational magnetic surfaces* (i.e. those on which q or $\iota\!\!\!-$ is the ratio of two integers) have special significance in the discussion of "flute-like" modes extensively discussed in Chap. 6, and in calculating the response to non-axisymmetric perturbations.

The generalised flute ordering requires wavelengths across magnetic field lines to be much shorter than the scale along field lines. For such modes in toroidal geometry, including interchange instabilities and so-called *ballooning modes*, it is useful to introduce a hybrid toroidal-poloidal coordinate variable

$$\alpha \equiv \zeta - q(s)\theta \tag{5.80}$$

that labels individual field lines on a flux surface as depicted in Fig. 5.5. Rewriting (5.78) in terms of α, we have the *Clebsch representation*

---

[13]Some authors reserve the terminology "magnetic coordinates" for straight-field-line coordinates, referring to the more general coordinates with arbitrary θ and ζ as "flux coordinates".

[14]The Kruskal=Shafranov condition was earlier identified in stability analysis for simpler cylindrical geometry—cf. Sect. 6.6.

[15]By analogy with $\hbar$ we use the notation $\iota\!\!\!-$ to denote $\iota/2\pi$, where ι is the increment in poloidal angle in radians after one toroidal circuit.

$$\mathbf{B} = \nabla\alpha \times \nabla\psi, \tag{5.81}$$

hence the constancy of α and ψ (and hence s) on a magnetic field line through the identities $\mathbf{B} \cdot \nabla\alpha = \mathbf{B} \cdot \nabla\psi = 0$. If we transform from (s, θ, ζ) coordinates to (α, s, θ) coordinates, then (5.79) is replaced by

$$\mathbf{B} \cdot \nabla = \frac{\psi'(s)}{J} \frac{\partial}{\partial\theta}, \tag{5.82}$$

where $\partial/\partial\theta$ denotes the partial derivative with α constant rather than ζ.

Although $s(\mathbf{r})$ is a single-valued function of $\mathbf{r}$, the angular coordinates are not. Thus the poloidal angle $\theta(\mathbf{r})$ is single-valued over multiple toroidal circuits, but increases by 2π over each poloidal circuit in the $\hat{\mathbf{e}}_\theta$ direction (analogous to the corresponding cylindrical coordinate), so α *decreases* by $2\pi q$; and the toroidal angle $\zeta(\mathbf{r})$ is also single-valued over multiple poloidal circuits, but increases by 2π over each toroidal circuit in the $\hat{\mathbf{e}}_\zeta$ direction, as does α.

Exercise

(Q1) Using the straight-field-line representation above and the tensor calculus techniques of Chap. 1 show:
(a) The poloidal and toroidal fluxes through surfaces of section $\theta = \text{const}$ and $\zeta = \text{const}$, respectively, are independent of the constants chosen and the precise choice of the θ and ζ coordinates.
(b) The safety factor is given by

$$q(s) = \frac{F(\psi)}{2\pi} \oint_\theta \frac{dl}{R|\nabla\psi|}. \tag{5.83}$$

5.12 Advected Ideal MHD Discontinuities

As in fluid mechanics (cf. Sect. 4.11), there are shock wave solutions describing discontinuities propagating with finite velocity relative to the MHD fluid. Indeed, there is a richer variety of cases in MHD, with three types of shock corresponding to the three branches of linear MHD waves treated in Sect. 5.9—cf. [6]. However, except for an Exercise generalising the Rankine–Hugoniot conditions of Sect. 4.11 to MHD shocks, in this section we consider a *different* class of discontinuities with zero propagated speed relative to the fluid. We call this class *advected discontinuities* because they are propagated purely by advection, so unlike shocks no matter passes through the interface on which the discontinuity occurs. We find that this class in turn decomposes into two types, *contact discontinuities* and *tangential discontinuities*.

Nevertheless, as for shock waves in Sect. 4.11 we consider infinitesimally thin transition regions where field variations become so large that one might expect discontinuities or even "infinities" to occur, since in ideal MHD there is also no natural length scale. Thus we again envisage interfaces that can be represented mathematically as surfaces, where now some or all in the set of principal ideal MHD fields $\{\rho(\mathbf{r},t), \mathbf{v}(\mathbf{r},t), p(\mathbf{r},t), \mathbf{B}(\mathbf{r},t)\}$ may change discontinuously. Once again we do not attempt to resolve the physical structure of the transition region, but analyse the discontinuities purely from the ideal MHD equations via generalised function theory—in particular, using the Heaviside step and Dirac delta functions discussed in Sect. 1.10, presuming that none of these principal fields has Dirac delta function behaviour at the interface but their derivatives may be singular there (although bounded elsewhere).[16] We represent the interface via an arbitrary smooth parameter $\varsigma(\mathbf{r},t)$, changing monotonically across the interface and such that the interface is on the level surface $\varsigma(\mathbf{r},t) = 0$ where $\varsigma(\mathbf{r},t)$ changes sign. Then we can write

$$\rho = \rho_- + (\rho_+ - \rho_-)H(\varsigma)\,, \tag{5.84}$$

$$\mathbf{v} = \mathbf{v}_- + (\mathbf{v}_+ - \mathbf{v}_-)H(\varsigma), \tag{5.85}$$

$$p = p_- + (p_+ - p_-)H(\varsigma), \tag{5.86}$$

$$\mathbf{B} = \mathbf{B}_- + (\mathbf{B}_+ - \mathbf{B}_-)H(\varsigma), \tag{5.87}$$

where the subscripts $+$ and $-$ denote the regions at either side of the transition layer, with the $\pm$ functions smoothly extended across it.[17] The Heaviside step function $H(\varsigma)$ selects the $+$ or $-$ branch according to which side the point $\mathbf{r}$ is on. The condition that the interface is advected by the flow is

$$\frac{d\varsigma}{dt} = 0 \ \text{ on } \varsigma = 0, \tag{5.88}$$

where $d\varsigma/dt$ at the interface can be written either as $(\partial_t + \mathbf{v}_- \cdot \nabla)\varsigma$ or $(\partial_t + \mathbf{v}_+ \cdot \nabla)\varsigma$.

Substituting (5.84)–(5.87) into the system of ideal MHD equations in Sect. 5.7, and setting the coefficients of the Dirac delta functions arising from the derivatives to zero, produces the relevant boundary conditions.[18] Thus (5.84) and (5.85) substituted into (5.33) yields

$$\left([\![\rho]\!] \frac{d\varsigma}{dt} + \bar{\rho}\, [\![\mathbf{v}]\!] \cdot \nabla\varsigma \right) \delta(\varsigma) = 0, \tag{5.89}$$

---

[16] A more general mathematical perspective is provided by the theory of matched asymptotic expansions—cf. [9, pp. 321–342]. This theory gives the leading order "outer region" behaviour—and when more physics is included as in Sect. 5.16, further resolves the discontinuity through an "inner region" expansion on smaller length scales and longer timescales.

[17] Dirac delta function behaviour is ruled out for the fundamental fields $\mathbf{v}$ and $\mathbf{B}$ on physical grounds (the kinetic and magnetic energies must be finite), such that both of these fields are in L^2–space. While it is sometimes useful to imagine a delta function mass density for mathematical convenience, there is no physical motivation for this and for simplicity we do not allow for that here.

[18] The product of a Heaviside step function and a delta function can usually be interpreted using $\delta(\cdot)H(\cdot) = \frac{1}{2}\delta(\cdot)$ that follows from $\delta(x)f(x) = f(0)\delta(x)$ with $f(x) = H(x)$ defined in (1.68), although we could work with conservation forms such as (5.41) to avoid having to use this formula.

where $[\![f]\!]$ again denotes the jump in any enclosed function f at the interface $\varsigma(\mathbf{r}, t) = 0$ (i.e. $[\![f]\!] \equiv f_+ - f_-$ evaluated at $\varsigma = 0$) and $\bar{\rho} = (\rho_+ + \rho_-)/2$ is the average density across the transition layer. On using the advection condition (5.88), condition (5.89) yields

$$\hat{\mathbf{n}} \cdot [\![\mathbf{v}]\!] = 0 \text{ on } \varsigma = 0, \tag{5.90}$$

where $\hat{\mathbf{n}} = \nabla\varsigma/|\nabla\varsigma|$ is the unit normal to the interface. As previously appreciated in Sect. 4.7 and elsewhere, the condition (5.90) implies non-cavitation at the interface, consistent with the advection assumption (5.88). The pressure evolution Eq. (5.35) is automatically satisfied for arbitrary $[\![p]\!]$ under (5.88) and (5.90), but the equations of motion and magnetic induction remain to be considered. This is readily achieved by also noting that the magnetic field condition $\nabla \cdot \mathbf{B} = 0$ applies globally such that

$$\nabla \cdot \mathbf{B} = [\![\mathbf{B}]\!] \cdot \nabla\varsigma \, \delta(\varsigma) \tag{5.91}$$

at any interface, whence

$$\hat{\mathbf{n}} \cdot [\![\mathbf{B}]\!] = 0 \text{ on } \varsigma = 0. \tag{5.92}$$

Thus the equation of motion in the form (5.39) or (5.41) immediately yields

$$\hat{\mathbf{n}} \cdot [\![p\mathsf{I} + \mathcal{T}]\!] = \left[\!\!\left[p + \frac{B^2}{2\mu_0} \right]\!\!\right] \hat{\mathbf{n}} - \frac{\hat{\mathbf{n}} \cdot \mathbf{B}}{\mu_0} [\![\mathbf{B}]\!] = 0 \text{ on } \varsigma = 0, \tag{5.93}$$

on equating the coefficient of $\delta(\varsigma)$ to zero; and the equation of induction in the form (5.46) likewise immediately yields

$$\hat{\mathbf{n}} \cdot \mathbf{B} \, [\![\mathbf{v}]\!] = 0 \text{ on } \varsigma = 0, \tag{5.94}$$

where $\mathbf{B}$ can represent either $\mathbf{B}_-$ or $\mathbf{B}_+$ because (5.92) implies $\hat{\mathbf{n}} \cdot \mathbf{B}$ is continuous.[19] The normal component of (5.93) yields the result that the total pressure (plasma pressure plus magnetic pressure) must be continuous—i.e.

$$\left[\!\!\left[p + \frac{B^2}{2\mu_0} \right]\!\!\right] = 0 \text{ on } \varsigma = 0, \tag{5.95}$$

and crossing (5.93) with $\hat{\mathbf{n}}$ yields the transverse component

$$\hat{\mathbf{n}} \cdot \mathbf{B} \, \hat{\mathbf{n}} \times [\![\mathbf{B}]\!] = 0 \text{ on } \varsigma = 0. \tag{5.96}$$

---

[19] The results (5.93) and (5.94) may also be derived from the original Eqs. (5.34) and (5.38) respectively, noting that $\partial_t \mathbf{B} = d\mathbf{B}/dt - \mathbf{v} \cdot \nabla \mathbf{B}$ in particular.

The condition (5.94) implies

$$\hat{\mathbf{n}} \cdot \mathbf{B}_- = \hat{\mathbf{n}} \cdot \mathbf{B}_+ = 0 \quad or \quad \hat{\mathbf{n}} \times [\![\mathbf{v}]\!] = 0 \quad \text{on } \varsigma = 0 \tag{5.97}$$

in addition to (5.90) above, and (5.96) implies

$$\hat{\mathbf{n}} \cdot \mathbf{B}_- = \hat{\mathbf{n}} \cdot \mathbf{B}_+ = 0 \quad or \quad \hat{\mathbf{n}} \times [\![\mathbf{B}]\!] = 0 \quad \text{on } \varsigma = 0, \tag{5.98}$$

with the "or" the logical inclusive disjunction.

Under the first alternative in each of (5.97) and (5.98), such that magnetic flux surfaces align with the interface, there may be a flow discontinuity ($\hat{\mathbf{n}} \times [\![\mathbf{v}]\!] \neq 0$)—i.e. a vortex sheet. The magnetic field could of course still be continuous across the transition layer ($[\![\mathbf{B}]\!] = 0$), when both alternatives in condition (5.98) are satisfied, and from (5.95) the plasma pressure is continuous ($[\![p]\!] = 0$). On the other hand, there could be a jump in the magnetic field $\mathbf{B}$ and an accompanying jump in the pressure p across the transition layer, in accordance with (5.95).[20]

Note that both alternatives in (5.97) and (5.98) may be satisfied. However, if the first alternative ($\hat{\mathbf{n}} \cdot \mathbf{B}_\pm = 0$) is not met, given (5.90) and (5.92) the second alternatives in (5.97) and (5.98) imply that $[\![\mathbf{v}]\!] = 0$ and $[\![\mathbf{B}]\!] = 0$ on $\varsigma = 0$, whereupon (5.95) again reduces to $[\![p]\!] = 0$ on $\varsigma = 0$. Thus in this case the only possible jump in the essential MHD fields envisaged at the interface under (5.84)–(5.87) is in the density ρ.

The possible jump in $\mathbf{B}$ under the first alternative in each of (5.97) and (5.98) corresponds to a strongly localised current flow in the narrow transition region illustrated in Fig. 5.6—cf. also [10, 11]. In analogy with the vortex sheet described by (3.64), such a singular current is called a *current sheet*. Thus on substituting (5.87) into the underlying pre-Maxwell equation (5.17), the second alternative in (5.98) is replaced by

$$\hat{\mathbf{n}} \times [\![\mathbf{B}]\!] = \mu_0 \mathbf{j}_* \quad \text{on } \varsigma = 0, \tag{5.99}$$

the sharply peaked component of the current density being idealised to a surface current $\mathbf{j}_*$ (a surface intensity) at the interface such that

$$\mathbf{j} = \mathbf{j}_- + (\mathbf{j}_+ - \mathbf{j}_-)H(\varsigma) + \mathbf{j}_* |\nabla\varsigma| \delta(\varsigma) \tag{5.100}$$

in the infinitesimally thin (delta function) transition layer limit.

The relative length and time scale of the actual transition layer generally determine whether or not to consider introducing any such surface intensity at a discontinuity. Given that electric charge separation may occur in plasmas on the Debye length scale, which must be much smaller than the width of the transition layer in MHD modelling, we might also envisage an electric field intensity $\mathbf{E}_*$ due to charge sheets

[20] The implications of this condition for the existence or otherwise of equilibrium solutions in non-symmetric systems are discussed in M. McGann, S.R. Hudson, R.L. Dewar and G. von Nessi (*Physics Letters A*, **374**, 3308–3314, 2010).

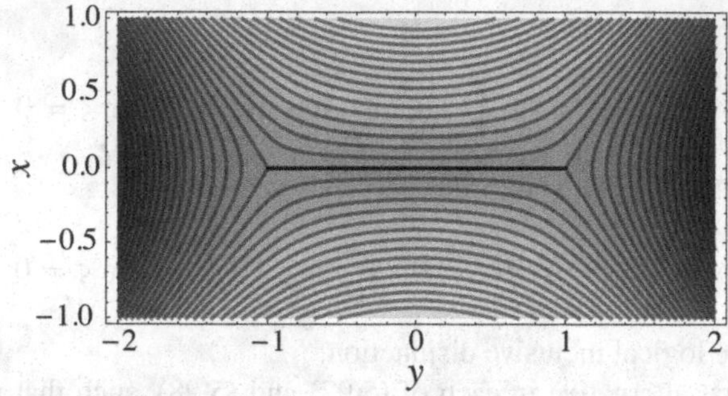

Fig. 5.6 Cross-section of a plasma with translational symmetry in the z-direction. A typical "Sweet-Parker" current sheet is indicated by the *horizontal black line*, and magnetic field lines by the contours of a flux function ψ, the magnitude of which is shown by shading—cf. also Exercises (Q1) and (Q2) below. Field lines change direction at the current sheet that is formed when they reconnect, which may occur during the nonlinear phase of a resistive tearing instability—cf. Sect. 6.10

of opposite sign (dipole sheets or *double layers*), causing a discontinuity in the electrostatic potential. While $\mathbf{v}$ and $\mathbf{B}$ are functions in L^2 space, the ideal MHD equation (5.36) giving $\mathbf{E} = -\mathbf{v} \times \mathbf{B}$ does not imply that the electric field $\mathbf{E}$ is square-integrable, because the product of two L^2 functions is not necessarily in L^2. While the possibility of double layers cannot be ruled out from an ideal MHD analysis, we shall defer consideration of such a possibility to Sect. 5.16, and here assume $\mathbf{E} = \mathbf{E}_- + (\mathbf{E}_+ - \mathbf{E}_-)H(\varsigma)$ with no delta function intensity in the transition region, leading to the condition

$$\hat{\mathbf{n}} \times [\![\mathbf{E}]\!] = \hat{\mathbf{n}} \cdot \mathbf{v}\, [\![\mathbf{B}]\!] \quad \text{on } \varsigma = 0 \tag{5.101}$$

from the field equation (5.37). The implied jump in the transverse components of $\mathbf{E}$ when $[\![\mathbf{B}]\!] \neq 0$ is certainly consistent with the ideal Eq. (5.36) applied outside the transition layer. Thus when $\hat{\mathbf{n}} \cdot \mathbf{B}_- = 0$ and $\hat{\mathbf{n}} \cdot \mathbf{B}_+ = 0$, from (5.90) and $\hat{\mathbf{n}} \times [\![\mathbf{E} + \mathbf{v} \times \mathbf{B}]\!] = 0$ we have (5.101). For the second alternative where $[\![\mathbf{v}]\!] = 0$ and $[\![\mathbf{B}]\!] = 0$ under (5.97) and (5.98), we have $[\![\mathbf{E}]\!] = -[\![\mathbf{v} \times \mathbf{B}]\!] = 0$ such that (5.101) is trivially satisfied.

In summary, the resulting ideal MHD boundary conditions at an advected interface are:

(1) the normal components of $\mathbf{v}$ and $\mathbf{B}$ must be continuous—i.e.

$$\hat{\mathbf{n}} \cdot [\![\mathbf{v}]\!] = \hat{\mathbf{n}} \cdot [\![\mathbf{B}]\!] = 0; \tag{5.102}$$

(2) the total pressure, kinetic plus magnetic, must be continuous—i.e.

$$\left[\!\!\left[p + \frac{B^2}{2\mu_0} \right]\!\!\right] = 0; \text{ and} \tag{5.103}$$

(3) the remaining conditions depend on what further continuity assumptions are made, and there are two cases.

(a) *Contact discontinuities*: The tangential components of both $\mathbf{v}$ and of $\mathbf{B}$ are continuous—i.e.

$$\hat{\mathbf{n}} \times [\![\mathbf{v}]\!] = \hat{\mathbf{n}} \times [\![\mathbf{B}]\!] = 0, \tag{5.104}$$

so together with (5.102) we have both $\mathbf{v}$ and $\mathbf{B}$ continuous. In particular, $\mathbf{B}$ need not be tangential to the interface ($\hat{\mathbf{n}} \cdot \mathbf{B}$ may be nonzero), and from (5.103) we also have that p is continuous. Thus only ρ may be discontinuous across the interface, so this case may be viewed as a form of entropy wave [6, p. 195].

(b) *Tangential discontinuities*: Either $\mathbf{v}$ or $\mathbf{B}$ or both are discontinuous, and $\mathbf{B}$ must be tangential to the interface ($\hat{\mathbf{n}} \cdot \mathbf{B}_\pm = 0$). Condition (5.103) now implies that p is discontinuous if the magnitude of the magnetic field $|\mathbf{B}|$ changes across the interface. Finally, in common with the contact discontinuity case (a), ρ may be discontinuous.

Exercises

(Q1) Revisit the Exercise in Sect. 5.10, leading to the Grad-Shafranov equation (5.73) for the static equilibrium of a plasma with translational symmetry in the z-direction. Suppose the plasma supports a discontinuity in $\mathbf{B}$ at a surface S, where $\psi(\mathbf{r})$ switches between two smooth flux functions $\psi_\pm(\mathbf{r})$, each continuously differentiable in the neighbourhood S. (Assume B_z and p are discontinuous across S.)
Express the boundary and jump conditions in terms of $\psi(\mathbf{r})$ and $F(\psi)$, and show that ψ must be constant on and continuous across S—i.e. $\psi_-(\mathbf{r}) = \psi_+(\mathbf{r}) =$ const for $\mathbf{r} \in S$.

(Q2) (Syrovatsky current sheet) Preparatory to using the complex variable method described in Sect. 3.6, define $\zeta = (x + iy)/\Delta$ and

$$w(\zeta) = \mathrm{sgn}(\Re\zeta)\left[\zeta\sqrt{\zeta^2+1} + \ln\left(\zeta + \sqrt{\zeta^2+1}\right)\right],$$

defined on the complex ζ-plane cut along the imaginary axis between $\zeta = \pm i$. Show that a flux function given by $\psi = (B_0\Delta)\,\Re\, w(\zeta)$ satisfies the equilibrium conditions in (Q1) above for a tangential discontinuity (current sheet) located on the y-axis between $y = \pm\Delta$. (Assume $B_z = 0$ and $p = 0$.)

(Q3) Neglecting gravity, deduce the generalised Rankine–Hugoniot equations for a plane shock from the ideal MHD equations, when the magnetic field **B** is parallel to the plane of the shock—viz.

$$[\![\rho v]\!] = 0, \quad \text{[cf. equation (4.126)]}$$

$$\left[\!\!\left[p + \rho v^2 + \frac{B^2}{2\mu_0} \right]\!\!\right] = 0,$$

$$\left[\!\!\left[\frac{v^2}{2} + \frac{\gamma}{\gamma - 1}\frac{p}{\rho} + \frac{B^2}{\mu_0 \rho} \right]\!\!\right] = 0,$$

Hence obtain the result

$$\frac{1}{\gamma - 1}\left[\!\!\left[\frac{p}{\rho}\right]\!\!\right] + \left[\!\!\left[\frac{B^2}{2\mu_0 \rho}\right]\!\!\right] + \left[\!\!\left[\frac{1}{\rho}\right]\!\!\right]\overline{\left(p + \frac{B^2}{2\mu_0}\right)} = 0,$$

where the bars denote the average of the quantities p and $1/\rho$ across the shock. (Hint: Initially solve for v^2 on either side of the shock, using the second equation and the square of the first.)

5.13 Vacuum Fields

A standard model used in designing and interpreting magnetically confined plasmas in fusion research assumes a highly conductive hot toroidal plasma volume is surrounded by a *vacuum region*, as depicted schematically in Figs. 5.2 and 5.7—e.g. as in tokamaks [12]. In this section, we develop this useful although somewhat idealised picture.

In a vacuum there is no current flow ($\mathbf{j} \equiv 0$), so (5.17) reduces to

$$\nabla \times \mathbf{B} = 0. \tag{5.105}$$

As in the case of irrotational flow where $\nabla \times \mathbf{v} = 0$, Eq. (5.105) implies the existence of a scalar potential χ (the magnetic potential in this context) such that

$$\mathbf{B} = \nabla \chi, \tag{5.106}$$

so from (5.1) χ satisfies the Laplace equation

$$\nabla^2 \chi = 0. \tag{5.107}$$

Thus the "subluminal" approximation (5.5) has reduced the equation for **B** from a wave equation to an elliptic partial differential equation in the spatial coordinates only (cf. the subsonic approximation discussed in Sect. 3.4). As in Sect. 3.5, χ can

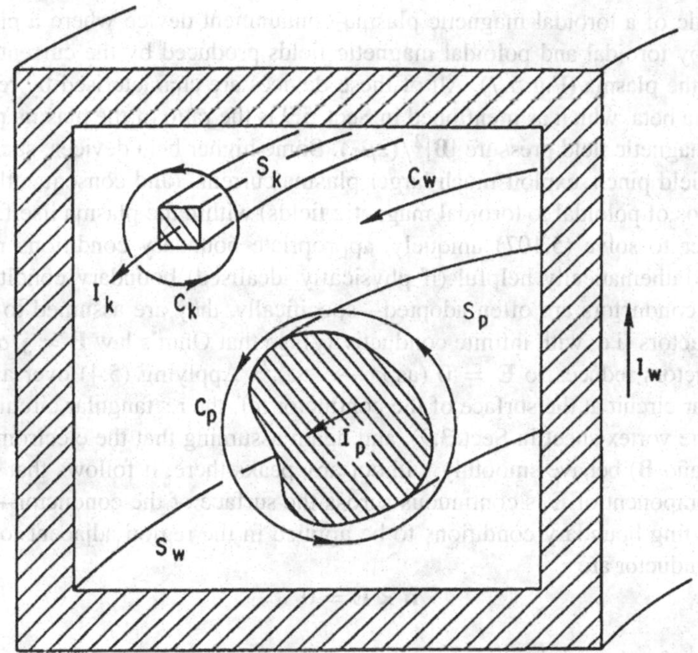

Fig. 5.7 Schematic of a general toroidal configuration (showing plasma region and conductor producing a poloidal magnetic field, and wall). Three sides of the wall may go to infinity, leaving a conductor along the major axis producing a toroidal magnetic field. The currents I_k, I_p and Ampère contours C_k, C_p are also shown

be multi-valued if the vacuum region is not simply connected. For a unique χ, the boundary conditions must be supplemented by specifying a sufficient number of circulation integrals [cf. (3.43)]

$$K_k \equiv \oint_{\Sigma_k} \mathbf{B} \cdot d\mathbf{r}, \tag{5.108}$$

where the $C_k = \partial \Sigma_k$ are irreducible circuits of the vacuum region taken anti-clockwise relative to the normals to the topologically distinct surfaces Σ_k that cut the vacuum domain.

Applying Stokes Theorem to (5.108) demonstrates that the circulations are related to the total currents I_k flowing through the surfaces Σ_k—i.e.

$$K_k = \mu_0 \int_{\Sigma_k} \mathbf{j} \cdot d\mathbf{S} = \mu_0 I_k. \tag{5.109}$$

These currents may be carried either by the plasma, by external conductors, or by a conducting wall or solenoid surrounding the vacuum and plasma—cf. Fig. 5.7, a schematic of a toroidal magnetic plasma-containment device where a plasma is confined by toroidal and poloidal magnetic fields produced by the currents in the coils and the plasma (Fig. 5.7). All of these devices are characterised by relatively low plasma beta, which as mentioned in Sect. 5.9 is the ratio of the plasma pressure p to the magnetic field pressure $|\mathbf{B}|^2/(2\mu_0)$. Some higher beta devices, such as the reversed field pinch, exploit much larger plasma currents (and consequently much larger ratios of poloidal to toroidal magnetic fields) within the plasma itself.

In order to solve (5.107) uniquely, appropriate boundary conditions must be applied. Mathematically helpful (if physically idealised) boundary conditions on electrical conductors are often adopted—specifically, they are assumed to be perfect conductors (i.e. with infinite conductivity), so that Ohm's law $\mathbf{E} = \mathbf{j}/\sigma$ within the conductors reduces to $\mathbf{E} = 0$ (as $\sigma \to \infty$).[21] Applying (5.4) over a narrow rectangular circuit at the surface of the conductor (cf. the rectangular circuit drawn through the vortex sheet in Sect. 3.9), and again assuming that the electromagnetic fields (**E** and **B**) behave smoothly without any peaks there, it follows that the tangential component of **E** is continuous across the surface of the conductor—i.e. the corresponding boundary conditions to be applied in the region adjacent to a fixed perfect conductor are

$$\hat{\mathbf{n}} \times \mathbf{E} = 0 \tag{5.110}$$

and $\hat{\mathbf{n}} \cdot \mathbf{v} = 0$, where $\hat{\mathbf{n}}$ denotes the unit normal at the surface of the conductor.

As mentioned in Sect. 5.12, under the sharp boundary model in ideal MHD the plasma is taken to be confined by magnetic pressure, corresponding to a tangential discontinuity such that

$$\hat{\mathbf{n}} \cdot \mathbf{B}^p = 0 \text{ and } \hat{\mathbf{n}} \cdot \mathbf{B}^v \equiv \hat{\mathbf{n}} \cdot \nabla\chi = 0 \text{ on } \varsigma = 0 \tag{5.111}$$

in which $\mathbf{B}^v$ and $\mathbf{B}^p$ denote the respective vacuum and plasma magnetic fields. Thus there is a magnetic field jump at the plasma-vacuum interface ($[\![\mathbf{B}]\!] \equiv \mathbf{B}^v - \mathbf{B}^p \neq 0$), and $[\![\rho]\!] \equiv -\rho$ and $[\![p]\!] \equiv -p$ simply refer to the plasma density and pressure if these fields are presumed to be zero in the vacuum. On the other hand, in passing we observe that a velocity field **v** may be defined in the vacuum to within an arbitrary parallel component $\hat{\mathbf{b}}_v \cdot \mathbf{v}$ (where $\hat{\mathbf{b}}_v = \mathbf{B}^v/|\mathbf{B}^v|$), by requiring that (5.36) hold there such that (5.85) in addition to (5.87) is meaningful.

Even if it is envisaged that the hot plasma is surrounded by a vacuum region rather than colder plasma, in the ideal MHD stability discussion of Chap. 6 it can be useful theoretically to assume that the plasma level surface $\varsigma = 0$ coincides with the surface of the conductor. In that case $\hat{\mathbf{n}}$ and $\nabla\varsigma$ are parallel at the surface, so we have (5.110) with $\hat{\mathbf{n}}$ replaced by $\nabla\varsigma$, since (5.4) applies in both the plasma and the conductor.Consequently, taking the dot product of (5.4) with $\nabla\varsigma$ and noting from

[21]Unless the coils are superconductors, this is only an acceptable approximation over times that are short compared with the resistive skin time δ^2/η, where δ is the thickness of the conductor and $\eta \equiv 1/(\mu_0\sigma)$ is the resistivity introduced in Sect. 2.12.

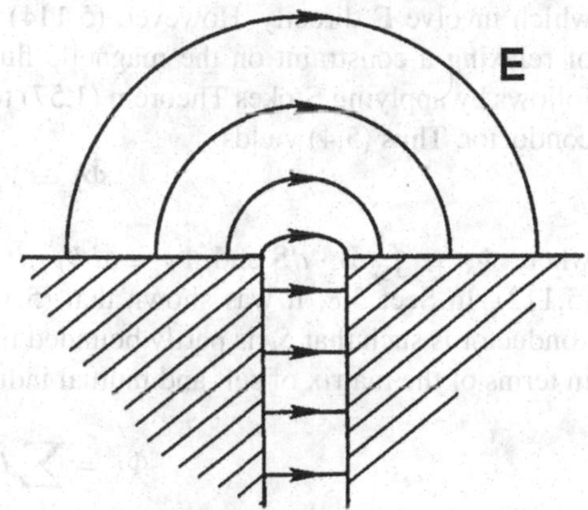

Fig. 5.8 Vacuum gap electric field: close-up of the electric field in the vicinity of an insulating gap, showing how the tangential component is essentially a delta function

(5.111) that $(\nabla\varsigma)\cdot\nabla\times\mathbf{E} = -\nabla\cdot(\nabla\varsigma\times\mathbf{E}) = 0$ on the surface of a perfect conductor, we obtain

$$\frac{\partial(\mathbf{B}\cdot\nabla\varsigma)}{\partial t} = 0 \tag{5.112}$$

if it is stationary (so $\partial_t\varsigma = 0$). Thus the boundary condition at the conductor is equivalent to

$$\hat{\mathbf{n}}\cdot\mathbf{B} = \text{constant in time } t, \tag{5.113}$$

and frequently $\mathbf{B}$ can be split into a steady background value $\mathbf{B}_0$ and a fluctuating time-dependent perturbation $\tilde{\mathbf{B}}$—i.e. the boundary condition at the stationary perfect conductor for the time-dependent component reduces to $\hat{\mathbf{n}}\cdot\tilde{\mathbf{B}} = 0$, so in an ideal MHD model we have simple Cauchy boundary conditions on the normal components of the velocity and magnetic field fluctuations.

However, we note that (5.110) must be modified if the conductor is connected to external circuitry such as a power supply and the current is changing with time. This is a consequence of the conductor being broken somewhere (Fig. 5.8), in order to run out leads to the external world or to prevent the copper shell shorting out the Ohmic heating transformer in the case of an old-fashioned tokamak. This break may be idealised as a gap of infinitesimal width. Although $\mathbf{E}$ is almost everywhere normal to the surface of the conductor, its tangential component is then infinite across the gap, so effectively (5.110) must be replaced by

$$\hat{\mathbf{n}}\times\mathbf{E} = -V_k\,\delta(\zeta)\,\hat{\mathbf{n}}\times\nabla\zeta \quad \text{(on a conductor, } k \text{ say)}, \tag{5.114}$$

where ζ is a tangential coordinate such that the break in the conductor corresponds to the surface $\zeta = 0$ and $\delta(\zeta)$ is the Dirac delta function. The quantity V_k is the "voltage across the gap" or electromotive force applied to the conductor k. Since there is no gap in the plasma, $V_k = 0$ when k is plasma. Although $\mathbf{E}$ is highly singular at the gap, this does not affect $\mathbf{B}$ since it is determined by (5.107)–(5.113), none of which involve **E** directly. However, (5.114) does have the important consequence of relaxing a constraint on the magnetic flux ϕ_k enclosed by a conductor, which follows by applying Stokes Theorem (1.57) to a circuit C_k around the surface of the conductor. Thus (5.4) yields

$$\dot{\Phi}_k = V_k, \tag{5.115}$$

where $\Phi_k \equiv \int_{S_k} \mathbf{B} \cdot d\mathbf{S}$ and $\dot{\Phi}_k \equiv d\Phi_k/dt$, independent of the choice of S_k—cf. (5.112). In Sect. 5.8, it was shown that (5.115) holds even if the topology of the conductor is such that S_k is partly bounded by an ideal plasma. Often, Φ_k is written in terms of the matrix of self and mutual inductances L_{kl}—i.e.

$$\Phi_k = \sum_l L_{kl} I_l, \tag{5.116}$$

where I_l are the currents flowing in the l'th conductor (or plasma). Then (5.114) yields

$$\sum_l \frac{d}{dt}(L_{kl} I_l) = V_k, \tag{5.117}$$

where the L_{kl} cannot be assumed independent of time due to the plasma motion.

5.14 Resistive MHD Model

We recall that Alfvén initially predicted the existence of MHD waves under the incompressible approximation. This led to early laboratory experiments using liquid mercury to investigate MHD waves—but as mentioned in Sect. 5.5, the resistivity of mercury is significant, so let us now again include the resistive equation of magnetic induction (5.30) in the mathematical model.

On neglecting gravity and assuming a slowly varying background flow in a quasi-uniform magnetic field, we recall that the equation of motion yielded the linearised perturbation Eq. (5.51), involving the frequency $\omega' = \omega - \mathbf{k} \cdot \mathbf{v}$ in the local rest frame. Since $\mathbf{B} \times (\mathbf{k} \times \mathbf{B}_1) = \mathbf{k}\,\mathbf{B} \cdot \mathbf{B}_1 - \mathbf{k} \cdot \mathbf{B}\,\mathbf{B}_1$, this perturbation equation may be rewritten

$$\rho\omega' \mathbf{v}_1 = \mathbf{k}(p_1 + \mu_0^{-1} \mathbf{B} \cdot \mathbf{B}_1) - \mu_0^{-1} \mathbf{k} \cdot \mathbf{B}\mathbf{B}_1. \tag{5.118}$$

The incompressible approximation $\nabla \cdot \mathbf{v}_1 = 0$ is appropriate for liquid mercury such that $\mathbf{k} \cdot \mathbf{v}_1 = 0$, and we also have $\mathbf{k} \cdot \mathbf{B}_1 = 0$ since $\nabla \cdot \mathbf{B}_1 = 0$. Thus $p_1 + \mu_0^{-1} \mathbf{B} \cdot \mathbf{B}_1 = 0$ from (5.118), and hence also

$$\rho\omega' \mathbf{v}_1 = -\mu_0^{-1} \mathbf{k} \cdot \mathbf{B}\,\mathbf{B}_1. \tag{5.119}$$

The linearised ideal magnetic induction perturbation equation (5.53) is replaced by the resistive form from (5.30) for an incompressible fluid—i.e.

$$(\omega' + i\eta k^2)\mathbf{B}_1 = \mathbf{k} \times (\mathbf{B} \times \mathbf{v}_1) = -\mathbf{k} \cdot \mathbf{B}\,\mathbf{v}_1. \tag{5.120}$$

Equations (5.119) and (5.120) immediately produce the resistive dispersion relation $\omega'^2 + i\eta k^2 \omega' - k_\parallel^2 c_A^2 = 0$ where $c_A = B/\sqrt{\mu_0 \rho}$ is the Alfvén speed as before, so the Alfvén wave dispersion relation $\omega'^2 = k_\parallel^2 c_A^2$ is recovered in the ideal MHD limit $\eta \to 0$. However, the roots $\omega' = -i\eta k^2 \pm \sqrt{k_\parallel^2 c_a^2 + \eta^2 k^4}/2$ of the resistive relation imply the phase speed of the Alfvén wave is significantly increased due to the positive definite resistive term under the square root, while the resistivity produces significant damping defined by the first term. These resistive effects therefore obscured Alfvén wave propagation in the laboratory experiments using mercury.

A compressible resistive MHD model is sometimes considered, when Eqs. (5.33)–(5.35) are supplemented by (5.30). Gravity may also be included—e.g. in a generalisation of the analysis of Sect. 4.7 to the interface between conducting fluids or plasmas in magnetic fields, including simplistic stability investigations where the gravity is presumed to simulate magnetic field curvature.

Let us also note that the term proportional to $\mathbf{k} \cdot \mathbf{B}$ in Eq. (5.120) becomes small in any neighbourhood where $\mathbf{k}$ becomes parallel to $\mathbf{B}$. Although this may not be so important when the resistivity η is large, as in the case of mercury in the discussion above, there can again be singular perturbations associated with higher derivatives introduced by the resistive term $\nabla \times (\eta \nabla \times \mathbf{B})$ in (5.30) to consider when η is small. Indeed, in analogy with boundary layers in fluids of small shear viscosity (high Reynolds number flows), the small resistivity may only be significant in a narrow region where $\mathbf{k} \cdot \mathbf{B} \simeq 0$, an observation particularly important in plasma stability theory—cf. Sect. 6.4 and subsequently. Thus in the next chapter we will first discuss ideal MHD stability on assuming the model outlined in Sect. 5.7, before proceeding to consider several important resistive instabilities that may arise. We will then incorporate plasma viscosity as another important non-ideal property in Sect. 6.11, since it reduces the growth rate of the singular resistive modes and produces enhanced plasma heating in the resulting viscoresistive layer—before finally discussing non-ideal instabilities due to the Hall effect, when the MHD model is modified as outlined in the next section.

5.15 Hall MHD Model

Hall MHD is widely interpreted to mean theory for a collisionless quasi-neutral plasma, where the generalised Ohm's law (2.98) is reduced to

$$\mathbf{E} + \mathbf{v} \times \mathbf{B} = \frac{1}{ne}(\mathbf{j} \times \mathbf{B} - \nabla p_e), \tag{5.121}$$

when both the inter-species and the electron pressure gradient are included. Some authors have omitted the electron pressure gradient term on the right-hand side of (5.121) such that

$$\mathbf{E} + \mathbf{v} \times \mathbf{B} = \frac{1}{ne} \mathbf{j} \times \mathbf{B}. \tag{5.122}$$

From either (5.122) or (5.121) when (2.99) applies, the consequent Hall equation of magnetic induction is

$$\begin{aligned} \frac{\partial \mathbf{B}}{\partial t} &= \nabla \times (\mathbf{v} \times \mathbf{B}) - \nabla \times \left(\frac{1}{ne} \mathbf{j} \times \mathbf{B} \right) \\ &= \nabla \times (\mathbf{v} \times \mathbf{B}) - \nabla \times \left(\frac{1}{\mu_0 ne} (\nabla \times \mathbf{B}) \times \mathbf{B} \right), \end{aligned} \tag{5.123}$$

on taking the curl and invoking (5.18) as before. Since

$$-\nabla \times \left(\frac{1}{ne} \mathbf{j} \times \mathbf{B} \right) = -\frac{1}{ne} \nabla \times (\mathbf{j} \times \mathbf{B}) + \frac{1}{n^2 e} \nabla n \times (\mathbf{j} \times \mathbf{B}), \tag{5.124}$$

the induction equation (5.123) includes a Hall term even if the plasma is homogeneous (n is constant), and a second contribution if it is not ($\nabla n \neq 0$). In the Exercise below, it emerges that the first term is responsible for socalled whistler waves, and the second term for Hall drift and shock-like waves. Alternatively, from (5.31) we have

$$-\nabla \times \left(\frac{1}{\mu_0 ne} (\nabla \times \mathbf{B}) \times \mathbf{B} \right) = \nabla \times \left[\frac{1}{\mu_0 ne} \left(\nabla_\perp \frac{B^2}{2} - B^2 \boldsymbol{\kappa} \right) \right], \tag{5.125}$$

where in particular we note the term involving the magnetic field curvature vector $\boldsymbol{\kappa}$ defined by (2.81). Moreover, the Hall term in (5.123) carries higher derivatives into the ideal magnetic induction equation (5.38) that are a potential source of singular behaviour, similar to but independent of the singular behaviour that may arise due to including the resistive term (or perhaps other terms) in the generalised Ohm's law (2.98).

In a quasi-neutral ion-electron plasma, the reduced form (5.122) is equivalent to $\mathbf{E} + \mathbf{v}_e \times \mathbf{B} \simeq 0$ because the electron velocity is $\mathbf{v}_e \simeq \mathbf{v} - \mathbf{j}/(ne)$ and $\mathbf{v} \simeq \mathbf{v}_i$ since $m_e \ll m_i$, as we noted in Sect. 2.10. Thus the electrons are "frozen-in" to the magnetic field but not the ions, which are therefore sometimes called "un-magnetised". It is also said that the Hall term decouples the ion and electron motion on ion inertial length scales, when collisionless Hall MHD is invoked to describe plasmas where the ion gyroradius is very large indeed. However, a Hall MHD model need not be collisionless. Thus in particular, when the term $\mathbf{j}/\sigma$ is retained in the generalised Ohm's law (2.98) in a resistive Hall model, the magnetic field is not "frozen-in" to either the ions or the electrons, and we recall the corresponding equation of magnetic induction is then (2.100)—i.e.

$$\frac{\partial \mathbf{B}}{\partial t} = \nabla \times (\mathbf{v} \times \mathbf{B}) - \nabla \times (\eta \nabla \times \mathbf{B}) - \nabla \times \left(\frac{1}{\mu_0 n e} (\nabla \times \mathbf{B}) \times \mathbf{B} \right), \quad (5.126)$$

involving the resistivity term with coefficient η due to particle collisions in addition to the Hall term as in (5.123). The relevant system of MHD equations again includes (5.33), (5.34) and either (5.35) or $\nabla \cdot \mathbf{v} = 0$. In Sect. 6.12, we discuss instabilities due to the Hall effect that arise in the collisionless and the collisional models—i.e. when either (5.123) or (5.126) is adopted, respectively.

Exercise

(Q1) Consider a Hall plasma of infinite extent permeated by a uniform magnetic field $\mathbf{B} = B\hat{\mathbf{j}}$ (B constant).

(a) If the plasma is uniform (number density n constant), such that the corresponding perturbed magnetic induction equation is

$$\frac{\partial \mathbf{B}_1}{\partial t} = -\frac{1}{\mu_0 n e} \nabla \times [(\nabla \times \mathbf{B}_1) \times \mathbf{B}] = -\frac{1}{\mu_0 n e} \mathbf{B} \cdot \nabla \nabla \times \mathbf{B}_1$$

on neglecting any velocity perturbation (given that the ion component is not "frozen-in" to the magnetic field), show that the frequency of plane waves with wave number k propagating in the direction of the magnetic field $\mathbf{B}$ is

$$\omega = \frac{k^2 B}{\mu_0 n e}.$$

(b) If the plasma is stratified perpendicular to the magnetic field, such that the number density is $n(z)$ say and the corresponding perturbed magnetic induction equation is well approximated by

$$\frac{\partial \mathbf{B}_1}{\partial t} = \frac{1}{\mu_0 n^2 e} \nabla n \times (\nabla \times \mathbf{B}_1 \times \mathbf{B}),$$

show that the frequency of plane waves with wave number k propagating transverse to the magnetic field $\mathbf{B}$ is

$$\omega = \frac{kB}{\mu_0 n e} \kappa \quad \left(\text{where } \kappa \equiv \frac{1}{n} \frac{dn}{dz} \text{ characterises the stratification}\right).$$

5.16 Advected Non-ideal MHD Discontinuities*

Advected discontinuities in the ideal MHD model were considered in Sect. 5.12, where under particular assumptions on field behaviour we found that there may be: (1) a contact discontinuity, when only the density may be discontinuous; or

(2) a tangential discontinuity, when not only the density but also either or both of the tangential velocity and magnetic fields—and hence the pressure—may be discontinuous. We mentioned that there is no natural small scale length in ideal MHD, and represented an assumed infinitesimally thin transition layer that may change spatially in negligible time as an interface on the kinematic level surface $\varsigma(\mathbf{r}, t) = 0$ where $\varsigma(\mathbf{r}, t)$ changes sign. Nevertheless, in principle we could proceed to consider many non-ideal plasma effects localised within the transition layer of finite width, analogous to the simpler inclusion of viscosity in the brief discussion of fluid shock structure in Sect. 4.12. On the other hand, when the ideal MHD model is abandoned and some non-ideal model is to be applied extensively throughout the plasma, for every non-ideal effect included in the model there is an associated characteristic length scale that is larger than the width of any conceivable transition region where other excluded non-ideal effects might be considered localised. Thus rather than attempt to resolve the physical structure within the associated transition region due to the excluded non-ideal effects, it may still be possible to envisage discontinuities in the particular extensive non-ideal MHD model adopted. On the other hand, unless the length scale over which some field variable varies significantly is rather smaller than the width of the transition layer associated with the particular non-ideal model, there may be no need to envisage a corresponding surface intensity.

In adopting a resistive MHD model, it is implicit that the resistive length scale is larger than the width of any conceivable transition region, with the resistive term in the equation of magnetic induction to be retained in the outer region. As in electromagnetic theory, it is then often assumed that no surface current $\mathbf{j}_*$ need be introduced—i.e. $\mathbf{j} = \mathbf{j}_- + (\mathbf{j}_+ - \mathbf{j}_-)H(\varsigma)$ may be assumed in addition to (5.84)–(5.87) at the interface, such that none in the extended set of MHD field quantities $\{\rho, p, \mathbf{v}, \mathbf{B}, \mathbf{j}\}$ has Dirac delta function behaviour at the interface although their derivatives may be singular there. Thus (5.92) is recovered, but (5.99) is replaced by

$$\hat{\mathbf{n}} \times [\![\mathbf{B}]\!] = 0 \quad \text{on } \varsigma = 0 \tag{5.127}$$

such that $[\![\mathbf{B}]\!] = 0$—i.e. the magnetic field $\mathbf{B}$ is continuous. The equation of motion in the form (5.39) then implies $[\![p]\!] = 0$—i.e. the pressure is also continuous.[22] Further, the condition from the third pre-Maxwell equation (5.18) is reduced to

$$\hat{\mathbf{n}} \times [\![\mathbf{E}]\!] = 0 \quad \text{on } \varsigma = 0. \tag{5.128}$$

[22] In passing, we note that this pressure continuity condition does not challenge the traditional notion of plasma confinement by a magnetic field, because any jump in the magnetic field envisaged under the ideal MHD model (when $\mathbf{j}_* \neq 0$) corresponds to a continuous but steep variation in the magnetic field under any resistive MHD model—i.e. the surface current in the ideal MHD boundary condition (5.99) occupies a thicker transition layer than that envisaged under condition (5.127), but which is nevertheless less than the macroscopic length scale of any field quantity outside it. The resistive diffusion time is also relatively long in most time-dependent scenarios—e.g. much longer than the time scale of the ideal and non-ideal MHD instabilities considered in the following chapter.

In addition to (5.90) from (5.33), for a plasma-plasma interface there are again conditions obtained from the relevant equation of magnetic induction. For example, if the resistive Hall equation of magnetic induction (5.126) rewritten as

$$\frac{d\mathbf{B}}{dt} = \mathbf{B}\cdot\nabla\mathbf{v} - \mathbf{B}\nabla\cdot\mathbf{v} - \nabla\times\left[\frac{\mathbf{j}}{\sigma} + \frac{1}{ne}\mathbf{j}\times\mathbf{B}\right] \tag{5.129}$$

is adopted, then

$$\hat{\mathbf{n}}\cdot\mathbf{B}\,[\![\mathbf{v}]\!] - \hat{\mathbf{n}}\times\left[\!\!\left[\frac{\mathbf{j}}{\sigma} + \frac{1}{ne}\mathbf{j}\times\mathbf{B}\right]\!\!\right] = 0 \quad \text{on } \varsigma = 0. \tag{5.130}$$

Thus *inter alia*, if either $\hat{\mathbf{n}}\cdot\mathbf{B} = 0$ or $[\![\mathbf{v}]\!] = 0$ holds, then the tangential component of $\mathbf{j}/\sigma + 1/(ne)\,\mathbf{j}\times\mathbf{B}$ must be continuous. If the Hall term is omitted, then condition (5.130) obviously reduces to

$$\hat{\mathbf{n}}\cdot\mathbf{B}\,[\![\mathbf{v}]\!] - \hat{\mathbf{n}}\times[\![\mathbf{j}/\sigma]\!] = 0 \quad \text{on } \varsigma = 0, \tag{5.131}$$

such that in the resistive MHD model of Sect. 5.14 the corresponding condition from the equation of magnetic induction reduces to continuity of the tangential component of $\mathbf{j}/\sigma$. (If neither $\hat{\mathbf{n}}\cdot\mathbf{B} = 0$ nor $[\![\mathbf{v}]\!] = 0$ hold, then of course the complete form of the relevant condition applies.)

On the other hand, in any extensive non-resistive MHD model there is no reference resistive length, and also the propensity for charged particle separation on the very small Debye length scale foreshadowed in Sect. 5.12. Thus even if it is realistic to assume there are similar transition regions where the resistivity is localised and the principal MHD fields $\{\rho, p, \mathbf{v}, \mathbf{B}\}$ are smooth and once again represented by (5.84)–(5.87), and that there are such expressions for the ion and electron pressures p_i and p_e individually, there may be an electric field intensity $\mathbf{E}_*$ to also include, similar to the current density $\mathbf{j}_*$ in (5.100) previously considered—i.e. the electric field is represented as

$$\mathbf{E} = \mathbf{E}_- + [\![\,\mathbf{E}\,]\!]\,H(\varsigma) + \mathbf{E}_*|\nabla\varsigma|\,\delta(\varsigma) \tag{5.132}$$

corresponding to significant charge separation and an associated double (dipole) layer at the interface, as foreshadowed in Sect. 5.12 and well known in electromagnetic theory (cf. p. 191 in Ref. [11]). For example, in the collisionless Hall MHD model there is then not only the condition

$$\hat{\mathbf{n}}\cdot[\![\,p\,\mathsf{I} + \mathcal{T}\,]\!] = 0 \quad \text{on } \varsigma = 0 \tag{5.133}$$

from the equation of motion (5.34) but also

$$(ne\,\mathbf{E})_* = -\hat{\mathbf{n}}\cdot[\![p_e\mathsf{I} + \mathcal{T}]\!] = [\![p_i]\!]\,\hat{\mathbf{n}} \quad \text{on } \varsigma = 0 \tag{5.134}$$

(where $p = p_i + p_e$) from (5.121), on recalling that $\mathbf{j} \times \mathbf{B} = \mu_0^{-1}(\nabla \times \mathbf{B}) \times \mathbf{B} = -\nabla \cdot \mathcal{T}$ where $\mathcal{T}$ denotes the magnetic stress tensor defined in Sect. 5.7. Condition (5.134) implies continuity of ion pressure ($[\![p_i]\!] = 0$) if and only if $(ne\,\mathbf{E})_* = 0$—and nonzero $\mathbf{E}_*$ implies

$$\hat{\mathbf{n}} \times [\![\mathbf{E}]\!] - \hat{\mathbf{n}} \times \nabla_2 \mathbf{E}_* = \hat{\mathbf{n}} \cdot \mathbf{v}\, [\![\mathbf{B}]\!] \text{ on } \varsigma = 0 \tag{5.135}$$

from (5.37), where ∇_2 denotes the two-dimensional vector differential operator in the plane of the level surface $\varsigma = 0$. There is also observational evidence for even more complicated physics, such as double discontinuities.

It seems that such physical complexities may have led some authors to avoid any discussion of a free boundary, and to consider only boundary conditions at a rigid wall that is usually assumed to be perfectly conducting, not only in MHD duct flow but also in analysing plasma stability—i.e. to adopt the conditions $\hat{\mathbf{n}} \cdot \mathbf{v} = 0$ and $\hat{\mathbf{n}} \cdot \mathbf{B} = 0$, and sometimes but not always $\hat{\mathbf{n}} \times \mathbf{E} = 0$, familiar from fluid mechanics and electromagnetic theory. However, any double layer or plasma sheath at the wall renders not only $\hat{\mathbf{n}} \times \mathbf{E} \neq 0$ from (5.135) but also

$$(\nabla \varsigma) \cdot \nabla \times \mathbf{E} = -\nabla \cdot (\nabla \varsigma \times \mathbf{E}) = -\nabla \cdot (\nabla \varsigma \times \nabla_2 \mathbf{E}_*),$$

when the relevant result (5.113) no longer follows. Moreover, there are some significant ideal and non-ideal MHD instabilities inherently associated with a free boundary, including important cases directly related to the classical Rayleigh–Taylor instability that was briefly discussed in Sect. 4.7.

Bibliography

1. H. Alfvén, *Cosmical Electrodynamics* (Oxford University Press, Oxford, 1950). (Pioneering work, including a discussion of MHD waves and solar physics)
2. D. Biskamp, *Nonlinear Magnetohydrodynamics* (Cambridge University Press, Cambridge, 1997). (Textbook with introductory chapters that may be read in conjunction with our presentation, prior to the discussion of nonlinear processes)
3. T.G. Cowling, *Magnetohydrodynamics* (Adam Hilger, Bristol, 1976). (Revision of an early concise text on MHD, with an emphasis on geophysical and astrophysical applications)
4. P.A. Davidson, *An Introduction to Magnetohydrodynamics* (Cambridge University Press, Cambridge, 2001). (Introductory textbook on MHD mentioned in the Preface)
5. J.P. Goedbloed, R. Keppens, S. Poedts, *Advanced Magnetohydrodynamics; with Applications to Laboratory and Astrophysical Plasmas* (Cambridge University Press, Cambridge, 2010). (Extends "Principles of Magnetohydrodynamics" to deal with streaming and toroidal plasmas, and also nonlinear dynamics)
6. J.P. Goedbloed, S. Poedts, *Principles of Magnetohydrodynamics* (Cambridge University Press, Cambridge, 2004). (First referenced in Chapter 2)
7. R.M. Kulsrud, *Plasma Physics for Astrophysicists* (Princeton University Press, Princeton, 2004). (The graduate textbook mentioned in the Preface, emphasising physical intuition in developing the analysis with astrophysical applications in mind)

8. G.E. Marsh, *Force-Free Magnetic Fields: Solutions, Topology and Applications* (World Scientific, Singapore, 1996). (A brief history of force-free magnetic fields before a discussion of field topology, helicity and multiply connected domains)
9. P.D. Miller, *Applied Asymptotic Analysis* (American Mathematical Society, Providence, 2006). (First referenced in Chapter 3)
10. E.N. Parker, *Spontaneous Current Sheets in Magnetic Fields with Applications to Stellar X-Rays* (Oxford University Press, Oxford, 1994). (Spontaneous tangential discontinuities (current sheets) in a magnetic field embedded in highly conducting plasma may explain the activity of external magnetic fields in astrophysics and the laboratory)
11. J.A. Stratton, *Electromagnetic Theory* (Adams Press, Chicago, 2007). (Reprint of the classic text originally published by McGraw-Hill in 1940, which covers fundamental electromagnetic theory in considerable depth)
12. J. Wesson, D.J. Campbell, *Tokamaks*, 3rd edn. (Oxford University Press, Oxford, 2004). (Surveys the important features of tokamak experiments, including equilibrium and confinement, heating, instabilities, surface interactions and diagnostics)

Chapter 6
MHD Stability Theory

As in the earlier discussion on waves, disturbances from an initial equilibrium are usually assumed small enough to justify linearisation of the perturbed equations in the mathematical model, as an important first step to define the evolution of a complex system in this chapter. Normal mode analysis involving Fourier forms for the disturbance is therefore often invoked when the initial state is one-dimensional (dependent on only one spatial variable), except that the time exponent is assumed to carry a real part, often with the additional imaginary part (when there is an accompanying oscillation). The sign of the real part then determines whether or not the disturbance grows or decays exponentially—i.e. whether or not the system is linearly unstable or stable, respectively. Identification of the stabilising and destabilising forces is particularly important in MHD stability analysis, in both laboratory and astrophysical applications. In the case of the ideal MHD model, a variational formulation (an energy principle) permits the analysis of more complicated geometries, such as in modern experiments in controlled thermonuclear fusion research. Important stabilising and destabilising forces are identified under this formulation, which is then applied to investigate the stability of cylindrical and toroidal configurations. More direct analysis is usually followed to investigate instability in non-ideal MHD models, but we demonstrate an extension of the variational formulation that includes viscosity. Although the main application we consider is magnetic confinement in thermonuclear research, the final section on Hall instability includes an aspect relevant to laser-driven fusion, and some topics of interest in astrophysics and solar physics are discussed there and elsewhere. The additional bibliography for this chapter once again provides suggested further reading.

6.1 Introduction

In Sect. 4.7, we discussed how a small perturbation of an initial state in a fluid may grow under gravity (Rayleigh–Taylor instability) or due to velocity shear (Kelvin–Helmholtz instability). Since the goal of magnetic confinement demands a design

R.J. Hosking and R.L. Dewar, *Fundamental Fluid Mechanics and Magnetohydrodynamics*, DOI 10.1007/978-981-287-600-3_6

that contains very hot plasma long enough for thermonuclear fusion to occur, the likelihood that certain perturbations grow significantly in that time presents an obvious obstacle [1, 2, 11]. A comprehensive theoretical investigation of the stability of any initial state of a dynamical system requires the consideration of all possible perturbations, but to some extent that task in relevant fusion research may be reduced by first considering the competition between stabilising forces such as magnetic tension and the various destabilising forces that may exist. We previously mentioned an early suggestion that gravity in Cartesian geometry might be regarded as a pseudo-centrifugal term simulating magnetic field curvature (cf. Sect. 5.7)—i.e. the gravitational acceleration g might be assumed to represent $v_{\rm th}^2/R_c$, where $v_{\rm th}$ is the representative thermal speed of the charged particles following the magnetic field lines with radius of curvature R_c. Although simplistic, this idea does provide some preliminary understanding prior to incorporating more realistic magnetic field geometry in the mathematical theory presented later in this chapter. Although we cannot attempt to provide a very thorough presentation of MHD stability in this book, we identify some important stabilising and destabilising mechanisms, and then extend our stability discussion from ideal MHD to non-ideal models as previously mentioned at the end of Sect. 5.14.

Over the years, various definitions of stability have been proposed. One definition is that a dynamical system is stable if and only if its undisturbed path of evolution is the limit of any disturbed path, a definition which extends the concept to time-dependent initial states. However, our purposes are well served by the earlier notion of a departure from an initially static state (cf. Sect. 5.10), or else from a steady (time-independent) state. The simplest stability theory assumes small perturbations of the initial state, such that the governing equations in the fundamental field variable perturbations are linear [3]. Sometimes the resulting initial value problem may be solved directly to decide whether the perturbation grows or decays in time (or possibly remains constant), but more commonly the Fourier form

$$f(\mathbf{r}, t) = f(\mathbf{r}) \exp(-i\omega t) \tag{6.1}$$

is again exploited, as also mentioned in the preamble above. Thus the initial value problem for the evolution of the small perturbations becomes an eigenvalue problem as before, where a system of differential equations is solved subject to relevant boundary conditions, which we discussed in Sect. 5.12 for ideal MHD and subsequently in Sect. 5.16 for some non-ideal models. However, we now anticipate that the eigenfrequency ω may have an imaginary part that renders a real exponent in time, as first encountered in Sect. 4.7. The perturbation then decays exponentially if the imaginary part of ω is negative, and the initial state is said to be *exponentially stable* when that is so for all allowable small perturbations. On the other hand, if the imaginary part of ω is positive for even one such perturbation, which therefore grows exponentially, the initial state is *exponentially unstable*. A perturbation is of course oscillatory whenever the real part of ω is nonzero, and if this perturbation grows the initial state is sometimes called overstable, as we mentioned at the end of Sect. 4.7. When the imaginary part of ω goes through zero for certain values of the system parameters, we may refer to a *marginally* (or neutrally) stable point that can often separate a stable from an unstable region in parameter space. This transition is often notable in developing stability theory—e.g. to suggest the appropriate flute ordering in toroidal geometry in Sect. 6.7, or wave number ranges where non-ideal instabilities are found to be significant such as in Sect. 6.9. There could be other than exponential instability, but that is not explored here.

An eigenvalue problem may be treated in at least two ways. The direct approach exploiting Fourier analysis often provides the solution for the perturbation functions satisfying the relevant system of governing differential equations and boundary conditions. Another way is to express the eigenvalue problem as a variational problem, involving a suitable functional. At the end of the eighteenth century, Laplace noted that any disturbance of a liquid in stable equilibrium under gravity increases the potential energy. In the second half of the nineteenth century, Lord Rayleigh showed that minimising the potential energy determines the eigenfrequencies of an elastic system, and after Hamilton the variational approach became a frequent feature in modern mathematical physics. A variational energy principle to investigate magnetohydrostatic equilibria under the ideal MHD model is discussed in Sect. 6.5. However, an initial state that is stable under the ideal MHD model may be rendered unstable when non-ideal plasma properties such as resistivity or Hall current are included, so further investigation is necessary. Variational formulations and analytical solutions are sometimes still possible. On the other hand, numerical simulation is often needed to investigate the stability of complex magnetic field designs, even in the linear ideal MHD model.

It is also well known that a linearly stable initial state may be unstable to larger perturbations, when linearisation of the perturbation equations is no longer valid. Analytical investigations of nonlinear behaviour are usually qualitative rather than quantitative, where phase diagrams or the construction of non-negative energy integrals or associated Lyapunov functions can assist [4]. Undergraduate textbooks sometimes refer to a ball under gravity on a surface, to illustrate linear and nonlinear stability considerations in an elementary way. Thus there is a stable point in a valley, about which the ball oscillates for small perturbations, but this point can be rendered unstable by a perturbation large enough to take the ball over a neighbouring hill through its peak, which is of course a linearly unstable point. Indeed, a sufficiently large perturbation could take the ball from the valley onto a neighbouring plane, which consists of marginally stable points according to linear theory but may likewise be unstable for such large perturbations. Friction might be expected to inhibit the perturbation, but it would need to be particularly strong to entirely prevent any of those outcomes. On the other hand, with or without friction a ball at the peak of a hill could conceivably access a neighbouring valley, so that a linearly unstable configuration may be stable to larger perturbations.

However, linear MHD stability theory has been emphasised historically—not only because linearity permits the powerful superposition techniques of Fourier analysis, but also because fast instabilities readily identified in the linear theory must be avoided altogether for a successful magnetic confinement design. We shall see that ideal MHD instabilities proceed on the time scale associated with the Alfvén speed, and that non-ideal instabilities can also proceed much faster than the resistive diffusion time.

We begin with a discussion of normal modes in a simple model in the next section, which leads to an elementary but informative analysis of ideal interchange instabilities in Sect. 6.3. Magnetic field shear introduced in Sect. 6.4 localises the instability; and in Sect. 6.5 we develop the powerful variational approach to ideal MHD stability analysis, prior to discussing ideal instabilities in cylindrical geometry in Sect. 6.6 and then toroidal geometry in Sect. 6.7. There is then a discussion of magnetorotational theory (a topic of astrophysical interest) in Sect. 6.8, before we proceed to consider resistive instabilities in Sects. 6.9 and 6.10. The role of plasma viscosity in damping instability on the one hand and enhancing the energy released in magnetic reconnexion on the other is discussed in Sect. 6.11, and finally we devote Sect. 6.12 to Hall instability.

6.2 Normal Modes in a Plasma Slab

Let us assume a magnetohydrostatic configuration defined by (5.63) when $\mathbf{v} = 0$, and again distinguish perturbation fields by the subscript 1 as in earlier chapters. In the ideal MHD model of Sect. 5.7, the corresponding first order linearised equations for small magnitude perturbations are

$$\frac{\partial \rho_1}{\partial t} + \mathbf{v}_1 \cdot \nabla \rho + \rho \nabla \cdot \mathbf{v}_1 = 0, \tag{6.2}$$

$$\rho \frac{\partial \mathbf{v}_1}{\partial t} + \nabla p_1 = \rho_1 \mathbf{g} + \mu_0^{-1} [(\nabla \times \mathbf{B}) \times \mathbf{B}_1 + (\nabla \times \mathbf{B}_1) \times \mathbf{B}], \tag{6.3}$$

$$\frac{\partial p_1}{\partial t} + \mathbf{v}_1 \cdot \nabla p + \gamma p_0 \nabla \cdot \mathbf{v}_1 = 0, \tag{6.4}$$

$$\frac{\partial \mathbf{B}_1}{\partial t} = \nabla \times (\mathbf{v}_1 \times \mathbf{B}), \tag{6.5}$$

with the propagated condition

$$\nabla \cdot \mathbf{B}_1 = 0. \tag{6.6}$$

These linear perturbation equations of course yield the ideal Alfvén and magnetosonic waves discussed in Sect. 5.9, but there can also be various MHD instabilities analogous to the Rayleigh–Taylor and Kelvin–Helmholtz instabilities discussed in Sect. 4.7. As mentioned in the previous section, we could investigate instability by considering such perturbation equations subject to appropriate initial and boundary conditions—i.e. as initial value problems. The solution of such an initial value problem in principle defines the time evolution of any assumed initial perturbation, and that approach has been followed in numerical simulation. However, the consideration of *normal modes* [Glossary Section "Normal Mode"] usually provides an efficient method for stability analysis.

A comprehensively analytic approach to determining the full spectrum of normal modes is usually only feasible for one-dimensional cases—i.e. when the unperturbed configuration is inhomogeneous with respect to a single spatial coordinate only—although some analytic simplification can be effected in more complicated geometries using suitably chosen asymptotic limits. For fuller normal mode analysis, large computer MHD codes are routinely used in the design of expensive fusion research facilities (to avoid building a potentially unstable system), but the insight gained from studying simple geometries and asymptotic limits is useful in the delicate numerical analysis required to write such codes. Thus in this chapter we first concentrate on one-dimensional slab and circular-cylinder models [Glossary Section "Slab Model"], and later an asymptotic short-wavelength approach to understanding more general cases.

For the slab geometry studied in this section, we adopt the gravitational confinement Cartesian coordinate convention [Glossary Section "Slab Model"]. As in Chap. 4, we can factor out the nontrivial z-dependence by separation of variables using the Fourier form

$$f(\mathbf{r}, t) = f(z) \exp[i(\mathbf{k} \cdot \mathbf{r} - \omega t)] \tag{6.7}$$

where $\mathbf{k} = k_x \hat{\mathbf{i}} + k_y \hat{\mathbf{j}}$ (ω and $\mathbf{k}$ are constant), representing a normal mode [Glossary Section "Normal Mode"] eigenfunction. As in the discussion of sound waves in Sect. 4.3, we note it is sometimes assumed that the disturbances have length scale rather smaller than the extent of the medium or the equilibrium scale, and an analogous eikonal ansatz may be introduced to represent such localised modes. However, larger scale instabilities can also be important, when appropriate plasma boundary conditions apply. In any case, the main objective is to derive the dispersion relation as before, except now as a relation between the generally complex frequency ω and the two-dimensional wave number vector $\mathbf{k}$.

6.3 Ideal Gravitational Interchange Instabilities

Hydrodynamic Rayleigh–Taylor instabilities arise whenever there is a decreasing pressure gradient (decreasing density) in the direction of any applied acceleration, but it is again useful to first discuss a sharp density discontinuity between two homogeneous regions in a vertical gravitational field (cf. Sect. 4.7). Thus let us consider a magnetohydrostatic configuration (no flow), where plasma under gravity in a uniform horizontal magnetic field has a density discontinuity at the plane interface $z = 0$ (an ideal MHD contact discontinuity, discussed in Sect. 5.12) such that

$$\rho = \rho_1 H(z) + \rho_2 H(-z), \quad \mathbf{g} = -g\,\hat{\mathbf{k}}, \quad \mathbf{v} = 0, \quad \mathbf{B} = B_x\,\hat{\mathbf{i}} + B_y\,\hat{\mathbf{j}}, \tag{6.8}$$

where ρ_1, ρ_2, $\mathbf{g}$ and now $\mathbf{B}$ are all constants and $H(z)$ again denotes the Heaviside step function. As before, the hydrostatic pressure is continuous but its gradient $dp/dz = -\rho g\,(z \neq 0)$ is discontinuous (with density) at the interface $z = 0$, since the magnetic field is assumed to be horizontal and identical on both sides of the interface in this simple case. Once again, assuming an infinitesimally thin transition region means in practice that the perturbation wavelengths must be significantly larger than its thickness, and there is no shock.

The "frozen-in" magnetic field concept discussed in Sect. 5.8, and the implied increase in magnetic tension when the interface is perturbed, suggests that the magnetic field should oppose the pressure-driven instabilities that always occur when $\rho_1 > \rho_2$ in the absence of the magnetic field ($\mathbf{B} = 0$). If incompressible disturbances are considered such that $\nabla \cdot \mathbf{v}_1 = 0$, where the subscript 1 denotes the perturbation quantity as before, from (6.2) the density is unperturbed everywhere outside the transition region at the interface in the present simple mathematical discussion. Thus as in Sect. 4.7, the gravity only enters the analysis in the pressure boundary condition at the perturbed interface, where there is the density discontinuity. On invoking (6.7), the residual linearised ideal MHD perturbation equations (6.3) and (6.5) reduce to

$$-i\omega\rho\mathbf{v}_1 + \nabla\Pi_1 = i\mu_0^{-1}\mathbf{k}\cdot\mathbf{B}\,\mathbf{B}_1, \tag{6.9}$$

$$-\omega\mathbf{B}_1 = \mathbf{k}\cdot\mathbf{B}\,\mathbf{v}_1, \tag{6.10}$$

where $\Pi_1 = p_1 + \mu_0^{-1}\mathbf{B}\cdot\mathbf{B}_1$ denotes the total pressure perturbation and $\mathbf{B}\cdot\nabla$ is replaced by $i\,\mathbf{k}\cdot\mathbf{B}$. This system reduces to the hydrodynamic case for precisely perpendicular propagation (where $k_\parallel = \mathbf{k}\cdot\mathbf{B}/B = 0$), so the initial uniform magnetic field cannot stabilise the corresponding particular Rayleigh–Taylor mode found in Sect. 4.7—and provides us with an early indication that $k_\parallel \simeq 0$ (the "*flute condition*", in nomenclature clarified in Sect. 6.6) can be quite significant in MHD stability theory.

Assuming $\rho\omega^2 \neq \mu_0^{-1}(\mathbf{k}\cdot\mathbf{B})^2$ to exclude stable Alfvén waves, the dispersion relation for MHD perturbations (when $\mathbf{B}_1 \neq 0$) follows for the resulting irrotational flow—i.e. we have $\nabla\times\mathbf{v}_1 = 0$ and $\nabla\times\mathbf{B}_1 = 0$, in addition to $\nabla\cdot\mathbf{v}_1 = 0$ and $\nabla\cdot\mathbf{B}_1 = 0$. The non-cavitation condition (4.65) met in the hydrodynamic discussion remains for the MHD contact discontinuity, rendering $[\![v_{1z}]\!] = 0$ in the linearised theory—but we now have the linearised condition of total pressure balance $[\![\Pi_1 + \zeta(dp/dz)]\!] = 0$ or $[\![\Pi_1]\!] + [\![\rho]\!]\,g\zeta = 0$ (cf. Sect. 5.12) to combine with the kinematic condition $-i\omega\zeta = v_{1z}$, which we apply to good approximation at $z = 0$ (cf. also Sect. 4.7).

The bounded velocity perturbation satisfying $\nabla\times\mathbf{v}_1 = 0$ and $\nabla\cdot\mathbf{v} = 0$ and the continuity condition $[\![v_{1z}]\!] = 0$ is $v_{1z} = C\exp[-k|z| + i(\mathbf{k}\cdot\mathbf{r} - \omega t)]$ where C is a constant, and the equation

$$\frac{d\Pi_1}{dz} = (-i\omega)^{-1}[\rho\omega^2 - \mu_0^{-1}(\mathbf{k}\cdot\mathbf{B})^2]\,v_{1z}$$

obtained from the z-components of (6.9) and (6.10) then produces the dispersion relation. Thus we obtain

$$[\![\Pi_1]\!] = \int_{0-}^{0+}\frac{d\Pi_1}{dz}\,dz = (i\omega k)^{-1}[(\rho_2 + \rho_1)\,\omega^2 - 2\mu_0^{-1}(\mathbf{k}\cdot\mathbf{B})^2]\,v_{1z}$$

at $z = 0$, to render

$$\omega^2 = \frac{(\rho_2 - \rho_1)gk + 2\mu_0^{-1}(\mathbf{k} \cdot \mathbf{B})^2}{\rho_2 + \rho_1} \tag{6.11}$$

as the generalisation of (4.74). The magnetic field term in (6.11) opposes the pressure-driven instability that would otherwise occur for $\rho_1 > \rho_2$, and the magnetic field is usually large enough such that ω^2 is non-negative except when $k_\parallel = \mathbf{k} \cdot \mathbf{B}/B \simeq 0$. Thus there are ideal MHD interchange instabilities when

$$\left(\frac{k_\parallel}{k}\right)^2 < \frac{\rho_1 - \rho_2}{\rho_2 + \rho_1} \frac{g}{kc_A^2} \tag{6.12}$$

even if the magnetic field $\mathbf{B} \neq 0$ is so large that $c_A = B/\sqrt{\mu_0 \rho}$ (the Alfvén speed) is dominant, as is typical in magnetic confinement systems. On the other hand, the essentially hydrodynamic case when $k_\parallel = 0$ involves no increase in magnetic energy, for the horizontal magnetic flux tubes associated with the higher density region readily interchange with their neighbours in the lower density region across the interface. Indeed, when $\rho_1 \gg \rho_2$ the corresponding interchange mode localised over a distance $O(k^{-1})$ has the classical Rayleigh–Taylor growth rate approaching $\sqrt{gk}$—cf. Sect. 4.7.

6.4 Magnetic Field Shear and Slab Modes

As the simple analysis for a density discontinuity in the previous section indicates, flute-like ($k_\parallel \simeq 0$) ideal MHD interchange instabilities can eliminate any density gradient at the rapid rate characterised by the Alfvén speed. However, it was suggested that MHD interchange instabilities would be inhibited when the direction of the magnetic field varies spatially—i.e. if there is *magnetic field shear*. The underlying physical notion was that the interchange of neighbouring flux tubes should then imply an increase in the magnetic energy of the system that reduces the energy available for their displacement, but it soon emerged that shear stabilisation is not effective everywhere.

Let us consider a plane slab of inhomogeneous plasma vertically stratified under gravity—i.e.

$$\nabla\rho = \frac{d\rho}{dz}\hat{\mathbf{k}}, \quad \mathbf{g} = -g\hat{\mathbf{k}}, \quad \mathbf{v} = 0, \tag{6.13}$$

with the magnetic field $\mathbf{B}(z)$ again in the horizontal plane ($B_z = 0$) but now varying in direction ($d\mathbf{B}/dz \neq 0$). The initial pressure balance then corresponds to

$$\frac{d}{dz}\left(p + \mu_0^{-1}B^2\right) + \rho(z)g = 0, \tag{6.14}$$

and on again invoking the Fourier form (6.7) the essential linearised perturbation equations for incompressible perturbations become

$$-i\omega\rho_1 + v_{1z}\frac{d\rho}{dz} = 0, \tag{6.15}$$

$$-i\rho\omega\mathbf{v}_1 + \nabla\Pi_1 = \rho_1\mathbf{g} + \mu_0^{-1}\left(i\mathbf{k}\cdot\mathbf{B}\,\mathbf{B}_1 + B_{1z}\frac{d\mathbf{B}}{dz}\right), \tag{6.16}$$

$$-\omega\mathbf{B}_1 = \mathbf{k}\cdot\mathbf{B}\,\mathbf{v}_1, \tag{6.17}$$

$$\nabla\cdot\mathbf{v}_1 = 0, \tag{6.18}$$

and the propagated condition $\nabla\cdot\mathbf{B}_1 = 0$, where $\Pi_1 = p_1 + \mu_0^{-1}\mathbf{B}\cdot\mathbf{B}_1$ is the total pressure perturbation as before. The gravity term $\rho_1\mathbf{g}$ now appears in the perturbation equation (6.16), introducing the density gradient $d\rho/dz$ to that equation on eliminating the perturbed density ρ_1 using (6.15). Elimination of Π_1 between the component equations of (6.16) then yields

$$\frac{d}{dz}\left(\rho\omega\frac{dv_{1z}}{dz}\right) - \rho\omega k^2 v_{1z} = \frac{k^2 g}{\omega}\frac{d\rho}{dz}v_{1z} + \frac{1}{\mu_0}\left[\frac{d^2(\mathbf{k}\cdot\mathbf{B})}{dz^2}B_{1z} - \mathbf{k}\cdot\mathbf{B}\left(\frac{d^2}{dz^2} - k^2\right)B_{1z}\right], \tag{6.19}$$

and then elimination of B_{1z} using the z-component of (6.17) produces the governing differential equation

$$\frac{d}{dz}\left(\left[\rho\omega^2 - \mu_0^{-1}(\mathbf{k}\cdot\mathbf{B})^2\right]\frac{dv_{1z}}{dz}\right) - k^2\left[\rho\omega^2 - \mu_0^{-1}(\mathbf{k}\cdot\mathbf{B})^2\right]v_{1z} - k^2 g\frac{d\rho}{dz}v_{1z} = 0. \tag{6.20}$$

Assuming the plasma is confined by horizontal rigid walls or the plasma is infinitely deep (such that v_{1z} vanishes at the plasma boundaries), on integrating (6.20) over the domain of z we obtain the corresponding Rayleigh variational form

$$\omega^2 = \frac{\int\rho\left[k_\parallel^2 c_A^2 |dv_{1z}/dz|^2 + k^2\left(k_\parallel^2 c_A^2 - g\kappa\right)|v_{1z}|^2\right]dz}{\int\rho\left(|dv_{1z}/dz|^2 + k^2|v_{1z}|^2\right)dz} \tag{6.21}$$

where $\kappa = \rho^{-1}\,d\rho/dz = d(\ln\rho)/dz$ is a measure of the density gradient, with $k_\parallel$ and c_A again denoting the parallel component of the wave number and the Alfvén speed, respectively. Thus in particular the plasma configuration is unstable to flute-like interchange modes *localised in the neighbourhood of the resonance surface where* $k_\parallel = 0$ ($z = z_s$ say), provided $g\kappa$ is sufficiently positive there—i.e. sufficient to overcome the positive definite stabilising contributions from the terms involving $k_\parallel^2 c_A^2$, in the integral in the numerator of (6.21) evaluated over that neighbourhood.

It is notable that the governing differential equation (6.20) is singular where

$$\rho\omega^2 = \mu_0^{-1}(\mathbf{k}\cdot\mathbf{B})^2, \tag{6.22}$$

which is the dispersion relation for Alfvén waves. The associated inertial term $\rho\omega^2$ is evidently significant near $k_\parallel = 0$, and this term shall be reconsidered later when non-ideal effects are incorporated.

Exercise

(Q1) When there is no field shear, the governing equation (6.20) reduces to

$$\frac{d}{dz}\left(\rho\frac{dv_{1z}}{dz}\right) - k^2\rho v_{1z} - \frac{k^2 g}{\omega^2}\frac{d\rho}{dz}v_{1z} = 0$$

—i.e. for interchange modes where $\mathbf{k}\cdot\mathbf{B} = 0$ everywhere. If the density $\rho(z) \to 0$ when $z \le 0$, show that the solution $v_{1z} = C\exp(-kz)$ in the half-space $z > 0$ defines modes with the maximum classical Rayleigh–Taylor growth rate $\sqrt{gk}$ (cf. also the discussion at the end of the previous section).

6.5 Ideal MHD Variational Principle

6.5.1 Linearised Equation of Motion

Let us now formulate ideal MHD stability analysis in another way, which readily allows identification of key sources of instability and has been used to investigate various plasma configurations. Intuitively, we anticipate that a physical system is unstable if it may move to a position of lower potential energy, as first envisaged in Sect. 5.10.3 and then in the simple case of the ball on a surface under gravity mentioned in Sect. 6.1. Since the ideal MHD model excludes dissipation, any decrease in potential energy produces an equal increase in the kinetic energy of the system. Mathematically, it emerges that a static plasma is ideal MHD stable if and only if a certain homogeneous quadratic form is positive definite. This form corresponds to the variation in the potential energy due to an arbitrary virtual displacement (consistent with the constraints).

The Lagrangian displacement $\boldsymbol{\xi}$ of a plasma fluid particle under the perturbation is defined by

$$\mathbf{r} = \mathbf{r}_0 + \boldsymbol{\xi}(\mathbf{r}_0, t), \tag{6.23}$$

where $\mathbf{r}_0$ is the original position of the particle prior to the virtual displacement. The corresponding Lagrangian velocity perturbation $\mathbf{v}_1(\mathbf{r}_0, t) = \partial_t\boldsymbol{\xi}$ equalsthe lowest order approximation to the Eulerian velocity for small $\boldsymbol{\xi}$ because

$$\mathbf{v}_1(\mathbf{r}, t) = \mathbf{v}_1(\mathbf{r}_0, t) + \boldsymbol{\xi} \cdot \nabla \mathbf{v}_1 + \cdots \tag{6.24}$$

($\boldsymbol{\xi} \cdot \nabla \mathbf{v}_1$ is second order), such that the linearised perturbation equations (6.2)–(6.5) of the ideal MHD model become

$$\frac{\partial \rho_1}{\partial t} + \frac{\partial \boldsymbol{\xi}}{\partial t} \cdot \nabla \rho + \rho \nabla \cdot \frac{\partial \boldsymbol{\xi}}{\partial t} = 0, \tag{6.25}$$

$$\rho \frac{\partial^2 \boldsymbol{\xi}}{\partial t^2} + \nabla p_1 = \rho_1 \mathbf{g} + \mu_0^{-1}[(\nabla \times \mathbf{B}) \times \mathbf{B}_1 + (\nabla \times \mathbf{B}_1) \times \mathbf{B}], \tag{6.26}$$

$$\frac{\partial p_1}{\partial t} + \frac{\partial \boldsymbol{\xi}}{\partial t} \cdot \nabla p + \gamma p \nabla \cdot \frac{\partial \boldsymbol{\xi}}{\partial t} = 0, \tag{6.27}$$

$$\frac{\partial \mathbf{B}_1}{\partial t} = \nabla \times \left(\frac{\partial \boldsymbol{\xi}}{\partial t} \times \mathbf{B} \right). \tag{6.28}$$

Integrating (6.25) and (6.27) with respect to time and then eliminating ρ_1 and p_1 from (6.26) yields

$$\rho \frac{\partial^2 \boldsymbol{\xi}}{\partial t^2} = \mathsf{F} \cdot \boldsymbol{\xi}, \tag{6.29}$$

involving the linear dyadic operator F such that

$$\begin{aligned} \mathsf{F} \cdot \boldsymbol{\xi} = \mu_0^{-1}[(\nabla \times \mathbf{B}_1) \times \mathbf{B} + (\nabla \times \mathbf{B}) \times \mathbf{B}_1] \\ + \nabla(\boldsymbol{\xi} \cdot \nabla p + \gamma p \nabla \cdot \boldsymbol{\xi}) - \mathbf{g} \nabla \cdot (\rho \boldsymbol{\xi}) \end{aligned} \tag{6.30}$$

where

$$\mathbf{B}_1 = \nabla \times (\boldsymbol{\xi} \times \mathbf{B}), \tag{6.31}$$

on integrating (6.28) with respect to time. In passing, we note that the propagated condition $\nabla \cdot \mathbf{B}_1 = 0$ follows trivially from (6.31).

6.5.2 Identification of ΔW and Self-adjointness

In simple geometry, Eq. (6.29) can be reduced to the solution of ordinary differential equations using separation of variables, but in more complicated geometries a variational approach is more powerful for approximation and numerical purposes. The variational approach is based on the kinetic energy

$$K[\boldsymbol{\xi}] = \frac{1}{2} \int_{\mathcal{P}} \rho \left(\frac{\partial \boldsymbol{\xi}}{\partial t} \right)^2 d\tau \tag{6.32}$$

and the change $\Delta W[\boldsymbol{\xi}] = W[\boldsymbol{\xi}] - W_0$ in the potential energy, where W and W_0 are the respective potential energies in the perturbed and equilibrium states given by (5.70) on neglecting gravity.[1]

Since there is no dissipation in ideal MHD, the sum $K + \Delta W$ of the kinetic and potential energy changes must be conserved, as can be seen by integrating (5.45) over the plasma volume $\mathcal{P}$.[2] Thus we must be able to obtain the equation of total energy conservation

$$\frac{\partial}{\partial t}\int_{\mathcal{P}}\left(\frac{1}{2}\rho\frac{\partial\boldsymbol{\xi}}{\partial t}\cdot\frac{\partial\boldsymbol{\xi}}{\partial t} + \Delta W[\boldsymbol{\xi}]\right) = 0, \tag{6.33}$$

as a first integral of (6.29). Dotting both sides of (6.29) with $\partial\boldsymbol{\xi}/\partial t$, and then integrating over $\mathcal{P}$ and rearranging appropriately, we have

$$\frac{\partial}{\partial t}\int_{\mathcal{P}}\frac{1}{2}\left[\rho\frac{\partial\boldsymbol{\xi}}{\partial t}\cdot\frac{\partial\boldsymbol{\xi}}{\partial t} - \boldsymbol{\xi}\cdot\mathbf{F}\cdot\boldsymbol{\xi}\right] = \frac{1}{2}\int_{\mathcal{P}}\left\{\frac{\partial\boldsymbol{\xi}}{\partial t}\cdot\mathbf{F}\cdot\boldsymbol{\xi} - \boldsymbol{\xi}\cdot\mathbf{F}\cdot\frac{\partial\boldsymbol{\xi}}{\partial t}\right\}d\tau. \tag{6.34}$$

Comparing (6.33) and (6.34), we first observe that they are the same form if and only if the right-hand side of (6.34) vanishes for all possible spatial functions $\boldsymbol{\xi}$ and $\partial\boldsymbol{\xi}/\partial t$. It is notable that these are *arbitrary* functions, because position and velocity may be specified independently as initial data for a second order differential equation such as (6.29), and any instant in time can be regarded as the initial instant. Thus $\mathbf{F}$ must be a *self-adjoint* operator, and indeed *Hermitian* because the coefficients of the operators on the right-hand side of (6.30) are real.[3] Although this can of course be established explicitly from (6.30) using integration by parts (cf. [6, pp. 465–467]), the above argument is not only much easier but also more fundamental.

---

[1] The remarkable article by Bernstein et al. (I. Bernstein, E.A. Frieman, M.D. Kruskal and R.M. Kulsrud, *Proceedings of the Royal Society London* A **244**, 17–40, 1958) first discussed the variational approach to stability analysis considered in this section. Although Bernstein et al. used the notation δW, we have designated this perturbation ΔW to avoid confusion with the first variation of W that vanishes at equilibrium, as we now consider the leading term involving the *second* variation in the increment to the potential energy [7]. We have also used square brackets, since W is a functional of $\boldsymbol{\xi}$—cf. the first footnote in Sect. 1.10.

[2] We will emphasise the model where the plasma extends to a fixed perfectly conducting boundary (the wall), so that it is the plasma contribution to the potential energy perturbation that we usually consider in this chapter, but Bernstein et al. (1958) discussed the configuration described in Sect. 5.13 where there is a vacuum region between the plasma edge and the wall (cf. Sect. 6.5.3). In the course of that extended analysis, the vector potential $\mathbf{A}$ replaces $\boldsymbol{\xi}$ as the field to be varied in the vacuum, unless $\boldsymbol{\xi}$ is defined through a special "*Newcomb gauge*" choice $\mathbf{A} = \boldsymbol{\xi}\times\mathbf{B}$ in order to unify the pressureless plasma and vacuum cases [6, Sect. 8.10].

[3] An operator F is said to be *self-adjoint* in some space of interest if $(\eta, F\xi) = (\xi, F\eta)$ for any η and ξ in the range of F, where the brackets denote the relevant function-space inner product—in our case, $(\mathbf{f}, \mathbf{g}) = \int \mathbf{f}\cdot\mathbf{g}\,d\tau$. The operator F is said to be *Hermitian* if $(\eta^*, F\xi) = (\xi^*, F\eta)^*$, where the superscript denotes the complex conjugate.

Having established how (6.33) and (6.34) can be regarded as the same equation, we can now identify $\Delta W[\boldsymbol{\xi}] = \Delta W[\boldsymbol{\xi}, \boldsymbol{\xi}]$, where the two-argument functional on the right-hand side is the Hermitian form defined for general complex $\boldsymbol{\eta}$ and $\boldsymbol{\xi}$ by

$$\Delta W[\boldsymbol{\eta}, \boldsymbol{\xi}] = -\frac{1}{2} \int_{\mathcal{P}} \boldsymbol{\eta}^* \cdot \mathsf{F} \cdot \boldsymbol{\xi} \, d\tau. \tag{6.35}$$

The negative of the right-hand side $\mathsf{F} \cdot \boldsymbol{\xi}$ in (6.29) evidently represents the restoring force density acting on the MHD fluid elements to resist the perturbation of the system away from equilibrium, so $\Delta W[\boldsymbol{\xi}, \boldsymbol{\xi}]$ may be interpreted as the total work done by the displacement $\epsilon\boldsymbol{\xi}$ against the force $-\epsilon\mathsf{F} \cdot \boldsymbol{\xi}$ as the time-like "switching parameter" ϵ runs from 0 to 1, providing a continuous connection between the equilibrium and perturbed states. In this pseudo-time evolution, $\partial(\epsilon\boldsymbol{\xi})/\partial\epsilon = \boldsymbol{\xi}$ plays the role of velocity and the factor 1/2 in (6.35) comes from the integral $\int_0^1 \epsilon \, d\epsilon$.

Identifying $\Delta W[\boldsymbol{\xi}]$ with an Hermitian form rather than a bilinear form has the advantage that it is real even for normal modes in the complex exponential form

$$\boldsymbol{\xi}(\mathbf{r}, t) = \boldsymbol{\xi}(\mathbf{r}) \exp(-i\omega t) \tag{6.36}$$

where $\omega \neq 0$, when Eq. (6.29) becomes

$$-\omega^2 \rho \boldsymbol{\xi} = \mathsf{F} \cdot \boldsymbol{\xi}. \tag{6.37}$$

For constant ρ Eq. (6.37) is an eigenvalue equation, with ω^2 the eigenvalue of $-\mathsf{F}/\rho$ and $\boldsymbol{\xi}(\mathbf{r})$ the eigenvector. If ρ is not constant, then (6.37) is a *generalised* eigenvalue equation,[4] which can be handled in a similar way to the ordinary eigenvalue equation by modifying the definition of inner product to include ρ as a weight function.

As in quantum mechanics, the (generalised) eigenvalue ω^2 is real because F is Hermitian, even though the eigenvector can be complex. This follows by taking a dot product of (6.37) with $\boldsymbol{\xi}^*$, integrating over the plasma volume to give

$$\omega^2 \int_{\mathcal{P}} \rho \, |\boldsymbol{\xi}|^2 \, d\tau = -\int_{\mathcal{P}} \boldsymbol{\xi}^* \cdot \mathsf{F}[\boldsymbol{\xi}] \, d\tau, \tag{6.38}$$

and then subtracting the similar result obtained from the complex conjugate of (6.37) and invoking the Hermitian property of F to obtain

$$(\omega^2 - \omega^{*2}) \int_{\mathcal{P}} \rho |\boldsymbol{\xi}|^2 \, d\tau = 0,$$

whence $\omega^2 - \omega^{*2} = 0$—i.e. ω^2 is real (though not necessarily positive). Thus the system is exponentially stable (pure oscillatory) when $\omega^2 > 0$, and exponentially unstable when $\omega^2 < 0$ (when ω is pure imaginary), with the transition from stability

[4] A generalised eigenvalue equation is of the form $Ax = \lambda Bx$.

to instability (*marginal stability*) at $\omega^2 = 0$. Another well known result due to the Hermitian property is that discrete normal modes are orthogonal, in this case with weight function ρ.

As with ΔW, we extend the kinetic energy K defined in (6.32) to an Hermitian form to allow complex normal modes in the form (6.36), by writing $K = \omega^2\mathcal{K}$ where the kinetic energy factor $\mathcal{K}$ is defined as the Hermitian form

$$\mathcal{K}[\boldsymbol{\xi}] = \frac{1}{2}\int_{\mathcal{P}} \left(\rho_\parallel|\xi_\parallel|^2 + \rho_\perp|\boldsymbol{\xi}_\perp|^2\right) d\tau. \tag{6.39}$$

For generality (cf. Sect. 6.7), in this definition we have allowed for MHD models when the density is anisotropic—i.e. $\boldsymbol{\rho} = \rho_\parallel\hat{\mathbf{b}}\hat{\mathbf{b}} + \rho_\perp\mathbf{I}_\perp$, involving parallel and perpendicular co to the anisotropic pressure tensor in Sect. 2.9.2. An example of the use of this generality is in the implementation of Freidberg's "collisionless MHD" model (cf. [6, pp. 32–38 and 260]), which corresponds to taking $\rho_\parallel = 0$.

However, in this section we are assuming $\rho_\parallel = \rho_\perp = \rho$. Dividing both sides of (6.38) by $\int\rho|\boldsymbol{\xi}|^2\,d\tau$ and comparing with (6.35) and (6.39) suggests the functional

$$\lambda[\boldsymbol{\xi}] = \frac{\Delta W[\boldsymbol{\xi}]}{\mathcal{K}[\boldsymbol{\xi}]} \tag{6.40}$$

known as the *Rayleigh quotient*, with (6.38) then the statement that $\lambda[\boldsymbol{\xi}]$ is a generalised eigenvalue ω^2 when $\boldsymbol{\xi}$ is a generalised eigenvector. The utility of the functional $\lambda[\boldsymbol{\xi}]$ derives from the principle that the first variation $\delta\lambda = (\delta\Delta W - \lambda\delta\mathcal{K})/\mathcal{K}$ vanishes for all $\delta\boldsymbol{\xi}$ if and only if $\boldsymbol{\xi}$ satisfies (6.37) with $\omega^2 = \lambda[\boldsymbol{\xi}]$. This principle allows approximate eigenvectors to be constructed by extremising (not necessarily minimising) $\lambda[\tilde{\boldsymbol{\xi}}]$ over a class of ansatz *trial functions* $\tilde{\boldsymbol{\xi}}$ in a Rayleigh–Ritz procedure, involving a superposition of basis functions with undetermined amplitudes useful for numerical work or an asymptotic approximation as in Sect. 6.7.

The Rayleigh quotient (6.40) provides a simple test of stability, given that the *infimum* $\min_{\boldsymbol{\xi}}(\lambda[\boldsymbol{\xi}])$ is bounded below but may be negative. Thus we can readily identify the *least stable* mode, including the fastest exponential instability when the infimum is less than zero. Since the denominator in (6.40) is obviously positive definite, to demonstrate instability it is sufficient to identify just one permissible displacement $\boldsymbol{\xi}$ that renders the numerator negative. On the other hand, the system is stable if and only if the numerator in (6.40) is positive definite for all permissible displacements. Indeed, we can interpret this stability criterion as the requirement that the form $\Delta W[\boldsymbol{\xi}^*, \boldsymbol{\xi}]$ be positive definite, upon introducing the $1/2$ factor to identify this form with the total perturbation in the potential energy (5.70) associated with the displacement away from equilibrium. Thus a configuration that is (locally) stable in the ideal MHD model corresponds to a minimum in the potential energy, as envisaged in Sect. 5.10.3—and in the simple case of the ball on a surface under gravity, where the ball is at the bottom of a valley.

6.5.3 Heuristic Form of ΔW

Various explicit forms of $\Delta W[\boldsymbol{\xi}, \boldsymbol{\xi}]$ have been derived. On setting (6.30) in (6.35) we have

$$\Delta W[\boldsymbol{\xi}, \boldsymbol{\xi}] = -\frac{1}{2}\int \boldsymbol{\xi}^* \cdot [\mu_0^{-1}(\nabla \times \mathbf{B}_1) \times \mathbf{B} + \mu_0^{-1}(\nabla \times \mathbf{B}) \times \mathbf{B}_1 + \nabla(\boldsymbol{\xi}\cdot\nabla p + \gamma p \nabla\cdot\boldsymbol{\xi}) - \mathbf{g}\nabla\cdot(\rho\boldsymbol{\xi})]\, d\tau, \quad (6.41)$$

where the integration is over the plasma volume. Then with the help of the identities

$$\boldsymbol{\xi}^* \cdot (\nabla \times \mathbf{B}_1) \times \mathbf{B} = \nabla\cdot[(\boldsymbol{\xi}^* \times \mathbf{B}) \times \mathbf{B}_1] - \mathbf{B}_1 \cdot \nabla \times (\boldsymbol{\xi}^* \times \mathbf{B}), \quad (6.42)$$

$$\boldsymbol{\xi}^* \cdot \nabla(\boldsymbol{\xi}\cdot\nabla p + \gamma p\nabla\cdot\boldsymbol{\xi}) = \nabla\cdot[(\boldsymbol{\xi}\cdot\nabla p + \gamma p\nabla\cdot\boldsymbol{\xi})\,\boldsymbol{\xi}^*] - (\boldsymbol{\xi}\cdot\nabla p + \gamma p\nabla\cdot\boldsymbol{\xi})\nabla\cdot\boldsymbol{\xi}^*, \quad (6.43)$$

from the Divergence Theorem (1.60) it immediately follows that

$$\Delta W[\boldsymbol{\xi}, \boldsymbol{\xi}] = \frac{1}{2}\int \Big[\mu_0^{-1}|\mathbf{B}_1|^2 - \boldsymbol{\xi}^*\cdot\mathbf{j}\times\mathbf{B}_1 + (\nabla\cdot\boldsymbol{\xi}^*)\,\boldsymbol{\xi}\cdot\nabla p + \gamma p\,|\nabla\cdot\boldsymbol{\xi}|^2 + \boldsymbol{\xi}^*\cdot\mathbf{g}\nabla\cdot(\rho\boldsymbol{\xi})\Big]\, d\tau - \frac{1}{2}\int [\,\mu_0^{-1}(\boldsymbol{\xi}^* \times \mathbf{B}) \times \mathbf{B}_1 + (\boldsymbol{\xi}\cdot\nabla p + \gamma p\nabla\cdot\boldsymbol{\xi})\,\boldsymbol{\xi}^*] \cdot d\mathbf{S} \quad (6.44)$$

where $\mathbf{B}_1(\boldsymbol{\xi}) = \nabla \times (\boldsymbol{\xi} \times \mathbf{B})$ on recalling (6.31). The surface integral vanishes if the plasma is unbounded ($\boldsymbol{\xi} \to 0$ suitably at infinity) or bounded by a rigid conductor ($\hat{\mathbf{n}} \cdot \boldsymbol{\xi} = 0$ and $\hat{\mathbf{n}} \cdot \mathbf{B} = 0$), but any negative term in the integrand of the volume integral is of course a potential source of instability. Although the third and fourth terms are zero for incompressible perturbations (where $\nabla\cdot\boldsymbol{\xi} = 0$), the second and last terms of the volume integral remain as obvious destabilising candidates. Indeed, it is the last term reduced to $\boldsymbol{\xi}^* \cdot \mathbf{g}\; \boldsymbol{\xi} \cdot \nabla\rho$ that drives the gravitational interchange instability of Sects. 6.3 and 6.4, when the respective density jump or density gradient is negative in the direction of the gravity $\mathbf{g}$.

Let us now neglect gravity (set $\mathbf{g} = 0$) and consider a tangential discontinuity in the classical magnetic confinement context (cf. Sect. 5.13), involving flux surfaces defined by $\nabla p = \mathbf{j} \times \mathbf{B}$ (cf. Sect. 5.10). The total potential energy variation given by (6.44) may then be expressed as—cf. Exercise (Q1):

$$\Delta W[\boldsymbol{\xi}, \boldsymbol{\xi}] = \Delta W_p + \Delta W_s + \Delta W_v, \quad (6.45)$$

in terms of the *plasma contribution*

$$\Delta W_p = \frac{1}{2}\int [\mu_0^{-1}|\mathbf{B}_1|^2 - \boldsymbol{\xi}_\perp^* \cdot \mathbf{j} \times \mathbf{B}_1 \\ + (\nabla \cdot \boldsymbol{\xi}_\perp^*)\,\boldsymbol{\xi}_\perp \cdot \nabla p + \gamma p\,|\nabla \cdot \boldsymbol{\xi}|^2]\, d\tau \tag{6.46}$$

involving $\mathbf{B}_1 = \nabla \times (\boldsymbol{\xi}_\perp \times \mathbf{B})$, the *surface contribution*

$$\Delta W_s = \frac{1}{2}\int |\hat{\mathbf{n}} \cdot \boldsymbol{\xi}_\perp|^2 \left[\!\left[\nabla \left(p + \frac{B^2}{2\mu_0} \right) \right]\!\right] \cdot d\mathbf{S}, \tag{6.47}$$

and the *vacuum contribution*

$$\Delta W_v = \int \frac{|\mathbf{B}_1^v|^2}{2\mu_0}\, d\tau^v. \tag{6.48}$$

Here $\boldsymbol{\xi} = \boldsymbol{\xi}_\perp + \xi_\parallel \hat{\mathbf{b}}$ where $\xi_\parallel = \boldsymbol{\xi} \cdot \hat{\mathbf{b}}$, the double brackets denote a jump discontinuity as before, and $\mathbf{B}_1^v$ is the perturbed magnetic field in the vacuum. The corresponding formulation with the gravity retained ($\mathbf{g} \neq 0$) was essentially given in the Bernstein et al. (1958) article previously cited in a footnote on p. 213.

6.5.4 *Physical Interpretation of Terms in the Plasma Contribution*

If the current density is separated into parallel and perpendicular components by writing $\mathbf{j} = \sigma \mathbf{B} + B^{-2}\mathbf{B} \times \nabla p$ where $\sigma \equiv \mathbf{j} \cdot \mathbf{B}/B^2$ (from the magnetohydrostatic equation $\nabla p = \mathbf{j} \times \mathbf{B}$ assumed here), and we also recall the magnetic field curvature vector $\boldsymbol{\kappa}$ defined in (2.81) and noted again in Sect. 5.7, the plasma contribution (6.46) may be re-expressed in a suitable form to provide physical understanding of the stabilising and destabilising terms—cf. Exercise (Q2):

$$\Delta W_p = \frac{1}{2}\int \Big[\mu_0^{-1}|\mathbf{B}_{1\perp}|^2 + \mu_0^{-1}B^{-2}\,|\mathbf{B}_1 \cdot \mathbf{B} - \mu_0\,\boldsymbol{\xi}_\perp \cdot \nabla p\,|^2 + \gamma p\,|\nabla \cdot \boldsymbol{\xi}\,|^2 \\ -2\,\boldsymbol{\xi}_\perp^* \cdot \boldsymbol{\kappa}\;\boldsymbol{\xi}_\perp \cdot \nabla p - \sigma\,\boldsymbol{\xi}_\perp^* \times \mathbf{B} \cdot \mathbf{B}_{1\perp} \Big]\, d\tau. \tag{6.49}$$

Thus the five terms in the integrand have physical interpretations as follows:

(1) $\mu_0^{-1}|\mathbf{B}_{1\perp}|^2$, magnetic field-line bending[5] term involving the perpendicular component $\mathbf{B}_{1\perp} = \mathbf{B}_1 - B_{1\parallel}\hat{\mathbf{b}}$;

---

[5]The extent to which the descriptions "field-line bending" and "field-line compression" can be rigorously justified is critiqued in Sect. 6.7.5.

(2) $\mu_0^{-1} B^{-2} |\mathbf{B}_1 \cdot \mathbf{B} - \mu_0 \boldsymbol{\xi}_\perp \cdot \nabla p|^2$, magnetic field-line compression term involving the parallel component $B_{1\parallel} = \mathbf{B}_1 \cdot \hat{\mathbf{b}}$;
(3) $\gamma p |\nabla \cdot \boldsymbol{\xi}|^2$, plasma fluid compression term;
(4) $-2 \boldsymbol{\xi}_\perp^* \cdot \boldsymbol{\kappa}\, \boldsymbol{\xi}_\perp \cdot \nabla p$, proportional to both the curvature vector and the pressure gradient, the curvature term (cf. the omitted gravity term $\boldsymbol{\xi}^* \cdot \mathbf{g}\, \boldsymbol{\xi} \cdot \nabla \rho$
(5) $-\sigma \boldsymbol{\xi}_\perp^* \times \mathbf{B} \cdot \mathbf{B}_{1\perp}$ proportional to the parallel current component, the kink term.

The first three of these terms are non-negative and therefore stabilising, and they are associated with plasma wave propagation as detailed below (cf. also Sect. 5.9). However, sufficiently negative contributions from either the fourth term or the fifth term drive instability. The degree to which displacements $\boldsymbol{\xi}(\mathbf{r})$ that isolate the field-line bending, magnetic field-line compression and curvature terms can be constructed (so as to make the interpretations above meaningful) is discussed in Sect. 6.7.5.

We now expand on the heuristic explanations of the five terms of (6.49) enumerated above[6]:

Term (1): This positive-definite stabilising term represents the magnetic energy variation associated with the Alfvén wave.

Term (2): This stabilising term may be rewritten, with the aid of the identity in Q2, in the alternative form

$$\mu_0^{-1} B^2 |\nabla \cdot \boldsymbol{\xi}_\perp + 2\boldsymbol{\xi}_\perp \cdot \boldsymbol{\kappa}|^2, \tag{6.50}$$

so approximates the energy variation associated with the fast magnetosonic wave.

Term (3): This stabilising term is the energy variation associated with the slow magnetosonic wave.

Term (4): This term may be destabilising (negative), driving magnetic interchanges in regions of unfavourable magnetic field curvature.

Term (5): This term can also be destabilising, driving current-driven or kink instabilities, in reference to their source or to the related distortion of current-carrying plasma columns.

Term (2) can be large in magnetised plasma, but for incompressible perpendicular perturbations (such that $\nabla \cdot \boldsymbol{\xi}_\perp = 0$) the field curvature remnant $4\mu_0^{-1} B^2 |\boldsymbol{\xi}_\perp \cdot \boldsymbol{\kappa}|^2$ may not completely offset the destabilising Term (4) and Term (5). Moreover, any parallel displacement component $\boldsymbol{\xi}_\parallel = \hat{\mathbf{b}}\hat{\mathbf{b}} \cdot \boldsymbol{\xi}$ contributes to Term (3) alone. In passing, we note that the fourth term only involves the perpendicular displacement component $\boldsymbol{\xi}_\perp$ but otherwise resembles the omitted gravity term, so magnetic interchange instabilities are similar to but distinct from the gravitationally driven interchanges discussed previously.

[6]We will make these interpretations more precise in Sect. 6.7 via a short-wavelength ordering.

6.5.5 Brief Note on the Eigenvalue Spectrum

The force operator F admits eigenvalue continua—i.e. not only discrete eigenvalues as discussed above. The spectrum of F (for constant ρ) depends upon whether or not $\mathsf{F}/\rho - \lambda\mathsf{I}$ can be inverted for complex λ. Indeed, the discrete spectra we assumed corresponds to λ such that $(\mathsf{F}/\rho - \lambda\mathsf{I}) \cdot \boldsymbol{\xi} = 0$ has a nontrivial solution—i.e. when $\mathsf{F}/\rho - \lambda\mathsf{I}$ is not invertible. The continua correspond to $\mathsf{F}/\rho - \lambda\mathsf{I}$ invertible but unbounded on some plasma surface [8]. However, to date it has been found that eigenvalue continua only occur on the positive real axis (i.e. such that $\omega^2 > 0$). There can be an accumulation of discrete negative eigenvalues near the origin (i.e. the point of marginal stability) when it is the lower bound of a continuum, but the normal mode analysis nevertheless determines whether or not there is exponential stability.

Exercises

(Q1) Derive the formulation (6.45)–(6.48), on setting $\mathbf{g} = 0$ in (6.41) and invoking the corresponding magnetohydrostatic equation $\nabla p = \mathbf{j} \times \mathbf{B}$.

(Q2) Show that $\mathbf{B}_1 \cdot \mathbf{B} = -2B^2\boldsymbol{\xi}_\perp \cdot \boldsymbol{\kappa} + \mu_0\, \boldsymbol{\xi}_\perp \cdot \nabla p - B^2\nabla\cdot\boldsymbol{\xi}_\perp$, and then derive the form of ΔW_p in (6.49) from (6.46).

6.6 Ideal Instabilities in Cylindrical Geometry

We recall from Sect. 5.10 that magnetohydrostatic configurations defined by $\nabla p = \mathbf{j} \times \mathbf{B}$ involve families of flux surfaces of particular interest for magnetic confinement. Let us now consider the ideal MHD magnetic interchange and kink instabilities identified in the previous section, assuming simply nested cylindrical flux surfaces. This is the configuration for plasma confinement in a current-carrying plasma cylinder, where there may be containment related to the well known mutual attraction of two conducting wires. If the plasma cylinder has circular cross-section, conventional cylindrical coordinates (r, θ, z) are suitable, with each cylindrical flux surface simply denoted by its radius. In the next section, we choose a generic curvilinear straight-field-line magnetic coordinate system to discuss toroidal flux surfaces, where each nested invariant magnetic surface is labelled by its enclosed poloidal flux. A toroidal configuration obviously introduces further magnetic field curvature, but the results obtained in simpler cylindrical geometry also considerably enhance an understanding of the ideal MHD instabilities identified in previous sections. Moreover, the analytic power of the ideal MHD variational principle for plasma stability analysis is evident in both non-Cartesian geometries.

In a cylindrical plasma configuration with pressure $p(r)$ and magnetic field $\mathbf{B}(r) = B_\theta(r)\hat{\mathbf{e}}_\theta + B_z(r)\hat{\mathbf{e}}_z$ corresponding to a current density $\mathbf{j}(r) = j_\theta(r)\hat{\mathbf{e}}_\theta + j_z(r)\hat{\mathbf{e}}_z$, the magnetohydrostatic pressure balance equation $\nabla p = \mathbf{j} \times \mathbf{B}$ becomes

$$\frac{dp}{dr} + \frac{B_\theta}{\mu_0 r}\frac{d}{dr}(rB_\theta) + \frac{B_z}{\mu_0}\frac{dB_z}{dr} = 0. \tag{6.51}$$

The current-carrying plasma region may be a column of radius a, but the plasma current may reverse or flow in an annular cross-section, and there could also be a surrounding current-free region. In any case, it is evident that the current-carrying plasma may be compressed by axial or azimuthal current flows, the fundamental idea behind pinch and θ-pinch operation. The components of the plasma displacement $\boldsymbol{\xi}(\mathbf{r}, t) = \xi_r\hat{\mathbf{e}}_r + \xi_\theta\hat{\mathbf{e}}_\theta + \xi_z\hat{\mathbf{e}}_z$ may be represented via normal modes of the form

$$f(r, \theta, z, t) = f(r)\exp[i(m\theta + k_z z - \omega t)]. \tag{6.52}$$

Indeed, the ends of a certain length L of the cylinder may also be identified such that $k_z = -2\pi n/L$ where n is an integer, so the modes are specified by the integer pair (m, n). Toroidal topology is therefore addressed, but not yet toroidal curvature.

However, it is convenient to adopt the dependent variable set[7]

$$\begin{aligned} \xi &= \hat{\mathbf{e}}_r \cdot \boldsymbol{\xi} = \xi_r, \\ \eta &= i\,\mathbf{k}\cdot\boldsymbol{\xi} = i\left(\frac{m}{r}\xi_\theta + k_z\xi_z\right) = \nabla\cdot\boldsymbol{\xi} - \frac{1}{r}\frac{d}{dr}(r\xi_r), \\ \zeta &= i\,(\boldsymbol{\xi}\times\mathbf{B})\cdot\hat{\mathbf{e}}_r = i\,(\xi_\theta B_z - \xi_z B_\theta). \end{aligned} \tag{6.53}$$

Thus for a fixed plasma boundary (when the surface term is zero) and omitting the gravity term, on excluding the case $m = n = 0$ and dropping the exponential factor we obtain from (6.44)

$$\begin{aligned} \Delta W_{m,n}[\xi, \eta, \zeta] = \frac{\pi}{2}\int r\Bigg[\Lambda\left(\xi, \frac{d\xi}{dr}\right) + \gamma p\left|\eta + \frac{1}{r}\frac{d}{dr}(r\xi)\right|^2 \\ + \mu_0^{-1}(m^2 + k_z r^2)\left|\zeta - \zeta_0(\xi, \frac{d\xi}{dr})\right|^2 / r^2\Bigg]\,dr \end{aligned} \tag{6.54}$$

per unit length of cylinder, where—cf. Exercise (Q1):

$$\begin{aligned} \Lambda\left(\xi, \frac{d\xi}{dr}\right) = \mu_0^{-1}(m^2 + k_z^2 r^2)^{-1}\left|(mB_\theta + k_z r B_z)\frac{d\xi}{dr} - (mB_\theta - k_z r B_z)\frac{\xi}{r}\right|^2 \\ + \mu_0^{-1}\left[(mB_\theta + k_z r B_z)^2 - 2\,B_\theta\frac{d}{dr}(rB_\theta)\right]\frac{|\xi|^2}{r^2} \end{aligned}$$

and

$$\zeta_0\left(\xi, \frac{d\xi}{dr}\right) = -(m^2 + k_z^2 r^2)^{-1} r\left[(mB_z - k_z r B_\theta)\frac{d\xi}{dr} + (mB_z + k_z r B_\theta)\frac{\xi}{r}\right].$$

[7] We follow the notation used in the elegant classical article by W.A. Newcomb (*Annals of Physics* **10**, 232–267, 1960).

It follows immediately that there is ideal MHD stability if

$$B_\theta \frac{d}{dr}(r B_\theta) \le 0 \; \forall r \text{ (i.e. on all cylindrical flux surfaces)}, \tag{6.55}$$

but this quite stringent sufficiency condition is not readily satisfied—although it is in the hard-core (inverse) pinch, where the plasma current of near annular cross-section is returned through a central conductor.

The non-negative terms in η and ζ in (6.54) vanish on setting

$$\eta = -\frac{1}{r}\frac{d}{dr}(r\xi) \text{ and } \zeta = \zeta_0\left(\xi, \frac{d\xi}{dr}\right),$$

such that the functional for further consideration is

$$\Delta W[\xi] = \mathrm{Min}_{\eta,\zeta}\, \Delta W_{m,n}(\xi, \eta, \zeta) = \frac{\pi}{2}\int r\Lambda\left(\xi, \frac{d\xi}{dr}\right) dr. \tag{6.56}$$

It is notable that this choice of η corresponds to an incompressible displacement ($\nabla\cdot\boldsymbol{\xi} = 0$), annulling the plasma fluid compression contribution in (6.49) associated with the slow magnetosonic wave. Under the boundary condition $\xi = 0$ at the fixed plasma boundary, on integrating by parts we obtain

$$\Delta W[\xi] = \frac{\pi}{2}\int \left[f(r)\left|\frac{d\xi}{dr}\right|^2 + g(r)|\xi|^2 \right] dr \tag{6.57}$$

where $f(r) = \mu_0^{-1} r (m^2 + k_z^2 r^2)^{-1} (m B_\theta + k_z r B_z)^2$,

and
$$\begin{aligned} g(r) &= (m^2 + k_z^2 r^2)^{-1}\, 2k_z^2 r^2 \frac{dp}{dr} \\ &+ \mu_0^{-1}(m^2 + k_z^2 r^2)^{-1}(m^2 + k_z^2 r^2 - 1)(m B_\theta + k_z r B_z)^2/r \\ &- \mu_0^{-1}(m^2 + k_z^2 r^2)^{-2}\, 2k_z^2 r\,(m^2 B_\theta^2 - k_z^2 r^2 B_z^2). \end{aligned}$$

The function $f(r)$ is non-negative, but it is zero at radii such that $\mathbf{k}\cdot\mathbf{B} = m B_\theta/r + k_z B_z = 0$—i.e. on a *(mode) rational flux surface*, which occurs at a radius r_s such that $q(r_s)$ is the rational fraction m/n. The safety factor $q(r_s)$, defined generally in (5.77), is given more explicitly in the cylindrical case by

$$q(r) = h(r)/L = 1/\,\iota\!\!\!{}^{-}(r) = \frac{2\pi r B_z(r)}{L B_\theta(r)}. \tag{6.58}$$

It characterises the pitch $h(r)$ and rotational transform $\iota\!\!\!{}^{-}(r)$ of the magnetic field lines on the flux surface of radius r (cf. also Sect. 5.11). Instability corresponds to contributions to the integrand in (6.57) from intervals where $g(r) < 0$ (such that

$\Delta W[\xi] < 0$), notably in the neighbourhood of the rational flux surfaces where $f(r)$ is small.

Normal modes satisfying the "flute condition" $\mathbf{k} \cdot \mathbf{B} = k_\parallel B \simeq 0$ for the approximate vanishing of $f(r)$ in a region of the plasma around a rational surface (cf. also Sects. 6.3, 6.4) and 6.7 are often called *flute-like modes*. This terminology reflects an architectural analogy illustrated in Fig. 6.1, as the associated wave perturbations $\boldsymbol{\xi}$ deform the rational surface into a shape resembling a fluted column where the waves (flutes) are approximately aligned with the helical magnetic field lines.

The $\{r_s\}$ are the resonance values in cylindrical geometry, previously noted in the elementary discussion of magnetic field shear in Cartesian geometry in Sect. 6.4. They are singular points of the Sturm–Liouville equation for the displacement extremal obtained from the Rayleigh quotient (6.40). If no such resonance value exists, it follows from Sturm–Liouville theory that the least eigenvalue is positive, so the system is stable. The $\{r_s\}$ are likewise singular points of the Euler–Lagrange equation in the plasma displacement ξ that minimises $\Delta W[\xi]$ in (6.57)—viz.

$$\frac{d}{dr}\left(f(r)\frac{d\xi}{dr}\right) - g(r)\,\xi = 0. \tag{6.59}$$

In the neighbourhood of r_s, on defining $x = r - r_s$ we have $f(r) = f''(r_s)x^2/2 + \mathrm{O}(x^3)$ with $f''(r_s) > 0$ provided $q'(r_s) \neq 0$; and $g(r) = g(r_s) + \mathrm{O}(x)$ with $g(r_s) \neq 0$

Fig. 6.1 Spiral fluted columns in the Great Colonnade at Apamea in Syria (Wikimedia Commons). Plasma waves with phase fronts aligned with the field lines on a magnetic surface are termed "flute modes" by analogy

provided $p'(r_s) \neq 0$—i.e. provided the magnetic field is not force-free at the singular surface. Thus near r_s the displacement satisfies—cf. Exercise (Q2):

$$x^2\frac{d^2\xi}{dx^2}+2x\frac{d\xi}{dx}+D_s\xi=0 \tag{6.60}$$

where the parameter $D_s = -2g(r_s)/f''(r_s)$. This classical Euler equation has a regular singularity at $x = 0$, and the familiar trial solution $\xi = |x|^\alpha$ yields the indicial equation $\alpha(\alpha-1)+2\alpha+D_s=0$ with roots $\alpha_{1,2}=(-1\pm\sqrt{1-4D_s})/2$ such that the solution may be irregular near the resonance value r_s. Thus if $D_s > 1/4$ the general solution of (6.60) for $x \neq 0$ is

$$\xi(x)=C_\xi|x|^{-1/2}\cos\left(\frac{1}{2}\sqrt{4D_s-1}\ln|x|+\phi\right)$$

where C_ξ and ϕ are constants—and this solution is highly oscillatory as $x \to 0$, violating a basic requirement of the classical calculus of variations used to obtain (6.60). However, reference back to the original form (6.57) shows that (in an interval where $g(r) < 0$) a localised displacement such that $\xi \simeq$ constant in the immediate neighbourhood of $x = 0$ is unstable—i.e. $\Delta W[\xi] < 0$. The $D_s < 1/4$ requirement for a regular solution renders the necessary condition for stability against ideal interchange modes in a negative pressure gradient, a result due to Suydam that may be expressed as

$$\frac{dp}{dr}+\frac{rB_z^2}{8\mu_0}\left(\frac{q'}{q}\right)^2>0 \tag{6.61}$$

everywhere, supplemented by

$$\frac{dp}{dr}+\frac{r}{8\mu_0}\left(\frac{dB_z}{dr}\right)^2>0 \tag{6.62}$$

at any zero of B_z for $m = 0$ modes. The second term in each case represents the stabilisation due to the magnetic field shear, as previously introduced in Cartesian geometry in Sect. 6.4, but which is now the variation in the magnetic field direction over the family of nested cylindrical flux surfaces (i.e. with radius r).[8]

In addition to the localised magnetic interchange modes just discussed, associated with the plasma pressure gradient and unfavourable magnetic field curvature, let us recall from Sect. 6.5 that there may also be current-driven instabilities. Indeed, the

[8]When (6.60) is valid its singular solutions can be represented by square integrable generalised functions with supports on either side of the singularity (i.e. on $x > 0$ or $x < 0$), where each single-sided generalised function corresponds to the classical solution on one side of the singularity.

function $g(r)$ is quadratic in the variable $u = -k_z r B_z/B_\theta = nq(r)$ and negative between the two real roots

$$u_{1,2} = E^{-1}\left(mA \pm \sqrt{4m^2k_z^2r^4 - 2k_z^2r^2(m^2 + k_z^2r^2)E\mu_0 r(dp/dr)/B_\theta^2}\right),$$

where $A = m^2(m^2-1)+(2m^2-1)k_z^2r^2+k_z^4r^4$ and $E = m^2(m^2-1)+(2m^2+1)$ $k_z^2r^2+k_z^4r^4$. By an appropriate choice of ξ, it follows that there are internal current-driven modes for the narrow range $m(1-4k_z^2)r_s^2E^{-1} < nq(r_s) < m$. However, it was soon confirmed that free boundary contributions to ΔW drive more dangerous external kink modes, which had been anticipated earlier on physical grounds. These current-driven kink modes produce large-scale displacements of the plasma cylinder.

Let us consider a plasma column of radial extent $0 \le r < a$ carrying the current $\mathbf{j}(r) = j_\theta(r)\,\hat{\mathbf{e}}_\theta + j_z(r)\,\hat{\mathbf{e}}_z$, surrounded by current-free (and therefore magnetically force-free) plasma in the second region $a < r < b$ extending to a co-axial conducting wall at $r = b$. The current-carrying plasma contribution must again be supplemented, because the interface at $r = a$ is not fixed. Equation (6.59) is recovered from the volume integral in (6.44) over the current-carrying plasma region when the gravity term is neglected, but integration by parts with the condition $\xi(a) = \xi_a$ at the displaced interface now yields the additional contribution

$$\Delta W_{\rm surf} = \frac{\pi}{2\mu_0}\frac{[mB_\theta(a)+k_zaB_z(a)]^2 - 2mB_\theta(a)[mB_\theta(a)+k_zaB_z(a)]}{m^2+k_z^2a^2}\,\xi_a. \tag{6.63}$$

On the other hand, the contribution (6.47) previously obtained for a plasma–vacuum interface is now zero because the magnetic field is continuous at the plasma-plasma interface $r = a$, but using $\nabla\cdot\mathbf{B} = 0$ and $\nabla\times\mathbf{B} = 0$ in the surface integral in (6.44) to eliminate the perturbation components $B_{1\theta}$ and B_{1z} in the surface integral produces the further additional contribution

$$\Delta W_{\rm region\,2} = \frac{\pi}{2\mu_0}\int_a^b \left[\frac{1}{m^2+k_z^2r^2}\left|\frac{d\eta}{dr}\right|^2 + \frac{|\eta|^2}{r^2}\right] dr$$

where $\eta(r) \equiv -irB_{1r} = (mB_\theta + k_zrB_z)\,\xi$. Minimising $\Delta W_{\rm region\,2}$ subject to the boundary conditions $\eta(a) = [mB_\theta(a)+k_zaB_z(a)]\xi_a$ and $\eta(b) = 0$ yields the relevant Euler–Lagrange equation

$$\frac{d}{dr}\left(\frac{r}{m^2+k_z^2r^2}\frac{d\eta}{dr}\right) - \frac{\eta}{r} = 0, \quad a < r < b. \tag{6.64}$$

In the "tokamak limit" $m^2 \gg k_z^2r^2$ ($|B_\theta|^2 \ll |B_z|^2$), this equation reduces to

$$r^2\frac{d^2\eta}{dr^2} + r\frac{d\eta}{dr} - m^2\eta = 0,$$

when the extremal satisfying the two boundary conditions is

$$\eta(r) = \left(1 - \frac{a^{2m}}{b^{2m}}\right)^{-1} \xi_a\, B_\theta(a)\, m \left(1 - \frac{nq(a)}{m}\right)\left(\frac{b^m}{r^m} - \frac{r^m}{b^m}\right)\frac{a^m}{b^m}$$

as $nq(a) = -k_z a B_z(a)/B_\theta(a)$. The minimum of this further additional contribution is therefore

$$\mathrm{Min}\{\Delta W_{\text{region 2}}\} = \frac{\pi}{2\mu_0}\left(1 - \frac{nq(a)}{m}\right)^2 m\,\lambda\,\xi_a^2, \tag{6.65}$$

where $\lambda \equiv (1 + a^{2m}/b^{2m})/(1 - a^{2m}/b^{2m})$. Hence the total variation per unit length to consider is

$$\Delta W = \Delta W_{\text{region 1}}$$
$$+ \frac{\pi}{2\mu_0}\xi_a^2 B_\theta(a)^2 \left[\left(1 - \frac{nq(a)}{m}\right)^2 (1 + m\lambda) - 2\left(1 - \frac{nq(a)}{m}\right)\right], \tag{6.66}$$

when the two contributions from (6.63) and (6.65) are combined. We recall the first term $\Delta W_{\text{region 1}}$ arising from the current-carrying region $0 < r < a$ can only be negative in the very narrow range of $q(r)$ values defined by $m(1 - 4k_z^2)r_s^2 E^{-1} < nq(r_s) < m$, and when $|B_\theta|^2 \ll |B_z|^2$ it is also relatively small. However, there can now be kink modes due to the second term in (6.66) if $q(a) < m/n$. The stability condition $q(a) > 1$, which was deduced by Kruskal and independently by Shafranov, corresponds to the fundamental mode $m = n = 1$.

It may seem odd to abandon the notion of a jump in the magnetic field at a plasma–vacuum interface envisaged for various magnetic confinement devices (cf. Sect. 5.13), which can be addressed using the separate energy contributions for the plasma, surface and vacuum (cf. Sect. 6.5). However, the replacement of the vacuum by a current-free plasma clearly has a physical basis, for there is often a hot current-carrying plasma core surrounded by a relatively cold current-free plasma. Freidberg reports that the ideal MHD stability for the external vacuum case is identical if there is no singular surface in the current-free plasma—but if there is, then the current-free plasma case is more stable, with the conducting wall effectively moved inward to the singular surface—cf. [6, pp. 264–266]. Reference may also be made to Chap. 9 in Ref. [8], for further discussion of the ideal MHD stability of cylindrical plasmas—e.g. the classical result that axisymmetric modes (where $m = 0$, including the degenerate mode $m = n = 0$) are all ideal MHD stable if there is stability when $m = 0$ and $k_z \to 0$.

Exercise

(Q1) Show that

$$\mathbf{B}_1 = i\left(\frac{m}{r}B_\theta + k_z B_z\right)\xi\,\hat{\mathbf{e}}_r + \left(k_z\zeta - \frac{d(\xi B_\theta)}{dr}\right)\hat{\mathbf{e}}_\theta - \left(\frac{1}{r}\frac{d(r\xi B_z)}{dr} + \frac{m}{r}\zeta\right)\hat{\mathbf{e}}_z.$$

Then noting (6.51), develop the integrand in the volume integral component of (6.44) in terms of ξ and ζ and hence verify (6.54).

6.7 Ideal Instabilities in Toroidal Geometry

Stability analysis in the toroidal geometry discussed in Chap. 5 is complicated by the fact that there is only one continuous spatial symmetry coordinate (the toroidal angle ζ in the axisymmetric case) or none (in the most general toroidal geometry). However, if we are considering the linear theory for an equilibrium state, the time-translation symmetry that allows normal mode analysis [Glossary Section "Normal Mode"] remains, but separation of variables cannot be used in toroidal geometry to factor an eigenfunction into a product of simple Fourier harmonics and a nontrivial function of only one variable—unlike the cylindrical case, where we have the form (6.52). In the axisymmetric toroidal case, the toroidal mode number n remains a "good quantum number", but the θ and s dependence must be represented by an infinite Fourier sum over the poloidal mode number m with s-dependent Fourier coefficients. In the general non-axisymmetric case, both the m and n Fourier harmonics become coupled and the normal mode spectrum is no longer simply classifiable, instead becoming "quantum chaotic" [13].

Nevertheless, as foreshadowed in Sect. 6.2, some analytic progress can be made using asymptotic methods—and in this section we develop a short-wavelength asymptotic ordering scheme, adapted to the special needs of MHD stability analysis. However, we cannot simply use the approach of Sect. 4.2, where the direction of $\mathbf{k}$ is arbitrary. To leading order in the slow-variation parameter ϵ (the ratio of wavelength to equilibrium scale length L), this is locally the plane wave case analysed in Sect. 5.9, and we recall from (5.61) that all three branches of the ideal MHD dispersion relation have $\omega^2 \geq 0$.[9] Thus the general approach is valid for approximating the high-frequency part of the MHD spectrum, but it cannot determine instability. The plane-wave dispersion relations do provide a clue as to where to look—viz. in the neighbourhood of *marginal stability* $\omega^2 = 0$, where small negative corrections can drive ω^2 negative. From (5.61a), we see that $\omega^2 = 0$ when $k_\parallel = 0$ for the Alfvén branch; and from (5.61b), for the slow magnetosonic branch $\omega^2 = 0$ when $\alpha = 0$, which *also* occurs only when $k_\parallel = 0$—cf. (5.62). On the other hand, ω^2 never vanishes on the fast magnetosonic branch, so unstable perturbations cannot include a fast magnetosonic component. For small $k_\parallel$, we may eliminate α from (5.61) and also simplify the eigenvectors of (5.58) to give the approximate dispersion relations and

[9] The prime on ω can be dropped, as we are considering only the stability of static equilibria ($\mathbf{v} = 0$ at equilibrium).

polarisations of the two branches (Alfvén and slow magnetosonic) that are important near marginal stability—viz. (cf. Exercise on pp. 174–175)

$$\rho\omega_{\mathrm{A}}^2 = \frac{(\mathbf{k}\cdot\mathbf{B})^2}{\mu_0}, \qquad \boldsymbol{\xi}_{\mathrm{A}} = \hat{\mathbf{k}}\times\hat{\mathbf{b}}, \tag{6.67a}$$

$$\rho\omega_{\mathrm{S}}^2 = \frac{\gamma p\,(\mathbf{k}\cdot\hat{\mathbf{b}})^2}{1+\frac{1}{2}\gamma\beta}, \qquad \boldsymbol{\xi}_{\mathrm{S}} = \left(1+\frac{\gamma\beta}{2}\right)\hat{\mathbf{b}} - \frac{\gamma\beta}{2}\frac{k_\parallel \mathbf{k}_\perp}{k_\perp^2}, \tag{6.67b}$$

where $\beta = 2\mu_0 p/B^2$ and the polarisation vectors $\boldsymbol{\xi}_{\mathrm{A}}$ and $\boldsymbol{\xi}_{\mathrm{S}}$ indicate direction only (they are dimensionless).

For a potentially unstable arbitrary perturbation $f(\mathbf{r}, t)$, these considerations are consistent with the following *generalised flute ordering*[10] with respect to the formal asymptotic expansion parameter $\epsilon \to 0$:

$$L \text{ and } |f|^{-1}\nabla_\parallel f \text{ are both } \mathrm{O}(1), \tag{6.68a}$$

$$|f|^{-1}\nabla_\perp f \quad \text{is} \quad \mathrm{O}(\epsilon^{-1}), \tag{6.68b}$$

$$B \text{ and } \boldsymbol{\xi}_\perp \text{ and } \xi_\parallel \text{ are all } \mathrm{O}(1), \tag{6.68c}$$

where $\nabla_\parallel = \hat{\mathbf{b}}\hat{\mathbf{b}}\cdot\nabla$ and $\nabla_\perp = \nabla - \nabla_\parallel$ as in (5.32). Note that f is not necessarily the single wave that guided the choice of ordering, for it may also be any perturbation that could in principle be represented as a superposition of Alfvén and slow magnetosonic waves. In the following discussion, we extend the wave terminology *longitudinal* to mean a vector field that is curl-free to leading order in ϵ, and the terminology *transverse* to mean a vector field that is divergence-free to leading order in ϵ. It is readily verified that these definitions are consistent in the eikonal representation (4.7), involving the usual application of these terms to wave fields with vector amplitudes (polarisations) parallel or transverse to $\mathbf{k}$, respectively.

This ordering will allow us to develop reduced forms of the potential energy ΔW given by (6.49)[11] in terms of one or two scalar fields, leading to the socalled *ballooning equation*. This equation describes flute-like modes with strong variation along the magnetic field lines (ballooning modes), and also modes with weak variation (Mercier modes) that are the toroidal generalisation of the interchange modes discussed in Sect. 6.6.

---

[10] The architectural analogy with fluted columns from the cylindrical case treated in the previous section (cf. Fig. 6.1) is carried over to the toroidal case, so waves satisfying the ordering $k_\parallel/k_\perp = \mathrm{O}(\epsilon)$ are termed "flute-like".

[11] In this section, only the plasma component of the energy variation under the fixed boundary assumption is considered, but we choose to drop the suffix p.

6.7.1 Reduced $\mathcal{K}$ and ΔW in the Generalised Flute Ordering

Since the Alfvén and slow magnetosonic frequencies are proportional to $k_{\parallel}$, the ordering (6.68a) is necessary for them to be finite in the generalised flute ordering. However, as previously mentioned there is no propagation direction such that the frequency of the fast magnetosonic wave is finite in this limit, so the fast magnetosonic branch must be excluded in the lowest order approximation. Consequently, although both parallel and perpendicular components of $\boldsymbol{\xi}$ are allowed under (6.67a) and (6.67b), the exclusion of the primarily longitudinal fast magnetosonic wave constrains $\boldsymbol{\xi}$ to be a transverse field and thus leads to the orderings

$$\boldsymbol{\xi} \text{ and } \nabla\cdot\boldsymbol{\xi} \text{ and } \nabla\cdot\boldsymbol{\xi}_{\perp} \text{ are all O(1)}, \tag{6.69a}$$

$$\mathbf{B}_{1\perp} \text{ and } B_{1\parallel} \text{ are both O(1)}. \tag{6.69b}$$

It is notable that these last quantities would be $O(\epsilon^{-1})$ if the fast magnetosonic mode were admitted—and as a final check, the orderings (6.68) and (6.69) render $\omega^2 = O(1)$ in the Rayleigh quotient variational form (6.40).

Under (6.67), the motion perpendicular to $\mathbf{B}$ is dominated by the Alfvén branch, which has transverse polarisation—but the slow magnetosonic wave provides a small perpendicular component through the second term in (6.67b), which has longitudinal polarisation. This structure is achieved through the representation

$$\boldsymbol{\xi} = \eta\mathbf{B} + \frac{\mathbf{B}\times\nabla\varphi}{B^2} - [\nabla\chi]^{(1)}, \tag{6.70}$$

where we flag subdominant $O(\epsilon)$ but nevertheless important terms through the notation $[\cdot]^{(1)}$. This representation uses $\eta = O(1)$ to represent the dominant component of the parallel displacement $\xi_{\parallel} = \eta\mathbf{B} + O(\epsilon^2)$, a stream-like function $\varphi = O(\epsilon)$ for the dominant transverse displacement, and a potential $\chi = O(\epsilon^2)$ to provide the subdominant longitudinal component. Thus the representation is complete and in principle exact, where the variable set $\{\eta, \varphi, \chi\}$ provides the three degrees of freedom needed to replace the three components of $\boldsymbol{\xi}$.

We note that η, φ and χ are assumed to satisfy the ordering (6.68b), with the $O(\epsilon^2)$ discrepancy between $\xi_{\parallel}$ and $\eta\mathbf{B}$ arising from $\hat{\mathbf{b}}\cdot\nabla\chi$, and that $|\boldsymbol{\xi}_{\perp}| = |\hat{\mathbf{b}}\times\nabla_{\perp}\varphi/B - \nabla_{\perp}\chi| = |\nabla\varphi|/B + O(\epsilon)$. Another consequence is that all of the terms in the divergence

$$\nabla\cdot\boldsymbol{\xi} = \mathbf{B}\cdot\nabla\eta + \nabla\times\left(\frac{\mathbf{B}}{B^2}\right)\cdot\nabla\varphi - \nabla^2\chi \tag{6.71}$$

from (6.70) are O(1) and so consistent with (6.69a), with $\nabla\cdot\boldsymbol{\xi}_{\perp}$ the sum of the second and third terms to leading order.

Using the representation (6.70), to lowest order the kinetic energy factor (6.39) becomes

$$\mathcal{K}^{(0)}[\eta, \varphi] = \frac{1}{2}\int_{\mathcal{P}} \left(\rho_{\|} B^2|\eta|^2 + \rho_{\perp}\frac{|\nabla\varphi|^2}{B^2}\right) d\tau, \tag{6.72}$$

where the superscript (0) is introduced to emphasise that this Hermitian form is $O(\epsilon^0)$.

The magnetic field perturbation $\mathbf{B}_1 = \nabla\times(\boldsymbol{\xi}\times\mathbf{B})$ at (6.31) may now be determined by first noting from (6.70) that

$$\boldsymbol{\xi}\times\mathbf{B} = \nabla\varphi + [\mathbf{B}\times\nabla\chi - \hat{\mathbf{b}}\hat{\mathbf{b}}\cdot\nabla\varphi]^{(1)}, \tag{6.73}$$

whence

$$\begin{aligned}\mathbf{B}_1 &= \frac{\mathbf{B}}{B^2}\times\nabla(\mathbf{B}\cdot\nabla\varphi) + \mathbf{B}\nabla^2\chi \\ &+ \left[\nabla\chi\cdot\nabla\mathbf{B} - \mathbf{B}\cdot\nabla\nabla\chi - \left(\nabla\times\frac{\mathbf{B}}{B^2}\right)\mathbf{B}\cdot\nabla\varphi\right]^{(1)},\end{aligned} \tag{6.74}$$

so that the perpendicular and parallel components of $\mathbf{B}_1$ are both O(1) as required by (6.69b).

Although χ does not contribute to $\boldsymbol{\xi}$ at leading order, we note that it *does* make leading-order contributions to $\mathbf{B}_1$ in (6.74) and to $\nabla\cdot\boldsymbol{\xi}$ in (6.71), and thus to ΔW in (6.49), which we can therefore write to leading order in ϵ as

$$\begin{aligned}\Delta W^{(0)} = \frac{1}{2}\int_{\mathcal{P}}\Bigg[&\frac{|\nabla(\mathbf{B}\cdot\nabla\varphi)|^2}{\mu_0 B^2} + \frac{|B^2\nabla\cdot\boldsymbol{\xi}_\perp + 2\boldsymbol{\kappa}\times\mathbf{B}\cdot\nabla\varphi|^2}{\mu_0 B^2} \\ &+ \gamma p\left|\nabla\cdot\boldsymbol{\xi}_\perp + \mathbf{B}\cdot\nabla\eta\right|^2 \\ &- \frac{(\boldsymbol{\kappa}\times\mathbf{B}\cdot\nabla\varphi^*)(\nabla p\times\mathbf{B}\cdot\nabla\varphi)}{B^4} \\ &- \frac{(\nabla p\times\mathbf{B}\cdot\nabla\varphi^*)(\boldsymbol{\kappa}\times\mathbf{B}\cdot\nabla\varphi)}{B^4}\Bigg] d\tau.\end{aligned} \tag{6.75}$$

The first, second and third terms in this form precisely correspond to Terms enumerated in Sect. 6.5.4—viz. (1) field-line bending, (2) field-line compression (6.50), and (3) fluid compression in (6.49), while Term (4) represents the curvature term symmetrised to demonstrate that $\Delta W^{(0)}$ is an Hermitian form. (The other terms are already clearly Hermitian, since the ∇ is an anti-Hermitian operator, as is $\mathbf{B}\cdot\nabla$ because $\nabla\cdot\mathbf{B} = 0$.) Term (5), the kink term, does not contribute at leading order and is therefore omitted.[12]

---

[12] On the symmetrisation of the curvature term and the question of whether or not to retain the kink term, see R.L. Dewar (Journal of Plasma Fusion Research **73**, 1123, 1997).

We now proceed to define a reduced Rayleigh quotient

$$\lambda^{(0)}[\xi_{\parallel}, \varphi, \chi] = \frac{\Delta W^{(0)}}{\mathcal{K}^{(0)}} \tag{6.76}$$

in lieu of (6.40), and use the Rayleigh variational principle to deduce a reduced eigenvalue problem for the lowest eigenvalue.

6.7.2 2-Field Reduction: Elimination of χ

In (6.75), χ is "hidden" in the field-line and fluid compression terms through $\nabla \cdot \boldsymbol{\xi}_{\perp}$ in (6.71). Rather than vary χ explicitly, we can treat $\nabla \cdot \boldsymbol{\xi}_{\perp}$ as the independent scalar field with respect to which $\lambda^{(0)}$ is minimised, with χ varied only implicitly.[13] Since χ does not appear in $\mathcal{K}^{(0)}$, we need to minimise only $\Delta W^{(0)}$—and its variation with respect to $\nabla \cdot \boldsymbol{\xi}_{\perp}$ vanishes if the Euler–Lagrange equation

$$\left(\frac{B^2}{\mu_0} + \gamma p\right) \nabla \cdot \boldsymbol{\xi}_{\perp} = -\frac{2\boldsymbol{\kappa} \times \mathbf{B} \cdot \nabla \varphi}{\mu_0} - \gamma p \mathbf{B} \cdot \nabla \eta \tag{6.77}$$

is satisfied. We can thus eliminate χ in (6.75) using (6.77), to reduce $\Delta W^{(0)}$ (and hence $\lambda^{(0)}$) to a functional in only η and φ—viz.

$$\begin{aligned} \Delta W^{(0)}[\eta, \varphi] = \frac{1}{2} \int_{\mathcal{P}} d^3x \Bigg[& \frac{|\nabla (\mathbf{B} \cdot \nabla \varphi)|^2}{\mu_0 B^2} \\ & + \gamma p \left(1 + \frac{\gamma \beta}{2}\right) |\nabla \cdot \boldsymbol{\xi}|^2 \\ & - \frac{(\nabla p \times \mathbf{B} \cdot \nabla_{\perp} \varphi^*)(\boldsymbol{\kappa} \times \mathbf{B} \cdot \nabla_{\perp} \varphi)}{B^4} + \text{c.c.} \Bigg], \end{aligned} \tag{6.78}$$

where the term in $\beta = 2\mu_0 p / B^2$ arises from the field-line compression term in (6.75). Since $\beta = o(1)$ in typical fusion experiments, the compressional effects are mainly dominated by the fluid term, so there is no significant error in setting the field-line compressibility term in (6.49) to zero from the outset. Indeed, the assumption of magnetic field-line incompressibility has been found to give accurate computational results for various toroidal devices, even for low-n modes—and decreased memory

[13] The existence of $\delta\chi$ for an arbitrary $\delta\nabla \cdot \boldsymbol{\xi}_{\perp}$ is justified by the invertibility of the Laplacian operator ∇^2 in (6.71).

requirements and increased computational speed under the two-field reduction of the Rayleigh quotient. In (6.78), we have

$$\nabla\cdot\boldsymbol{\xi} = \frac{1}{1+\frac{1}{2}\gamma\beta}\left[\mathbf{B}\cdot\nabla\eta - \frac{2\boldsymbol{\kappa}\times\mathbf{B}\cdot\nabla\varphi}{B^2}\right], \tag{6.79}$$

the reduced form after elimination of χ from (6.71).

6.7.3 1-Field Reduction: Elimination of $\xi_\parallel$

We observe that η only appears in the stabilising combined fluid and magnetic-field compression term in (6.78)—cf. (6.79). Consequently, the stationarity of $\lambda^{(0)}(\eta, \varphi)$ under arbitrary variations of η^* yields the Euler–Lagrange equation

$$-\hat{\mathbf{b}}\cdot\nabla\frac{\gamma p}{1+\frac{1}{2}\gamma\beta}\left[\mathbf{B}\cdot\nabla\eta - \frac{2\boldsymbol{\kappa}\times\mathbf{B}\cdot\nabla_\perp\varphi}{B^2}\right] = \omega^2\rho_\parallel B\eta \tag{6.80}$$

recalling that $\boldsymbol{\rho} = \rho_\parallel\hat{\mathbf{b}}\hat{\mathbf{b}} + \rho_\perp\mathbf{I}_\perp$, or on gathering terms involving η on the left:

$$D_S(\omega^2)\eta = -\frac{1}{B^2}\mathbf{B}\cdot\nabla\left(\frac{2\gamma p}{1+\frac{1}{2}\gamma\beta}\frac{\boldsymbol{\kappa}\times\mathbf{B}\cdot\nabla_\perp\varphi}{B^2}\right), \tag{6.81}$$

where the *slow magnetosonic wave operator*

$$D_S(\omega^2) = \frac{1}{B^2}\mathbf{B}\cdot\nabla\frac{\gamma p}{1+\frac{1}{2}\gamma\beta}\mathbf{B}\cdot\nabla + \omega^2\rho_\parallel \tag{6.82}$$

with the operators $\mathbf{B}\cdot\nabla$ acting on everything to their right. This slow wave magnetosonic operator may be shown to be singular for some $\omega^2 \geq 0$ on each magnetic surface, and produces the *slow magnetosonic continuum* in the MHD spectrum. However, in this section we only consider the unstable case $\omega^2 < 0$, and in principle solve (6.81) in terms of the inverse operator D_S^{-1} to give η as a functional of φ—thus reducing both $\mathcal{K}^{(0)}$ and $\Delta W^{(0)}$ to functionals in φ only.

This 1-field reduction might be mainly of academic or numerical interest, if it were not for the special case $\rho_\parallel = 0$[14] corresponding to the "collisionless MHD" model that Freidberg argues may be more appropriate to describe the physics of high-temperature plasmas than conventional MHD—cf. [6, pp. 32–38 and 260]. On setting $\rho_\parallel = 0$ in (6.80) and comparing with (6.79), we see that $\hat{\mathbf{b}}\cdot\nabla\nabla\cdot\boldsymbol{\xi} = 0$,

[14] As discussed in the article cited in the footnote after (6.75), $\rho_\parallel \to 0$ is a rather singular limit because $\nabla\cdot\boldsymbol{\xi} \to 0$ requires the resonant Fourier component of $\xi_\parallel = B\eta$ to diverge on a rational magnetic surface. However, we do not need to know $\xi_\parallel$ explicitly, so we ignore this issue here.

which is satisfied when $\nabla \cdot \boldsymbol{\xi} = 0$ everywhere—i.e. on assuming *incompressibility*. Consequently, on omitting the compression term from (6.78) we have the 1-field collisionless MHD model stability functional

$$\Delta W^{(0)} = \frac{1}{2} \int_{\mathcal{P}} \left[\frac{|\nabla (\mathbf{B} \cdot \nabla \varphi)|^2}{\mu_0 B^2} - \frac{(\boldsymbol{\kappa} \times \mathbf{B} \cdot \nabla \varphi^*)(\nabla p \times \mathbf{B} \cdot \nabla \varphi)}{B^4} - \frac{(\nabla p \times \mathbf{B} \cdot \nabla \varphi^*)(\boldsymbol{\kappa} \times \mathbf{B} \cdot \nabla \varphi)}{B^4} \right] d\tau. \quad (6.83)$$

Since $\mathcal{K}^{(0)} = \frac{1}{2} \int \rho_\perp |\nabla_\perp \varphi|^2 / B^2 \, d\tau$ now depends only on φ, so does $\lambda^{(0)}$. Varying φ^* in the Rayleigh principle $\delta \Delta W^{(0)} = \omega^2 \delta \mathcal{K}^{(0)}$ then gives the generalised eigenvalue equation for normal modes

$$\mathbf{B} \cdot \nabla \left[\nabla \cdot \frac{1}{\mu_0 B^2} \nabla (\mathbf{B} \cdot \nabla \varphi) \right] + \nabla \cdot \left[\frac{\boldsymbol{\kappa} \times \mathbf{B}}{B^4} \nabla p \times \mathbf{B} \cdot \nabla \varphi \right] + \nabla \cdot \left[\frac{\nabla p \times \mathbf{B}}{B^4} \boldsymbol{\kappa} \times \mathbf{B} \cdot \nabla \varphi \right] + \omega^2 \nabla \cdot \frac{\rho_\perp}{B^2} \nabla \varphi = 0. \quad (6.84)$$

In Sect. 6.6 on cylindrical geometry, we noted that radially localised ideal interchange instabilities occur in the neighbourhood of rational magnetic surfaces—i.e. on singular surfaces of radius r_s where $q(r_s) = m/n$ (and $\mathbf{B} \cdot \nabla$ terms are zero). Equation (6.84) would provide an alternative starting point to treat such high-n cylindrical modes, but it is mainly used for the toroidal case where the curvature of any magnetic field line depends upon the poloidal angle θ, coupling different Fourier components m.

The toroidal curvature on the inside of a toroidal flux surface (closest to the Z-axis) is favourable (i.e. stabilising for interchanges) but that on the outside of the surface (furthest from the Z-axis) is unfavourable, so that the plasma along a given field line experiences alternating regions of favourable and unfavourable magnetic field curvature. This suggested that there could be configurations with stabilising "magnetic wells", where the *average* curvature around the torus would be favourable enough to avoid interchanges. However, while this is true for cylindrical configurations, it was soon recognised that in toroidal geometries additional pressure-driven perturbations can develop with even larger amplitude, localised in those regions of unfavourable magnetic field curvature—viz. socalled *ballooning* modes, with flux surface deformations that resemble aneurysms or blowouts in weak regions of pneumatic tyres. Both interchange and ballooning modes in toroidal geometry can be analysed using the techniques that we sketch in the next subsection.

6.7.4 Ballooning Equation

The Euler–Lagrange equation (6.84) is a partial differential equation that is not separable in general geometry. The numerical solution of this equation (to calculate normal mode frequencies and growth rates) is simpler than starting with (6.49), but nevertheless also takes considerable effort. Instead, (6.84) can be solved asymptotically—cf. [11, pp. 263–268 and Chap. 8]. However, once again we cannot simply use the approach of Sect. 4.2 where the direction of **k** is arbitrary, since we need a more specialised approach that respects the generalised flute ordering (6.68) by constraining **k** to be locally perpendicular to **B**. This constraint can be enforced by transforming to (α, s, θ) coordinates, where the field-line label α (5.80) is such that $\mathbf{B} = \nabla\alpha \times \nabla\psi$ (5.81), and then choosing the representation

$$\varphi(s, \theta, \zeta) = \bar{\varphi}(\theta|\alpha, s) \exp[i S(\alpha, s) - i\omega t] \qquad (6.85)$$

where $\bar{\varphi}(\theta|\alpha, s)$ is assumed to vary on the equilibrium scale but the phase variation is rapid. Specifically, $\mathbf{k} = \nabla S$ is ordered large as $O(\epsilon^{-1})$, and the frequency ω is taken to be O(1). Adopting this eikonal ansatz (6.85), to lowest order we obtain the *ballooning equation*

$$\mathbf{B}\cdot\nabla\left(\frac{\mathbf{k}^2}{\mu_0 B^2}\mathbf{B}\cdot\nabla\bar{\varphi}\right) + \frac{2\mathbf{k}\cdot\nabla p \times \mathbf{B}\boldsymbol{\kappa}\times\mathbf{B}\cdot\mathbf{k}}{B^4}\bar{\varphi} + \omega_k^2\,\frac{\rho_\perp \mathbf{k}^2}{B^2}\bar{\varphi} = 0, \qquad (6.86)$$

an *ordinary* differential equation to be solved as a generalised eigenvalue equation by integration along each magnetic field line α = const, s = const—with θ the independent variable in (6.86), as is indicated in (6.85) by separating the variable and constant arguments of $\bar{\varphi}$ with a | divider symbol. Since this eigenvalue problem is local to a field line, rather than global over the whole plasma like (6.84), it does not immediately give us a normal mode. We therefore use the notation ω_k^2 to distinguish the local eigenvalues on each field line from the global normal mode eigenvalue ω^2, and discuss how such local solutions can be combined to form a normal mode below.

The eikonal $S(\alpha, s)$ is a constant on each field line, so $\mathbf{k}\cdot\mathbf{B} = 0$ automatically. Explicitly, the wave vector is $\mathbf{k} = k_\alpha\nabla\alpha + k_s\nabla s$, where $k_\alpha = \partial S/\partial\alpha$ and $k_s = \partial S/\partial s$ are also constant along magnetic field lines. While k_α and θ_k are constant local to a field line, **k** itself is far from constant because $\nabla\alpha$ and ∇s vary along a field line. In terms of the curvilinear basis set $\{\mathbf{e}^s(s,\theta,\zeta), \mathbf{e}^\theta(s,\theta,\zeta).\mathbf{e}^\zeta(s,\theta,\zeta)\}$ for the straight-field-line coordinates (cf. Sect. 5.11) that are periodic in θ and ζ, in (α, s, θ)-coordinates we have the basis vectors $\nabla\alpha = \mathbf{e}^\zeta - q(s)\mathbf{e}^\theta - q'(s)\theta\,\mathbf{e}^s$ and $\nabla s = \mathbf{e}^s(s, \theta, \alpha + q\theta)$ that are functions of $(s, \theta, \alpha + q\theta)$. We note two remarkable properties of these vectors: (a) for irrational q, ∇s is not periodic but *quasi-periodic* in θ—i.e. its "waveform" never exactly repeats along a field line[15]; and (b) $\nabla\alpha$ has

[15]This vector quasi-periodicity is not actually manifest in axisymmetric systems, because only scalars appear in the coefficients of the ballooning equation that are not functions of ζ, and so are periodic in θ. However, it has profound implications in non-axisymmetric systems—cf. P. Cuthbert and R.L. Dewar (*Physics of Plasmas* **7**, 2302, 2000).

a "secular" term $-q'\theta\,\mathbf{e}^s$, linear in θ. Replacing $\nabla\alpha$ and ∇s with the designated explicit forms above, we therefore obtain

$$\mathbf{k} = k_\alpha\left[\mathbf{e}^\zeta - q(s)\mathbf{e}^\theta - q'(s)(\theta-\theta_k)\mathbf{e}^s\right], \tag{6.87}$$

involving the angle-like constant $\theta_k = k_s/q'(s)k_\alpha$.[16]

As a consequence of the secular terms, the solution to (6.86) cannot be periodic but is more or less localised (depending on q') in the vicinity $\theta \approx \theta_k$. Thus the ordinary differential equation (6.86) may be solved numerically under the boundary conditions $\bar{\varphi}(\theta) \to 0$ as $\theta \to \pm\infty$, to give the local eigenvalue $\omega_k^2(\alpha, s, \theta_k)$, depending only on the ratio k_s/k_α in ω_k because k_α^2 may be factored out of (6.86). The set of all solutions for different (α, s, θ_k) forms a dispersion relation with two degrees of freedom (cf. Chap. 4), defining a component of a normal mode through the solution of the eikonal equation $\omega_k^2(\alpha, s, k_\alpha, k_s) = \omega_k^2(\alpha, s, \theta_k) = \omega^2$, regarded as a partial differential equation for S with ray equations in $(\alpha, s, k_\alpha, k_s)$ phase space as its characteristics.

Before proceeding to consider the global normal mode problem, let us first address a paradoxical aspect of the ballooning equation—viz. how its solutions that are not periodic in θ can produce a physical normal mode. In order to allow non-periodic boundary conditions, we implicitly lifted the problem to the $(\theta, \zeta) \in \mathbb{R}\times\mathbb{R}$ covering space of a given torus $s = \text{const}$—i.e. we abandoned the topological identification of angles under 2π increments. However, not all of the periodicity information is lost in doing so, for apart from the secular terms in $(\theta - \theta_k)$ the scalar coefficients in the ballooning equation remain periodic. As elsewhere in this book, we consider only the axisymmetric case so that ζ is an "ignorable" coordinate here—i.e. apart from the independent variable θ, the ballooning eigenfunctions $\bar{\varphi}$ depend only on the parameters s and θ_k. As θ_k occurs in the ballooning equation only through the secular terms in $(\theta-\theta_k)$, it can readily be shown that all functions $\bar{\varphi}(\theta + 2\pi l|s, \theta_k + 2\pi l)$ for integer l obey the same equation as a given ballooning eigenfunction $\bar{\varphi}(\theta|s, \theta_k)$, so the solutions of (6.86) are unique up to a constant normalising factor for given s, θ_k and ω_k^2. Thus on using a normalisation that is invariant under 2π shifts in θ, we identify these functions $\bar{\varphi}(\theta + 2\pi l|s, \theta_k + 2\pi l) = \bar{\varphi}(\theta|s, \theta_k)$. Equivalently, on replacing θ by $\theta - 2\pi l$ we see that an infinity of ballooning eigenfunctions with the same ω_k^2 but different θ_k can be generated, simply by translating a given eigenfunction by integer multiples of 2π in θ using the *shift identity*

$$\bar{\varphi}(\theta|s, \theta_k + 2\pi l) = \bar{\varphi}(\theta - 2\pi l|s, \theta_k). \tag{6.88}$$

A corollary is that ω_k^2 is a periodic function of θ_k.

On setting the toroidal mode number $k_\alpha = n$, we can separate the eikonal as $S = n\alpha + \int_{s_<}^{s} k_s(s')\,ds'$, where $k_s(s)$ is a solution of $\omega_k^2(s, k_s/nq') = \omega^2$ and $s_<$

---

[16]This constant $k_s/q'(s)k_\alpha$ is often denoted by θ_0, but the notation θ_k from R.L. Dewar, J. Manickam, R.C. Grimm and M.S. Chance (*Nuclear Fusion* **21**, 493, 1981) seems more appropriate, to indicate its relation to **k**.

(a *turning point*) is the smallest value of s for which a solution exists. We can now resolve the ballooning paradox by generating a periodic solution through an infinite superposition of degenerate solutions in the form of (6.85) with different θ_k (and hence k_s), but brought back to the same θ_k (and k_s) by using the shift identity (6.88), to obtain the *ballooning representation*

$$\bar{\varphi}_k(s,\theta,\zeta) = \exp\left(in\zeta + i\int_{s_<}^{s} k_s(s')\,ds' - i\omega t\right) \times \sum_{l=-\infty}^{\infty} \bar{\varphi}(\theta - 2\pi l|s,\theta_k)\exp[-inq(\theta - 2\pi l)]. \tag{6.89}$$

The sum converges unconditionally provided $\bar{\varphi}(\theta)$ decays at infinity faster than $1/|\theta|$, which is assured in the *ballooning unstable* case $\omega_k^2 < 0$, when the decay is exponential. Assuming convergence, periodicity in θ readily follows by replacing θ with $\theta + 2\pi$ on both sides and making the change of dummy summation index $l = l' + 1$, clarifying that the right-hand side has been left unchanged.

In the marginal case $\omega_k^2 = 0$, where asymptotic analysis of the $|\theta| \to \infty$ limit shows that the decay proceeds as an inverse fractional power of $|\theta|$, the convergence issue is much more delicate and non-analytic behaviour develops at the rational surfaces s_m where $q(s_m) = m/n$. It may be shown that the condition that the singularity at a rational surface be *square integrable* coincides with the Mercier criterion for stability to localised interchange modes (an extension of the Suydam criterion (6.61) in cylindrical geometry to toroidal geometry), which involves an average over each magnetic flux surface [6, 9].

However, (6.89) represents a travelling rather than a standing wave, and no normal mode. Indeed typically, and certainly near the maximum growth rate, there is a range $s_< \le s \le s_>$ well within the plasma where there are *two* solutions $\theta_k = \theta_k^{\pm}(s)$ within the range $\theta_k \in (-\pi, \pi)$ of the 1-degree-of-freedom eikonal equation $\omega_k^2(s, \theta_k) = \omega^2$. Correspondingly, there are two travelling wave branches $k_s^{\pm} = nq'\theta_k^{\pm}(s)$ that convert from one to the other on reflection at the turning points $s_<$ and $s_>$. The eikonal formalism breaks down near these points, but asymptotic matching to a local Airy function asymptotic approximation can be used to show there is a phase change of $\pi/2$ at each reflection, making a net change of π after two reflections—cf. [11, pp. 300–304]. The requirement that this doubly reflected wave be the same as the original unreflected wave provides the *quantisation condition* for normal modes—viz.

$$n \oint \theta_k(s)q'(s)ds = 2\pi\left(N + \frac{1}{2}\right) \tag{6.90}$$

where N is a positive integer, so the most unstable normal mode for given n corresponds to $N = 0$. The loop integral symbol $\oint$ signifies the sum of the phase integrals over the original and reflected waves, which is indeed an integral around the contour $\omega_k^2(s,\theta_k) = \omega^2$ surrounding the minimum of ω_k^2 in the s, θ_k-plane.[17]

6.7.5 Geometric Interpretations

Another feature of the asymptotic ordering approach is that it allows us to explore, whether and in what sense, some of the terms in the form for ΔW given in (6.49) are distinct in a rigorously geometric way. Rather than restrict ourselves to the more specialised generalised flute ordering, in the following discussion we simply assume a *short-wavelength ordering* such that all components of $\mathbf{k}$ (including $k_\parallel$) are $\mathrm{O}(\epsilon^{-1})$. More precisely, we replace (6.68a) and (6.68b) with the orderings

$$L = \mathrm{O}(1)\,, \quad |f|^{-1}\nabla f = \mathrm{O}(\epsilon^{-1}), \tag{6.91}$$

while retaining the ordering (6.68c). As we retain only linear order in displacements $\boldsymbol{\xi}$, their ordering in ϵ is arbitrary so we still take $\boldsymbol{\xi} = \mathrm{O}(1)$, but do not impose the orderings (6.69a) and (6.69b). In particular, let us reconsider the field-line bending and compression terms, to make precise the definition of a displacement field $\boldsymbol{\xi}$ that either (a) bends but does not compress or (b) compresses but does not bend the magnetic field lines. If the geometric identifications of "field-line bending term" and "field-line compression term" are precise, the second term of ΔW_p should vanish under displacements of type (a), while the first term should vanish under displacements of type (b).

Let us now introduce the concept of a *Lagrangian variation* operator Δ on a field quantity, when the plasma is displaced from its original background position to its perturbed state as $\boldsymbol{\xi}$ is "turned on" at fixed time t, such that

$$\Delta \equiv \delta + \boldsymbol{\xi}\cdot\nabla \tag{6.92}$$

where δ produces the more usual *Eulerian variation*—i.e. the variation of a field quantity at a fixed point in space. Thus the Lagrangian variation is the change seen by a fluid element as it is perturbed from its background to its final position. It is a property of ideal MHD that the Lagrangian variation of the density, pressure and magnetic field can all be expressed in terms of the strain dyadic $\nabla\boldsymbol{\xi}$. Moreover, these fields are all $\mathrm{O}(\epsilon^{-1})$ in the short-wavelength ordering (6.91). Lagrangian variations also have the advantage that they vanish for uniform translations of the system, which should not change any energy terms. We now define Lagrangian measures for magnetic field-line compression and magnetic field-line bending variations with the

[17] Although $N = 0$ is the case least appropriate for the asymptotic analysis, it is known that the analogous quantisation condition in the case of the quantum oscillator is exact, and numerical comparisons in the R.L. Dewar et al. (1981) article cited in the previous footnote have found good agreement with full normal mode calculations.

dimension of magnetic field. We now proceed to reconsider the displacement fields of type (a) and (b) defined above.

Magnetic Field-Line Compression

As magnetic field-line density is measured by B, we use $\Delta B = (\Delta B^2)/2B$ as the natural measure for magnetic field-line compression. This may be calculated as follows

$$\begin{aligned}\Delta B &= B^{-1}\left[\mathbf{B}\cdot\mathbf{B}_1 + \boldsymbol{\xi}\cdot\nabla\left(\frac{B^2}{2}\right)\right] \\ &= B^{-1}\left[\mathbf{B}\cdot\mathbf{B}_1 - \mu_0\boldsymbol{\xi}_\perp\cdot\nabla p + \mathbf{B}\cdot(\nabla\mathbf{B})\cdot\boldsymbol{\xi}\right],\end{aligned} \tag{6.93}$$

where reference to (6.31) led us to $\mathbf{B}_1 = \nabla\times(\boldsymbol{\xi}\times\mathbf{B}) = \mathrm{O}(\epsilon^{-1})$ for the Eulerian variation $\delta\mathbf{B}$, and the identity $\boldsymbol{\xi}\cdot(\nabla B^2/2) = \mathbf{B}\cdot(\nabla\mathbf{B})\cdot\boldsymbol{\xi} - \mu_0\boldsymbol{\xi}_\perp\cdot\nabla p$ obtained by dotting the magnetohydrostatic condition $\mu_0\nabla p = (\nabla\times\mathbf{B})\times\mathbf{B}$ with $\boldsymbol{\xi}$ is used to eliminate the subdominant O(1) term $\boldsymbol{\xi}\cdot(\nabla B^2/2)$. This provides a form that is directly comparable with Term (2) of (6.49), with which it agrees *if* the equilibrium magnetic field is constant along field lines, $\mathbf{B}\cdot\nabla\mathbf{B} = 0$ as in Cartesian (slab) geometry. As our motivation for the generalised flute mode orderings in Sect. 6.7 was in fact based on plane waves in a uniform magnetic field, the description of Term (2)—i.e. $|\mathbf{B}\cdot\mathbf{B}_1 - \mu_0\boldsymbol{\xi}_\perp\cdot\nabla p|^2/\mu_0 B$—as the "field-line compression term" is justified for heuristic purposes, but the omission of the subdominant term $\mathbf{B}\cdot(\nabla\mathbf{B})\cdot\boldsymbol{\xi}$ means that we cannot regard such a description as exact for general equilibria.

Magnetic Field-Line Bending

For a measure of magnetic field-line bending we calculate the *Eulerian* change of the unit tangent vector $\hat{\mathbf{b}} = \mathbf{B}/B$ (i.e. the change of the magnetic field direction caused by a plasma displacement), and multiply by B in order to give the variation the dimensions of magnetic field:

$$B\,\delta\hat{\mathbf{b}} = \mathbf{B}_1 - \hat{\mathbf{b}}\,\delta B = \mathbf{B}_1 - \hat{\mathbf{b}}\hat{\mathbf{b}}\cdot\mathbf{B}_1. \tag{6.94}$$

This is precisely the $\mathbf{B}_{1\perp}$ used in Term (1) of (6.49), but it differs from the corresponding Lagrangian definition $B\Delta\hat{\mathbf{b}}$ by the subdominant term $B\boldsymbol{\xi}\cdot\nabla\hat{\mathbf{b}}$ that vanishes only for an equilibrium magnetic field of constant direction. If we accept that the Lagrangian definition is the more fundamental, then we again conclude that the heuristic identification of Term (1) as the "field-line bending term" is reasonable although not exact.

6.8 Magnetorotational Theory

MHD stability theory has interesting applications in astrophysics. For example, in the early twentieth century Jeans suggested that the galactic spiral structures observed by Hubble could be material arms shed from the galactic centre and wound into spiral forms by galactic rotation. However, he also recognised that a material arm could break up into clumps under self-gravitation, the phenomenon that became known as Jeans instability. Chandrasekhar and Fermi then represented a material arm as an infinite self-gravitating plasma cylinder of radius 250 parsecs with density 2×10^{-24} grams per cubic centimetre, to show that incompressible transverse perturbations are insignificant on the galactic time scale in the presence of a uniform axial magnetic field above 7×10^{-6} gauss according to ideal MHD stability theory. Thus the wavelength and e-folding time of the fastest instability at 2.7×10^3 parsecs and 1.0×10^8 years in the absence of a magnetic field became 3.4×10^5 parsecs and 1.1×10^{10} years (the approximate age of our Galaxy) in a uniform axial magnetic field of 7×10^{-6} gauss, and there was some observational support at the time for a magnetic field of this magnitude. Their work also addressed other consequences of self-gravitation, such as star formation.

Many astrophysicists have since considered the stability of thin discs. One motivation for this was the rapid stellar rotation in our Galaxy eventually observed, indicating that the large-scale spiral pattern would become so wound up as to be indistinguishable in a rotation period of about 10^8 years (much less than the estimated age of the Universe) if the arms largely consisted of the same material. Since strong *differential* rotation was also observed, Lin and Shu proceeded to consider the hydrodynamic stability of an infinitesimally thin pressureless (cold) differentially rotating disc to explore the maintenance of the spiral pattern as a density wave, a concept originally due to B. Lindblad and others. Two essential assumptions in the Lin and Shu analysis are self-gravitation and that the motion is always in the plane of the thin disc. They assumed a quasi-stationary spiral structure (QSSS) as a working hypothesis, but Jeans instability was again an issue. In particular, Toomre found that short-wavelength axisymmetric perturbations remain unstable in a thin disc of stars with a large-scale Schwarzschild velocity distribution unless their root-mean-square radial speed is at least $O(G\sigma_0/\kappa)$, where σ_0 denotes the stellar surface density (κ is the socalled epicyclic frequency dependent on the stellar angular velocity Ω defined below and G is Newton's universal constant of gravitation). Consequently, Lin and Shu included random stellar motion by multiplying the surface density in the dispersion relation for the stars by a reduction factor

$$\mathcal{F}_\nu(x) = \frac{1-\nu^2}{x}\left(1 - \frac{\nu\pi}{\sin(\nu\pi)}\,\frac{1}{2\pi}\int_{-\pi}^{\pi} \mathrm{e}^{-x(1+\cos s)}\cos(\nu s)\,ds\right),$$

where $\nu = (\omega - m\Omega)/\kappa$ is a frequency ratio involving the mode eigenfrequency ω and azimuthal wave number m, and $x = \alpha^2 \left\langle c_{\hat{\omega}}^2 \right\rangle / \kappa^2$ (α is the dimensionless radial wave number and $\left\langle c_{\hat{\omega}}^2 \right\rangle$ denotes the mean-square value of the stellar radial speed). However, Toomre then found that the group velocity of the density waves seemed sufficient to obliterate any spiral pattern in a thin stellar disc within a few rotation periods—and since tight spirals also did not appear as instabilities in many-body computer simulations, he subsequently proposed another mechanism called swing amplification that generated large-scale spiral patterns as tidal transients in a thin disc of stars.

The focus on the stellar component in the work of those authors reflected the view that the stars constitute most of the disc material in our Galaxy—but its spiral pattern is primarily outlined by the youngest and brightest stars that appear to be born at density maxima in the interstellar gas, suggesting that there is a continual renewal process primarily associated with the gas rather than the stellar component. Further, the interstellar gas dynamics must be influenced by any prevailing magnetic field, and this is true whether or not the equation of state (3.3) is included in the mathematical model—i.e. whether or not the gas is so tenuous that its pressure may be neglected. In this section, we derive an ideal MHD criterion for the stabilisation of an infinitesimally thin pressureless plasma disc against the short wavelength self-gravitational perturbations Lin and Shu considered in their hydrodynamic analysis, assuming an initial magnetic field that is perpendicular at the plane of the disc. This MHD criterion is not only applicable in the spiral pattern context, but also wherever the underlying mathematical model is appropriate—e.g. a magnetised plasma disc has also been considered in modelling a quasar or galactic nucleus.

Many astrophysicists have also considered accretion or protostellar discs, to model planetary and star formation for example. In particular, Balbus and Hawley recognised the relevance of a magnetorotational instability due to a weak magnetic field in a thin Keplerian disc, where self-gravitation does not play an important role. This interesting instability has no hydrodynamic counterpart, and undermines an otherwise stable angular velocity distribution in the strong central gravitational field in the disc. Originally found by Velikov and Chandrasekhar, the motivation that led to its rediscovery together with some intuitive insight is discussed in Ref. [10], and we present some simplified further analysis below.

6.8.1 *Planar Short-Wavelength Self-gravitational Modes in a Thin Disc*

In an infinitesimally thin plasma disc, the mass and electric current distributions may be written

$$\rho(r,\theta,z,t) = \sigma(r,\theta,t)\,\delta(z) \qquad \text{and} \qquad \mathbf{j}\,(r,\theta,z,t) = \mathbf{J}(r,\theta,t)\,\delta(z),$$

where $\delta(z)$ denotes the Dirac delta function, so the continuity equation (5.33) yields

$$\frac{\partial\sigma}{\partial t} + \frac{1}{r}\frac{\partial}{\partial r}[r\sigma v_r(r,\theta,t)] + \frac{1}{r}\frac{\partial}{\partial\theta}[\sigma v_\theta(r,\theta,t)] = 0 \tag{6.95}$$

on the plane $z = 0$, where $\sigma(r, \theta, t)$ is the surface density; and ignoring the pressure term, the equation of motion (2.56) becomes

$$\sigma \frac{d\mathbf{v}}{dt} = \sigma \nabla \phi + \mathbf{J} \times \mathbf{B}, \tag{6.96}$$

where ϕ denotes the gravitational potential (such that $\mathbf{g} = \nabla \phi$) and $\mathbf{B}(r, \theta, 0, t)$ is the magnetic field at $z = 0$ due to some planar flow. For ideal MHD motion, there is also the equation of magnetic induction (5.38)—i.e.

$$\frac{\partial \mathbf{B}}{\partial t} = \nabla \times (\mathbf{v} \times \mathbf{B}), \tag{6.97}$$

again applicable on $z = 0$. In addition, throughout all space we have the Poisson equation for the gravitational potential

$$\nabla^2 \phi(r, \theta, z, t) = -4\pi G \sigma \delta(z), \tag{6.98}$$

where

$$\nabla^2 = \frac{\partial^2}{\partial r^2} + \frac{1}{r}\frac{\partial}{\partial r} + \frac{1}{r^2}\frac{\partial^2}{\partial \theta^2} + \frac{\partial^2}{\partial z^2}. \tag{6.99}$$

Finally, we have the electromagnetic equations (5.16) and (5.17)—viz. $\nabla \cdot \mathbf{B} = 0$ and now

$$\nabla \times \mathbf{B} = \mu_0 \mathbf{J}\, \delta(z). \tag{6.100}$$

The Fourier form is $f_1(r) \exp[i(m\theta - \omega t)]$ for planar azimuthal perturbations, where as before m is an integer but both $f_1(r)$ and ω are generally complex. Moreover, the form $\sigma_1(r) \exp[i(m\theta - \omega t)]$ for the surface density perturbation may be rendered

$$\sigma_1(r, \theta, t) = a(r) \exp(\omega_I t) \cos[m\theta - \omega_R t + b(r)]$$

on writing $\sigma_1(r) = a(r) \exp[ib(r)]$, where $a(r)$ varies slowly while $b(r)$ varies rapidly with r in an eikonal ("local") short-wavelength approximation. The perturbed density distribution maxima then defines a spiral pattern of form $m(\theta - \theta_0) = b(r) - b(r_0)$ in which (r_0, θ_0) is any chosen reference point, rotating with angular velocity ω_R—and with an angle of inclination $\vartheta_{\rm i}$ given by

$$\tan \vartheta_{\rm i} = m \left[r \frac{d\, b(r)}{d\, r} \right]^{-1},$$

and trailing arms for $m \geq 0$ if $db(r)/dr > 0$ (or leading arms if $db(r)/dr < 0$).

Furthermore, the solution for the three-dimensional self-gravitational potential perturbation over $0 < r_0 < r < \infty$ satisfying the perturbation equation obtained from (6.98) and bounded as $r \to \infty$ may be expressed as

$$\phi_1(r, \theta, z, t) = C H_m(\alpha r)\, e^{-\alpha|z|} \exp[i(m\theta - \omega t)], \tag{6.101}$$

where C is a constant and $H_m(\alpha r)$ is an mth order Hankel function. Thus the surface density perturbation in the plasma disc is

$$\sigma_1(r) = \frac{\alpha}{2\pi G}\, \phi_1(r), \tag{6.102}$$

if we assume the stellar background and any other nearby material is passive (unperturbed) on the time-scale of interest. It is notable that

$$H_m(\alpha r) \simeq \sqrt{\frac{2}{\pi \alpha r}} \exp\left[\pm i\left(\alpha r - \frac{1}{2}m\pi - \frac{1}{4}\pi\right)\right] \tag{6.103}$$

for $\alpha r \gg 1$, where the $+$ and $-$ correspond to the respective Hankel functions $H_m^{(1)}$ and $H_m^{(2)}$. Indeed, in the short radial wavelength eikonal approximation near $r = R$ say, where $H_m(\alpha r)$ is taken to vary much more rapidly with r than any initial quantity (and then may be treated as constant), the perturbation forms are $f_1(r, \theta, t) \sim \exp[i(\omega t - m\theta \pm \alpha r)]$ on $z = 0$ and Eq. (6.95) produces

$$i(\omega - m\Omega)\sigma_1 = \sigma_0 \frac{1}{r}\left(\frac{d}{dr}(r v_{1r}) + i m v_{1\theta}\right) \simeq \sigma_0 \frac{d v_{1r}}{dr} = \pm i \sigma_0\, \alpha v_{1r}, \tag{6.104}$$

assuming that $v_{1r} \sim v_{1\theta}$. Near the radius $r = R$ (where $\alpha r \gg 1$), we also have

$$\left.\frac{\partial \phi_1}{\partial r}\right|_{z=0} = \pm 2\pi i G \sigma_1 \quad \text{and} \quad \left.\frac{\partial \phi_1}{\partial \theta}\right|_{z=0} = O\left(\frac{\sigma_1}{\alpha}\right). \tag{6.105}$$

Let us now consider an initial magnetic field that is perpendicular at the plane of the disc—i.e. such that $\mathbf{B} = B_z(r)\hat{\mathbf{e}}_z$ at $z = 0$. This magnetic field may be quite weak, as emerges below, and it may originate either internally or externally. Thus it may correspond to a poloidal magnetic field generated by an azimuthal electric current distribution $\mathbf{j}(r, z) = J_\theta(r)\,\delta(z)\hat{\mathbf{e}}_\theta$ in the disc, or it could be due to a seed field emanating from elsewhere. The analysis below relates to the first case, but the ideal MHD stability criterion obtained is precisely the same in either case—as one would expect for the eikonal ("local") perturbations considered.

From (6.100) and $\nabla \cdot \mathbf{B} = 0$, the components of the magnetic field perturbation corresponding to planar current perturbations J_{1r} and $J_{1\theta}$ are

$$B_{1r}(r, \theta, z, t) = \mu_0 J_{1\theta}\, \text{sgn}(z) \exp(-\alpha|z|) \tag{6.106}$$

$$B_{1\theta}(r, \theta, z, t) = -\mu_0 J_{1r}\, \text{sgn}(z) \exp(-\alpha|z|) \tag{6.107}$$

$$B_{1z}(r, \theta, z, t) = \mu_0 \left[\frac{\partial (r J_{1\theta})}{\partial r} - \frac{\partial J_{1r}}{\partial \theta}\right] \frac{1}{\alpha} \exp(-\alpha|z|). \tag{6.108}$$

Thus if we ignore any accompanying background disturbances, the linearised perturbation equations for the infinitesimally thin plasma disc at $z = 0$ with initial surface density σ are (presuming $v_{1r} \sim v_{1\theta}$):

$$\begin{aligned}
i(\omega - m\Omega)\sigma_1 &= \pm i\sigma\alpha v_{1r} \\
-i(\omega - m\Omega)v_{1r} - 2\Omega v_{1\theta} &= \pm 2\pi i\, G\sigma_1 + \frac{B_z}{\sigma} J_{1\theta} + \frac{J_\theta}{\sigma} B_{1z} - \frac{J_\theta B_z}{\sigma^2}\sigma_1 \\
-i(\omega - m\Omega)v_{1\theta} + \frac{\kappa^2}{2\Omega} v_{1r} &= -\frac{B_z}{\sigma} J_{1r} \\
i(\omega - m\Omega)B_{1z} &= \pm i\alpha B_z J_{1z} \\
B_{1z} &= \pm i\alpha\mu_0 J_{1\theta} \\
\pm\alpha J_{1r} - (m/R)J_{1\theta} &= 0,
\end{aligned}$$

from (6.96) and (6.105), (6.97), (6.107) and $\nabla\cdot\mathbf{J} = 0$. Here

$$\kappa^2 = 4\Omega^2(R)\left(1 + \frac{r}{2\Omega}\frac{d\Omega}{dr}\right)\bigg|_{r=R} = \left(4\Omega^2 + r\frac{d\Omega^2}{dr}\right)\bigg|_{r=R} \tag{6.109}$$

defines the epicyclic frequency κ at any radius R. (In our Galaxy the epicyclic frequency κ is often written in terms of socalled local Oort constants, and varies only slowly with radius.)

Elimination between these perturbation equations yields the dispersion relation

$$(\omega - m\Omega)^2 = \kappa^2 - \left(2\pi G\sigma - \frac{B_z^2}{\mu_0\sigma}\right)\alpha, \tag{6.110}$$

so the vertical magnetic field strength B_0 stabilises the localised short-wavelength perturbations according to the MHD stability criterion

$$B_z > \sqrt{2\pi G\mu_0\sigma^2}. \tag{6.111}$$

For example, if we assume a surface density $\sigma = \mathrm{O}(10^{-3})\,\mathrm{gm\,cm^{-2}}$, the critical magnitude of the magnetic field $\sqrt{2\pi G\mu_0\sigma^2}$ for marginal stability is only $\mathrm{O}(10^{-6})$ gauss—i.e. a similar magnitude to that previously found sufficient to prevent the breakup of the plasma cylinder model for a material galactic arm on the galactic time scale. Any accompanying perturbation in the stellar background or neighbourhood may increase this order of magnitude estimate, but the plasma density σ_0 probably remains the major factor in defining the critical magnetic field under the stability criterion (6.111). Thus short-wavelength disturbances that produce tight spiral wave patterns may be stabilised by a relatively small magnetic field, independent of any finite plasma pressure effect analogous to the stabilisation of a thin disc of stars due to their random motion. Finally, we note that the MHD stability criterion (6.111) is also applicable to magnetised disc equilibria even if there is little or no rotation, as in some quasar modelling.

6.8.2 Magnetorotational Instabilities in a Keplerian Disc

Let us now consider a differentially rotating Keplerian disc of small but finite thickness, where self-gravitation is negligible relative to a strong central gravitational force. Let us continue to assume that the plasma is subject to a magnetic field $\mathbf{B} = B_z\hat{\mathbf{e}}_z$, which may now be weak such that the equilibrium condition becomes $r\Omega^2(r) \simeq GM/r^2$—i.e. the angular momentum is largely balanced by the central gravitational force. We denote the projected surface density by $\sigma(r)$ and the azimuthal angular speed by $\Omega(r)$ as before, but here invoke the ideal compressible model of Sect. 5.7 and consider *axial* eikonal ("local") perturbations of Fourier form $f_1(r)\exp[i(kz - \omega t)]$. Thus for $f_1(r)$ varying slowly with r and $kr \gg 1$, from (5.33)–(5.35) and (5.38) the linearised perturbation equations (in our usual subscript notation) are

$$-\omega\frac{\sigma_1}{\sigma} + kv_{1z} = 0 \tag{6.112}$$

$$i\omega v_{1r} + 2\Omega v_{1\theta} + i\frac{kB_z}{\mu_0\sigma}B_{1r} = 0 \tag{6.113}$$

$$-i\omega v_{1\theta} + \frac{\kappa^2}{2\Omega}v_{1r} - i\frac{kB_z}{\mu_0\sigma}B_{1\theta} = 0 \tag{6.114}$$

$$-\omega v_{1z} + k\frac{p_1}{\sigma} = 0 \tag{6.115}$$

$$\frac{p_1}{p} - \gamma\frac{\sigma_1}{\sigma} = 0 \tag{6.116}$$

$$\omega B_{1r} + kB_z v_{1r} = 0 \tag{6.117}$$

$$i\omega B_{1\theta} + ikB_z v_{1\theta} + \frac{d\Omega}{d\ln R}B_{1r} = 0 \tag{6.118}$$

$$B_{1z} = 0, \tag{6.119}$$

where κ is again the epicyclic frequency given by (6.109). There is decoupling in this simplified system for an axial magnetic field, such that the subset in $\{v_{1r}, v_{1\theta}, B_{1r}, B_{1\theta}\}$ renders the dispersion relation we seek as the condition for the corresponding nontrivial solution—viz.

$$\det\begin{pmatrix} i\omega & 2\Omega & ikB_z/(\mu_0\sigma) & 0 \\ \kappa^2/(2\Omega) & -i\omega & 0 & -ikB_z/(\mu_0\sigma) \\ kB_z & 0 & \omega & 0 \\ 0 & ikB_z & d\Omega/d\ln R & i\omega \end{pmatrix} = 0.$$

where $c_A = \sqrt{B_z^2/(\mu_0\sigma)}$ is the Alfvén speed, or

$$\omega^4 - \left(2k^2c_A^2 + \kappa^2\right)\omega^2 + k^2c_A^2\left(k^2c_A^2 + \frac{d\,\Omega^2(r)}{d\ln r}\right) = 0. \tag{6.120}$$

Consequently, if $d\,\Omega^2(r)/d\ln r < 0$ there is an exponentially growing instability ($\omega^2 < 0$) for wave numbers satisfying

$$k^2c_A^2 < -\frac{d\,\Omega^2(r)}{d\ln r}. \tag{6.121}$$

Moreover, on setting $c_A = 0$ it follows that this magnetorotational instability has no hydrodynamic counterpart—cf. also the Exercise below.

Since $\Omega^2(r) = GM/r^3$ implies $d\Omega^2(r)/d\ln r \equiv r d\Omega^2(r)/dr = -3\Omega^2(r) < 0$ at the arbitrary radius r in the Keplerian disc, the unstable wave numbers k are more restricted for larger magnetic field magnitude B_z, according to the criterion (6.121). However, the maximum growth rate $3\Omega(r)/4$ that occurs at the wave number k where $k^2c_A^2 = 15\Omega(r)^2/16$ is so large that the perturbation amplifies more than 100 times each rotation period. Numerical simulations have shown that turbulence due to this instability strongly enhances radial transport of angular momentum, possibly explaining various astrophysical phenomena such as star formation and intense X-ray sources associated with black holes or neutron stars.

Exercise

(Q1) Consider a rotating reference frame with origin located at the unperturbed position $\mathbf{R}(t)$ of an arbitrary element moving in a circular orbit of radius R, in a Keplerian plasma disc distribution. Thus if $\boldsymbol{\Omega} = \Omega(R)\,\hat{\mathbf{e}}_z$ denotes the angular velocity about the disc centre, the pressureless equation of motion is (cf. the discussion of rotating reference frames in Sect. 3.14):

$$\sigma\left(\frac{d\mathbf{v}}{dt} + 2\boldsymbol{\Omega}\times\mathbf{v} - \Omega^2\boldsymbol{\rho}\right) = -\frac{GM}{\rho^2}\,\hat{\mathbf{e}}_r,$$

where $\mathbf{v}(\mathbf{r}, t)$ is the velocity of the element of density σ relative to the origin of the rotating frame, $\boldsymbol{\rho} = \mathbf{R} + \mathbf{r}$ is its position relative to the disc centre, and $-GM/\rho^2\hat{\mathbf{e}}_r$ is the gravitational force due to the localised mass M at $\rho = 0$.
(a) Noting that $\Omega^2(R+\xi_r)[(R+\xi_r)\hat{\mathbf{e}}_r + \xi_\theta\hat{\mathbf{e}}_\theta] \simeq \Omega^2(R)R\,\hat{\mathbf{e}}_r + \xi_r\, d\Omega^2/dr\,|_R\,\hat{\mathbf{e}}_r$ under a small planar displacement $\boldsymbol{\xi} = \xi_r\hat{\mathbf{e}}_r + \xi_\theta\hat{\mathbf{e}}_\theta$ ($|\boldsymbol{\xi}| \ll R$), derive the appropriate linearised equations for axial perturbations of form $f(t)\exp(ikz)$, on writing the perturbation velocity $\mathbf{v}_1 = \dot{\boldsymbol{\xi}}$ (where the dot denotes differentiation with respect to the time t).

(b) Deduce the hydrodynamic dispersion relation (at $r = R$)

$$\omega^2 = 4\Omega^2(R) + R\frac{d\,\Omega^2(r)}{dr}\bigg|_{r=R} = \kappa^2,$$

where κ is the epicyclic frequency defined by Eq. (6.109).
(c) Show that

$$\kappa^2 = \frac{1}{r^3}\frac{d\,(r^2\Omega(r))^2}{dr}\bigg|_{r=R},$$

such that the disc is hydrodynamically stable when the angular momentum $r^2\Omega(r)$ increases with radius.

6.9 Resistive-g Instabilities

We now recall that there are various non-ideal features that may be introduced into the MHD model. In particular, if the resistive term is retained in the equation of magnetic induction (2.100) and we assume the resistivity coefficient is unperturbed, the ideal linearised perturbation equation (6.17) is replaced by the resistive form

$$-\,i\omega\mathbf{B}_1 = i\mathbf{k}\cdot\mathbf{B}\,\mathbf{v}_1 - \nabla\times(\eta\nabla\times\mathbf{B}_1). \tag{6.122}$$

As foreshadowed in Sect. 5.14, this resistive modification may be significant even if the resistivity coefficient is small ($\eta \ll 1$)—viz. for modes where once again $\mathbf{k}$ is almost parallel to $\mathbf{B}$, so the resistive term $\nabla\times(\eta\nabla\times\mathbf{B}_1)$ in (6.122) becomes comparable with $i\mathbf{k}\cdot\mathbf{B}\,\mathbf{v}_1$. It emerges that the relaxation of the "frozen-in field" constraint (due to the nonzero resistivity) leads to new resistive modes.

Let us reconsider the slab model assuming (6.13) in this section, in an initial discussion of resistive interchange ("resistive-g") instabilities. For a vertical density gradient and horizontal magnetic field, Eqs. (6.15) and (6.16) that allow for magnetic field shear again yield (6.19). However, from (6.122) the equation of magnetic induction for resistive perturbations is

$$-\,i\omega B_{1z} = i\mathbf{k}\cdot\mathbf{B}\,v_{1z} + \eta\left(\frac{d^2}{dz^2} - k^2\right)B_{1z}, \tag{6.123}$$

on assuming constant but nonzero resistivity η.

6.9.1 Uniform Magnetic Field

In the case of a *uniform* horizontal magnetic field (cf. Sect. 6.3), we might consider plane wave propagation where the z-dependence is entirely ignored, when (6.19) and (6.123) produce a nontrivial solution provided

$$\begin{vmatrix} \rho\omega + \dfrac{g}{\omega}\dfrac{d\rho}{dz} & \mu_0^{-1}\mathbf{k}\cdot\mathbf{B} \\ \mathbf{k}\cdot\mathbf{B} & \omega + i\eta k^2 \end{vmatrix} = 0 \tag{6.124}$$

or

$$\omega^3 + i\eta k^2\omega^2 - (k_\parallel^2 c_\mathrm{A}^2 - g\kappa)\omega + i\eta k^2 g\kappa = 0, \tag{6.125}$$

where as before $\kappa = \rho^{-1}d\rho/dz$, $k_\parallel = \mathbf{k}\cdot\mathbf{B}/B$ and $c_\mathrm{A} = B/\sqrt{\mu_0\rho}$ is the Alfvén speed. In the absence of resistivity ($\eta = 0$), the dispersion relation (6.125) reduces to the ideal MHD form $\omega^2 = k_\parallel^2 c_\mathrm{A}^2 - g\kappa$, which implies stability when $k_\parallel > \sqrt{g\kappa}/c_\mathrm{A}$. However, when $k_\parallel > \sqrt{g\kappa}/c_\mathrm{A}$ and the resistivity η is small but finite, there is a new root such that

$$\sigma \equiv -i\omega = \frac{\eta k^2 g\kappa}{k_\parallel^2 c_\mathrm{A}^2 - g\kappa}. \tag{6.126}$$

Thus at short wavelengths where the configuration is stable according to the ideal MHD stability criterion, there are new resistive interchange ("resistive-g") modes for small but finite η, with significant real growth rates σ where $k_\parallel \to k_\mathrm{crit} \equiv \sqrt{g\kappa}/c_\mathrm{A}$ from above.

6.9.2 Sheared Magnetic Field

In the ideal MHD analysis for a sheared magnetic field in Sect. 6.4, the governing differential equation (6.20) is singular at the resonance surface $z = z_s$ where $k_\parallel = \mathbf{k}\cdot\mathbf{B}/B = 0$. When η is small but finite, (6.123) reduces to the ideal form (6.17) except near $z = z_s$, where the resistive term should be retained. Thus in the presence of magnetic field shear, the resonance region near $k_\parallel = 0$ may now be referred to as the resistive region. Let us also recall the fluid mechanics analogy when the shear viscosity coefficient is small, where the higher order viscous term must be retained in boundary layers although the flow may be treated as ideal elsewhere (cf. Sect. 3.8).

Consider a sheared magnetic field $\mathbf{B} = (\hat{\mathbf{i}}z/L_S + \hat{\mathbf{j}})B_0$, where L_S is the representative shear length and B_0 measures the field strength, and consider flute-like modes where $\mathbf{k}\cdot\mathbf{B} \simeq (kB_0/L_S)z$ in the resistive region (narrow for small η) near $z = 0$ where $k = k_y$ and $k_z = 0$. Thus the first term on the right-hand side of (6.19) still does not arise, leaving the focus of the present analysis on the gravitational term. Let us seek a shock-like solution in the narrow resistive region, where the field variables in the coupled equations (6.19) and (6.123) are presumed to vary rapidly, such that the higher order derivative terms dominate all others independent of ω or the space variable z. Under the ad hoc assumption that the resistive growth rate is real, it is again convenient to write $\sigma = -i\omega$ and introduce the scaled variables

$$w = \frac{z}{\delta}\,,\ \delta^4 = \frac{\eta\sigma L_S^2}{k^2 c_A^2}\,,\ \mathcal{G} = \frac{g\kappa L_S^2}{c_A^2}\,,\ a = \frac{g\kappa k^2\delta^2}{\sigma^2} \text{ and } V = \frac{kB_0\delta}{\sigma L_S} v_{1z}$$

into the perturbation equations (6.19) and (6.123), to obtain after some algebra

$$\left(\frac{d^2}{dw^2} + a - w^2\right) V = iwB_{1z} \tag{6.127}$$

$$\frac{d^2 B_{1z}}{dw^2} = \frac{\mathcal{G}}{a}\,(B_{1z} - iwV). \tag{6.128}$$

The parameter δ is a measure of the resistive region thickness (where $\delta \to 0$ as the resistivity $\eta \to 0$), $\mathcal{G}$ is the gravity parameter, and the growth rate σ is implicit in the eigenvalue a.

When $\mathcal{G} > 1$, ideal instabilities grow on the fast Alfvén time scale defined by (6.22) such that the ratio $\mathcal{G}/a \to \infty$ as $\eta \to 0$, so $B_{1z} - iwV \simeq 0$ and (6.127) corresponds to (6.20). However, this ratio $\mathcal{G}/a$ in (6.128) is finite as $\eta \to 0$ for the resistive instabilities that arise with growth rate $\sigma = \mathrm{O}(\eta^{1/3})$ as discussed below. Nevertheless, since $|w| \to \infty$ as $\eta \to 0$ the resistive equations (6.127) and (6.128) do reduce to

$$B_{1z} - iwV \simeq 0\,,\quad \frac{d^2}{dw^2}B_{1z} + \frac{\mathcal{G}}{w^2}B_{1z} \simeq 0 \tag{6.129}$$

near the resistive region boundaries (i.e. where $|z| \simeq \delta$ or $|w| \simeq 1$), equivalent to the ideal equations. Thus there is an "inner" dissipative region and an "outer" ideal region, as in classical Prandtl boundary layer theory (cf. Sect. 3.8). Moreover, if our discussion is restricted to $\mathcal{G} \ll 1$ such that the configuration is strongly stable under the ideal MHD criterion, the dominant approximate solution for $|w| \simeq 1$ is $B_{1z} \simeq |w|^{\mathcal{G}}$. We may therefore adopt $B_{1z} \simeq C$ (a constant) as a first approximation in the resistive region, and proceed to solve

$$\left(\frac{d^2}{dw^2} + a - w^2\right) V = iCw \tag{6.130}$$

on $-\infty < w < \infty$, treating w as a "stretched" variable as $\eta \to 0$ in the terminology of asymptotic theory. In passing, let us note that the higher order term on the left-hand side of (6.128) defines the fluctuation in the magnetic field perturbation across the resistive region, where the magnitude of B_{1z} is approximately constant. In contrast, the magnitude of the associated velocity field perturbation varies substantially across the resistive region, such that $B_{1z} - iwV \simeq 0$ near the resistive region boundaries.

On taking the Fourier transform

$$\widehat{V}(\alpha) = \int_{-\infty}^{\infty} \mathrm{e}^{i\alpha w}\, V(w)\, dw,$$

(6.130) becomes

$$\left(\frac{d^2}{d\alpha^2} + a - \alpha^2\right)\widehat{V}(\alpha) = -2\pi C\delta'(\alpha), \tag{6.131}$$

where the right-hand side involves the derivative of the Dirac delta function. The corresponding homogeneous equation is valid almost everywhere (except at $\alpha = 0$), and readily solved by setting $\widehat{V}(\alpha) = \mathrm{e}^{-\alpha^2/2}\, W(\alpha)$ such that

$$\frac{d^2}{d\alpha^2}W(\alpha) - 2\alpha\frac{dW}{d\alpha} + (a-1)W = 0.$$

The general solution of this equation is $W(\alpha) = c_1 H_\nu(\alpha) + c_2\,\mathrm{e}^{\alpha^2} H_{-\nu-1}(\alpha)$, where $H_\nu(\alpha)$ is the Hermite function of degree $\nu = (a-1)/2$. However, we require $c_2 = 0$ and ν an integer to ensure that $\widehat{V}(\alpha)$ is bounded as $|\alpha| \to \infty$. Thus the solutions are $V(\alpha) = c^{(\pm)}\mathrm{e}^{-\alpha^2/2} H_n(\alpha)$, where the $+$ and $-$ superscripts specify the constants in the respective sub-intervals $\alpha > 0$ and $\alpha < 0$, involving the Hermite polynomials. Moreover, integrating (6.131) over $(0-, \alpha)$ where $\alpha > 0$ produces

$$\frac{d\widehat{V}}{d\alpha} - \left.\frac{d\widehat{V}}{d\alpha}\right|_{0-} = -2\pi C\delta(\alpha)$$

such that

$$\left.\frac{d\widehat{V}}{d\alpha}\right|_{0+} - \left.\frac{d\widehat{V}}{d\alpha}\right|_{0-} = 0 \quad\text{and}\quad \widehat{V}(0+) - \widehat{V}(0-) = -2\pi C \neq 0.$$

The solution compatible with this discontinuity at $\alpha = 0$ is such that $\widehat{V}(0) \neq 0$ and $\widehat{V}(0+) \simeq -\widehat{V}(0-)$, which requires $\nu = 2n$ (i.e. ν must be an even integer) because all odd Hermite polynomials are zero at the origin. The fastest growing mode corresponds to $n = 0$, when

$$1 = \frac{g\kappa k^2\delta^2}{\sigma^2} = \frac{g\kappa k^2}{\sigma^2}\left(\frac{\eta\sigma L_S{}^2}{k^2 c_{\mathrm{A}}{}^2}\right)^{1/2}.$$

Thus the growth rate

$$\sigma = \eta^{1/3}\left(\frac{g\kappa k L_S}{c_{\mathrm{A}}}\right)^{2/3} \tag{6.132}$$

is proportional to $\eta^{1/3}$ (asymptotically as $\eta \to 0$) as anticipated, and reduced by the magnetic field shear (proportional to $L_S^{2/3}$). It may also be verified immediately that the resistive region thickness is

$$\delta = \eta^{1/3} \left(\frac{g\kappa L_S{}^4}{k^2 c_A{}^4} \right)^{1/6} . \tag{6.133}$$

It is notable that the eigenvalue a is determined from the "inner" resistive region solution alone. This corresponds to local destabilisation, as the "frozen-in" constraint is removed in this region due to the nonzero resistivity. Nevertheless, outside the resistive region both of the field perturbations vary slowly across the entire plasma slab (the perturbations are not localised), and the response is said to be electromagnetic because $B_{1z} \neq 0$ almost everywhere (except at highly conducting slab boundaries, for example). There is a localised electrostatic mode where $B_{1z} \simeq C = 0$ everywhere, when the discontinuity requirement detailed above does not apply and therefore ν may also be an odd integer. However, the fastest growing electrostatic mode again corresponds to $n = 0$, so its growth rate is also given by (6.132).

6.10 Resistive Tearing Instabilities

In the previous section, it emerged that the growth rate of both local and global resistive-g modes in a sheared magnetic field is proportional to $\eta^{1/3}$ as $\eta \to 0$. Unlike the ideal MHD pressure-driven interchange modes under gravity we discussed in Sects. 6.3 and 6.4, the resistive-g modes are not prevented by magnetic field shear, although their growth rate is reduced. Thus on introducing the characteristic number $S = \tau_R/\tau_H$, the ratio of the hydromagnetic time $\tau_H = L/c_A$ and the resistive diffusion time $\tau_R = L^2/\eta$ (where L is a suitable characteristic length), the dimensionless growth rate of the resistive-g modes is $\hat{\sigma} = \sigma\tau_R \sim S^{2/3}$ where $S \gg 1$. This *Lundquist number* S (sometimes less appropriately called the magnetic Reynolds number) is typically $O(10^3)$ or greater in magnetic confinement configurations, and usually much larger in astrophysics.

There are two other resistive modes with dimensionless growth rates $\hat{\sigma} \sim S^{\zeta}$ where $0 < \zeta < 1$ and $S \gg 1$. The more important is the large-scale resistive counterpart to the ideal kink instability called *tearing*, due to the pattern of significant disruption it causes driven by the release of magnetic field tension following magnetic reconnexion in the resistive region. Matching of an "outer" ideal solution to the "inner" resistive solution is required to define the growth rate of this characteristically non-local tearing mode, analogous to matching the outer ideal solution with the solution in the viscous boundary layer in fluid mechanics (cf. Sect. 3.8). In the previous section, the magnetic field shear was assumed linear in z and it was found to reduce the growth rate of the resistive-g modes. In contrast, the long-wavelength resistive tearing modes arise in regions of strong magnetic field shear associated with significant current flow, such that the first term in the square brackets on the right-hand side of (6.19) previously omitted becomes dominant. Magnetic reconnexion was first suggested as the probable mechanism for solar flares [5], although it has emerged that the resistive plasma model is inadequate in that context (cf. Sects. 6.11 and 6.12). However, magnetic reconnexion due to some non-ideal modification is relevant much more generally, including plasma relaxation as mentioned in the brief commentary near the end of Sect. 5.10. The pioneering resistive stability analysis due to Furth and others, briefly represented below (cf. also [12]), allows for arbitrary field shear and also introduces a resistivity gradient that drives the third resistive mode known as *rippling*.

The Cartesian slab model is again adopted—but it is assumed that the plasma, with not only the density but also the resistivity dependent on z (to allow for rippling), is permeated by the more general horizontal magnetic field $\mathbf{B}(z) = B_x(z)\hat{\mathbf{i}} + B_y(z)\hat{\mathbf{j}}$ and bounded by perfectly conducting walls at $z = \pm z_w$. Let us introduce the dimensionless variables

$$\begin{aligned} \zeta &= \frac{z}{L}, \quad \psi = \frac{B_{1z}}{B_0}, \qquad V = ik_x v_{1z}\tau_R, \\ \alpha &= kL, \quad F = \frac{\mathbf{k}\cdot\mathbf{B}}{kB_0}, \qquad \hat{\sigma} = \sigma\tau_R, \\ \tau_R &= \frac{L^2}{\eta}, \quad \tau_H = \frac{L}{c_{\mathrm{A}}}, \qquad S = \frac{\tau_R}{\tau_H} \end{aligned} \tag{6.134}$$

with L a characteristic length in the vertical z direction and $k = \sqrt{k_x^2 + k_y^2}$ the usual wave number, which are similar but not identical to those in the original analysis of Furth and others. From (6.19) and (6.122), the linearised perturbation equations for the vertical field components may then be written

$$(\rho V')' - \alpha^2\left(\rho - \frac{S^2 G}{\hat{\sigma}^2}\right)V = \frac{\alpha^2 S^2 F}{\hat{\sigma}}\left[\psi'' - \left(\alpha^2 + \frac{F''}{F}\right)\psi\right], \tag{6.135}$$

$$\psi'' - \alpha^2\left(1 + \frac{\hat{\sigma}}{\eta\alpha^2}\right)\psi = \frac{F}{\eta}\left(1 + \frac{\eta' F'}{\hat{\sigma} F}\right)V, \tag{6.136}$$

where the derivatives are taken with respect to the dimensionless spatial variable ζ, ρ is the normalised density, η is the normalised resistivity, G (proportional to $g\kappa$) represents the gravity that we recall may crudely simulate the magnetic field curvature, and the characteristic Lundquist number $S \gg 1$. The first term in the square bracket on the right-hand side of (6.19) produces the additional current-dependent term in (6.135) with coefficient F''/F, which was lost in the resistive-g analysis given in the previous section. Thus there is again a fourth order system of ordinary differential equations to consider that reduces to the corresponding second order ideal perturbation equation in the limit $S \to \infty$, except in the resistive region near the resonance surface where $F = 0$—i.e. near where the wave vector $\mathbf{k}$ is perpendicular to the magnetic field $\mathbf{B}$.

The previous assumption that the growth rate is real can be proven to extend to any resistive instability admitted by the generalised model (6.135) and (6.136). Normalising the resistivity so that $\eta F' = 1$ and noting that $V = 0$ and $\psi = 0$ at the rigid perfectly conducting boundaries, on multiplying (6.135) and (6.136) by appropriate complex conjugates and then integrating one obtains

$$\int d\zeta \left\{ \frac{\hat{\sigma}^2}{|\hat{\sigma}|^2 \alpha^2 S^2} \left[\rho |V'|^2 + \alpha^2 \left(\rho - \frac{S^2 G}{\hat{\sigma}^2} \right) |V|^2 \right] + \frac{\hat{\sigma} F' - F''/F}{|\hat{\sigma} F' - F''/F|} \left| \psi'' - \left(\alpha^2 + \frac{F''}{F} \right) \psi \right|^2 + |\psi'|^2 + \left(\alpha^2 + \frac{F''}{F} \right) |\psi|^2 \right\} = 0. \qquad (6.137)$$

Taking the imaginary part of (6.137) implies $\Re(\hat{\sigma}) \leq 0$ if $\Im(\hat{\sigma}) \neq 0$, so there are no overstable modes, validating our assumption introduced in Sect. 6.9 that any unstable perturbation has real growth rate.

Adopting the stretched variable $\theta = [\zeta - \zeta_s + \eta'/(2\hat{\sigma})]/\epsilon$ (where $\epsilon \ll 1$), we obtain the scaled forms of the field equations (6.135) and (6.136) for the narrow resistive layer—viz.

$$\frac{d^2 U}{d\theta^2} + \left(\Lambda - \frac{1}{4}\theta^2 \right) U = (\theta - \delta)\psi, \qquad (6.138)$$

$$\frac{d^2 \psi}{d\theta^2} - \epsilon^2 \alpha^2 \psi = \epsilon \Omega [4\psi + (\theta + \delta_1) U] \qquad (6.139)$$

in the notation of Furth and others, where

$$\epsilon = \left(\frac{\hat{\sigma}\eta\rho}{4\alpha^2 S^2 (F')^2} \right)^{1/4}, \qquad U = \frac{4\epsilon F'}{\hat{\sigma}} V,$$

$$\Lambda = \frac{(\eta')^2}{16\epsilon^2\hat{\sigma}^2} + \frac{S^2\epsilon^2\alpha^2 G}{\hat{\sigma}^2 \rho} - \epsilon^2\alpha^2, \; \delta = \left(\frac{F''}{F'} + \frac{\eta'}{2\eta} \right) \Big/ (4\Omega), \qquad (6.140)$$

$$\Omega = \frac{\epsilon\hat{\sigma}}{4\eta}, \qquad \delta_1 = \frac{\eta'}{8\eta\Omega}.$$

In the limit $S \to \infty$ such that $\epsilon \to 0$, from (6.139) we may infer that $\psi \simeq$ constant over the resistive region—i.e. $B_{1z} \simeq$ constant as before. When the resistivity gradient term is ignored, and the field shear is relatively small such that the term proportional to F'' is also negligible as in Sect. 6.9, then $\delta \simeq 0$ and (6.138) reduces to

$$\frac{d^2 U}{d\theta^2} + \left(\Lambda - \frac{1}{4}\theta^2 \right) U = \text{const.}\ \theta \qquad (6.141)$$

consistent with (6.130) where $w = \theta/\sqrt{2}$. In the present notation, the fastest short wavelength ($\alpha \gg 1$) resistive-g mode driven by the gravity term (when

$\Lambda \simeq S^2\epsilon^2\alpha^2 G/(\hat{\sigma}^2\rho) = 1/2$) has the growth rate

$$\hat{\sigma} = \eta^{1/3}\left(\frac{S\alpha G}{\rho^{1/2}|F'|}\right)^{2/3} \tag{6.142}$$

consistent with (6.132), where we first observed that the growth is inhibited but not eliminated by the magnetic field shear ($F' \neq 0$ here).

As anticipated, tearing modes are characteristically non-localised and relatively long wavelength ($\alpha < 1$), and arise when the resistivity gradient and gravity terms in Λ are relatively small. The limiting ideal equation (in the limit $S \to \infty$) for the outer region is evidently

$$\psi'' - \left(\alpha^2 + \frac{F''}{F}\right)\psi = 0 \tag{6.143}$$

from (6.135), with an asymptotic expansion solution

$$\xi \sim \exp[-\alpha(\zeta - \zeta_s)](\psi_0 + \alpha\psi_1 + \cdots), \quad \alpha \ll 1, \tag{6.144}$$

such that $\psi_0 \sim |F|$ and at first order

$$\psi_1'' - \frac{F''}{F}\psi_1 = 2\psi_0 \,\mathrm{sgn}(\zeta - \zeta_s), \tag{6.145}$$

so $\psi_1 \sim F_\pm^2/F'$ near ζ_s. The matching requirement is effectively met in this context by equating the jump in the logarithmic derivative across the resistive region

$$\Delta = \left.\frac{d\ln\psi}{d\zeta}\right|_{\zeta-}^{\zeta+}$$

evaluated in the outer region with the appropriate form for Δ in the inner region, because this represents the desired dispersion relation. Thus we have from the outer region

$$\Delta_{\text{outer}} \simeq (F')^2\,(F_+^{-2} + F_-^{-2})\,/\,\alpha,$$

where $F_\pm$ denote the values of F at the respective boundaries $\zeta_s\pm$. For the inner region, Furth et al. chose to expand in terms of the orthonormal Hermite polynomials to render (for $\psi \simeq$ constant)

$$\begin{aligned}\Delta_{\text{inner}} &= 2^{7/2}\Omega\sum_{m=0}^{\infty}\frac{\Gamma(m+\frac{1}{2})}{\Gamma(m+1)}\left[\frac{\Lambda - \frac{1}{2}}{\Lambda - (2m+\frac{3}{2})} - \frac{\delta\delta_1/4}{\Lambda - (2m+\frac{1}{2})}\right]\\ &= 2^{7/2}\pi\Omega\left[\frac{\Gamma(\frac{3}{4} - \frac{1}{2}\Lambda)}{\Gamma(\frac{1}{4} - \frac{1}{2}\Lambda)} + \frac{\delta\delta_1}{8}\frac{\Gamma(\frac{1}{4} - \frac{1}{2}\Lambda)}{\Gamma(\frac{3}{4} - \frac{1}{2}\Lambda)}\right]\\ &\simeq 12\Omega,\end{aligned}$$

on noting that $\delta_1 \simeq 0$ and $\Lambda \ll 1$ when resistivity gradient and gravity contributions are negligible. Equating Δ_{outer} to Δ_{inner} produces the dispersion relation

$$\hat{\sigma} = (F')^2 \left(\frac{2S\eta^{3/2}}{9\alpha\rho^{1/2}} \right)^{2/5} (F_+^{-2} + F_-^{-2})^{4/5} \tag{6.146}$$

for the tearing mode, driven by the strong magnetic field shear (with growth rate proportional to the square of $F' \neq 0$).

The resistive rippling mode corresponds to dominance of the resistivity gradient in Λ, with maximum growth rate when $\Lambda \simeq (\eta')^2/(16\epsilon^2\hat{\sigma}^2) = 1/2$—viz.

$$\hat{\sigma} = \left[\frac{(\eta')^2 \alpha S |F'|}{4\eta^{1/2}\rho^{1/2}} \right]^{2/5}. \tag{6.147}$$

The resistivity gradient term is larger than the gravity term in Λ when

$$\hat{\sigma} < \frac{(\eta')^2 (F')^2}{4\eta |G|}, \tag{6.148}$$

such that resistive-g instabilities typically have the faster growth rate when $G > 0$ and the magnetic field shear (represented by F') is not too large, but rippling may become more evident at plasma–vacuum boundaries for example (where the resistivity gradient η' is rather large).

6.11 Effect of Plasma Viscosity on Stability

Even if a well designed configuration avoids ideal MHD instabilities, we now know that the resistive-g (or pressure-driven resistive interchange) and resistive tearing (the current-driven resistive kink) modes can produce large-scale plasma displacements. Indeed, although the growth rate of the resistive-g instability is moderated somewhat in the presence of magnetic field shear, the tearing instability is driven by the magnetic field shear and reduces the current density gradient as the magnetic field lines in the resistive region break and reconnect. Moreover, although no resistive instability grows on the fast Alfvén time-scale of ideal MHD modes, they all develop at a rate much faster than classical resistive diffusion (characterised by the resistive time τ_R). However, the continuing strong interest in magnetic confinement prompted many theoretical investigations of resistive instabilities in cylindrical and toroidal geometry—again usually by normal mode analysis or various supplementary numerical simulations in both linear and nonlinear regimes, rather than by a variational approach that had proven so powerful in extending the ideal MHD stability theory to non-Cartesian geometries, as discussed in Sects. 6.5–6.7. Variational techniques to analyse dissipative and nonlinear systems can give useful qualitative insights, but not necessarily all the explicit quantitative outcomes often sought. In brief, in addition to smaller scale disturbances it became clear that major instabilities to counter in magnetic confinement designs include: (1) ideal and resistive interchanges (or ballooning modes in toroidal configurations) driven by the pressure gradient, wherever there is unfavourable magnetic field curvature; and (2) ideal and resistive kink (tearing) instabilities driven by a current term involving the magnetic field shear in the MHD equation of motion.

Various experimental magnetic confinement programmes have been partly successful. Tearing modes in tokamaks, where the toroidal magnetic surfaces are created by a large applied longitudinal magnetic field and a longitudinal current, can be controlled by a rather stringent upper limit on the magnitude of that current [14]. High current, high magnetic field shear devices such as the reverse-field-pinch allow higher β (i.e. a desirably higher ratio of plasma pressure to magnetic pressure for a thermonuclear reactor) and do promise plasma confinement at safety factor values much less than 1. Robinson found ideal and tearing mode stable configurations in such devices for central β values approaching 20 %, but the resistive-g mode was seen to be a continuing threat. Other instabilities have been identified over the years too, and not all of them are relatively localised like the resistive rippling mode mentioned above. The need to control persistent instability in any attempt at magnetic confinement, and of course to better understand the physics in the laboratory and elsewhere, have motivated the inclusion of further non-ideal effects in the plasma model. In this section, we discuss the inclusion of plasma viscosity (cf. Sects. 2.9 and 2.10), which provides (1) partial stabilisation of ideal and resistive modes and (2) enhanced energy release due to magnetic field reconnexion, of particular interest in solar physics.

6.11.1 Magnetoviscous Stabilisation of Ideal and Resistive Instabilities

Let us again consider a magnetohydrostatic configuration defined by (5.63) when $\mathbf{v} = 0$, but now introduce plasma viscosity so the equation of motion in the set of linearised perturbation equations becomes

$$\rho\frac{\partial \mathbf{v}_1}{\partial t} + \nabla p_1 + \nabla\cdot\mathbf{t}_1 = \mu_0^{-1}[(\nabla\times\mathbf{B})\times\mathbf{B}_1 + (\nabla\times\mathbf{B}_1)\times\mathbf{B}] + \rho_1\mathbf{g}, \qquad (6.149)$$

and as before adopt the resistive equation of magnetic induction

$$\frac{\partial \mathbf{B}_1}{\partial t} = \nabla\times(\mathbf{v}_1\times\mathbf{B}) - \nabla\times(\eta\nabla\times\mathbf{B}_1). \qquad (6.150)$$

Introducing the Lagrangian displacement vector $\boldsymbol{\xi}$ and magnetic vector $\mathbf{R}$, defined by

$$\frac{\partial \boldsymbol{\xi}}{\partial t} = \mathbf{v}_1(\mathbf{r}_0, t) \quad \text{and} \quad \frac{\partial \mathbf{R}}{\partial t} = \mathbf{B}_1(\mathbf{r}_0, t) \tag{6.151}$$

relative to any initial reference position $\mathbf{r}_0$, the system of linearised perturbation equations reduces to

$$P\ddot{\underline{\boldsymbol{\xi}}} + K\dot{\underline{\boldsymbol{\xi}}} + D\underline{\boldsymbol{\xi}} = 0. \tag{6.152}$$

Here the dot once again denotes time differentiation, the underlined symbol $\underline{\boldsymbol{\xi}}$ denotes the column six-vector $(\boldsymbol{\xi}, \mathbf{R})^T$, and the coefficient matrices are

$$P = \begin{bmatrix} \rho & 0 \\ 0 & 0 \end{bmatrix}, \quad K = \begin{bmatrix} L_0 & 0 \\ 0 & \mu_0^{-1} L_2 \end{bmatrix}, \quad D = \begin{bmatrix} L_1 L_3 + L_4 & -L_1 L_2 \\ -\mu_0^{-1} L_2 L_3 & \mu_0^{-1} L_2^2 \end{bmatrix} \tag{6.153}$$

with the implicit linear operators

$$\begin{aligned}
L_0 \dot{\boldsymbol{\xi}} &= \nabla \cdot \mathbf{t}_1(\dot{\boldsymbol{\xi}}), \\
L_1(\mathbf{R}) &= \mu_0^{-1}[(\nabla \times \mathbf{B}) \times \mathbf{R} + (\nabla \times \mathbf{R}) \times \mathbf{B}], \\
L_2 \mathbf{R} &= \nabla \times (\eta \nabla \times \mathbf{R}), \\
L_3 \boldsymbol{\xi} &= \nabla \times (\boldsymbol{\xi} \times \mathbf{B}), \\
L_4 \boldsymbol{\xi} &= \mathbf{g} \nabla \cdot (\rho_0 \boldsymbol{\xi}) - \nabla(\gamma p \nabla \cdot \boldsymbol{\xi} + \boldsymbol{\xi} \cdot \nabla p).
\end{aligned} \tag{6.154}$$

Equation (6.152) is a generalised form for the dissipative system, with plasma viscosity and resistivity combined in the dissipative coefficient matrix K and the resistivity also rendering the otherwise ideal MHD coefficient matrix D nondiagonal. Thus when the resistivity η is zero, (6.152) reduces to the plasma viscosity modification of the earlier ideal MHD form (6.29) in the displacement three-vector $\boldsymbol{\xi}$. One may readily identify the relevant driving terms in the implicit linear operators, for the now familiar ideal or resistive instabilities.

Let us again consider an inner product over the solution space, but now for the 6-vector entities. For real exponential growth rates ($\sim e^{\sigma t}$), from (6.152) we have the quadratic relation

$$\sigma^2 + 2\kappa\sigma + \alpha = 0 \tag{6.155}$$

with coefficients

$$2\kappa = \frac{\int_V \underline{\boldsymbol{\xi}}^* \cdot K \underline{\boldsymbol{\xi}} \, d\tau}{\int_V \underline{\boldsymbol{\xi}}^* \cdot P \underline{\boldsymbol{\xi}} \, d\tau} \quad \text{and} \quad \alpha = \frac{\int_V \underline{\boldsymbol{\xi}}^* \cdot D \underline{\boldsymbol{\xi}} \, d\tau}{\int_V \underline{\boldsymbol{\xi}}^* \cdot P \underline{\boldsymbol{\xi}} \, d\tau}, \tag{6.156}$$

where again the asterisk denotes the complex conjugate and the integration is taken over the plasma volume V. Equation (6.155) is similar to the eigenvalue equation for an harmonic oscillator with damping coefficient κ. However, the Rayleigh quotient

α is usually no restoring force here, for it is the source of instabilities driven by the unfavourable magnetic field contributions from the linear operator L_1 to the matrix D. Thus for $\kappa > 0$ the necessary and sufficient condition for instability is $\Re\{\sqrt{\kappa^2 - \alpha}\} > \kappa$, which reduces to $\alpha < 0$ for real α. Nevertheless, the growth rate of any instability is reduced from $\sqrt{|\alpha|}$ when the coefficient κ is zero to $|\alpha|/(2\kappa)$ for large positive κ—i.e. the instability remains, but it is significantly damped.

We have seen in Sect. 2.9 that the plasma pressure tensor involves an expansion $\mathbf{t} = \mathbf{t}_{\|} + \mathbf{t}_{g} + \mathbf{t}_{\perp} + \cdots$ in magnetised plasma where $\omega_c\tau \gg 1$, except in the near neighbourhood of any magnetic null. Further, if we adopt the simple form for the deformation tensor

$$\mathbf{s} \equiv \{\nabla \mathbf{v}\} = \frac{1}{2}[\nabla \mathbf{v} + (\nabla \mathbf{v})^T] - \frac{1}{3}\nabla \cdot \mathbf{v}\,\mathbf{I}$$

corresponding to the result (2.80) in Sect. 2.10 valid for sufficiently small τ, from (2.71) the parallel ion viscosity component entering (6.149) is

$$\mathbf{t}_{\|} = -3\mu\left[\hat{\mathbf{b}} \cdot \nabla(\mathbf{v} \cdot \hat{\mathbf{b}}) - \mathbf{v} \cdot (\hat{\mathbf{b}} \cdot \nabla \hat{\mathbf{b}}) - \frac{1}{3}\nabla \cdot \mathbf{v}\right]\left(\hat{\mathbf{b}}\hat{\mathbf{b}} - \frac{1}{3}\mathbf{I}\right) \tag{6.157}$$

where $\hat{\mathbf{b}} = \mathbf{B}/B$. Except in Cartesian geometry, the second term in the square brackets in (6.157) proportional to magnetic field curvature ensures that parallel viscosity enters the theory, although the first and third terms are both small in the neighbourhood of the resonance surfaces for the most damaging incompressible modes previously considered (where the parallel wave number $k_{\|} \simeq 0$ and $\nabla \cdot \mathbf{v} \simeq 0$). In particular, if it is again assumed that the plasma is bounded by a rigid perfectly conducting wall, the boundary conditions include: (1) vanishing $\hat{\mathbf{n}} \cdot \boldsymbol{\xi}$ (and possibly other components of $\boldsymbol{\xi}$ in the presence of plasma viscosity); (2) vanishing $\hat{\mathbf{n}} \cdot \mathbf{R}$ (or vanishing $\mathbf{R}$ if the wall is "at infinity"); and (3) vanishing $\hat{\mathbf{n}} \times \mathbf{E}$, so that $\hat{\mathbf{n}} \times (\nabla \times \mathbf{R})$ vanishes. For such Cauchy-type conditions the dissipative numerator is

$$\begin{aligned}\int_V \underline{\boldsymbol{\xi}}^* \cdot K\,\underline{\boldsymbol{\xi}}\,d\tau &= \int_V \boldsymbol{\xi}^* \cdot L_0\,\boldsymbol{\xi}\,d\tau + \mu_0^{-1}\int_V \mathbf{R}^* \cdot L_2\mathbf{R}\,d\tau \\ &= \int_V \left[3\mu\left|\left(\hat{\mathbf{b}}\hat{\mathbf{b}} - \frac{1}{3}\mathbf{I}\right) : \nabla\boldsymbol{\xi}\right|^2 + \mu_0^{-1}\eta|\nabla \times \mathbf{R}|^2\right],\end{aligned} \tag{6.158}$$

so that $\kappa > 0$. Linear stability calculations for cylindrical flux surfaces showed that parallel ion viscosity could significantly reduce the growth rate of the resistive-g mode in tearing mode stable magnetic field configurations identified for the reverse-field pinch, except for the axisymmetric ($m = 0$) mode where the nonlinear response was followed. Critical local pressure gradients dependent on the magnetic field shear were determined, implying a critical pressure profile with an average β value of around 10 %. However, since not only the pressure p but also the ion-ion collision time τ is expected to rapidly increase as the plasma temperature increases, there are additional terms in the modified form (2.82) of $t_{\|,\|}$ that may become important. As previously mentioned, in simulations Eq. (2.79) could be invoked directly to render the scalar field $t_{\|,\|}$ in the parallel viscosity component $\mathbf{t}_{\|} = (3/2)t_{\|,\|}\{\hat{\mathbf{b}}\hat{\mathbf{b}}\}$.

Parallel ion viscosity had earlier been overlooked—partly because the classical shear viscosity appropriate for fluids but not for magnetised plasma (except in the near neighbourhood of a magnetic null) was adopted by many authors; partly because gyrokinetic theory had only produced the gyroviscous (cross) component $\mathbf{t}_g$ given by (2.72) as previously mentioned in Sect. 2.9; and also partly because the analysis was undertaken for the plasma slab—i.e. in Cartesian coordinates, where there is no geometric magnetic field curvature term as in (6.157), so the dominant contribution in the resonance (or resistive) region from the leading parallel viscosity component $\mathbf{t}_{\|}$ was lost but one or both of the two lower order components $\mathbf{t}_g$ and $\mathbf{t}_{\perp}$ in the expansion (2.70) were included. There may be exceptions in non-Cartesian geometries where the magnetic field curvature term is locally negligible, when one or both of these additional components do become more important than the parallel viscosity. However, it does seem that stabilisation has generally been found—whether in the non-dissipative context of ideal modes where the gyroviscous component $\mathbf{t}_g$ only is included, or when either or both of the gyroviscous or perpendicular viscosity components $\mathbf{t}_g$ and $\mathbf{t}_{\perp}$ are retained in resistive theory. This is of course entirely consistent with the universal result in the above analysis when $\kappa > 0$.

Other non-ideal effects have been included in extensive numerical calculations, where both the initial linear and subsequent nonlinear mode behaviour may be followed, and there have been some further related theoretical developments. For example, when electron diamagnetic effects are included in the generalised Ohm's law (2.96) together with ion viscosity, the tearing instability splits into two branches—viz. the socalled drift-tearing (or reconnecting) mode in addition to the resistive kink. There is stabilisation of both of these branches, with the drift-tearing branch being completely stabilised even if collisions are relatively rare, although the resistive kink mode may remain unstable very close to the ideal MHD stability boundary. More work to incorporate magnetic field curvature or other physical features such as electron trapping is being undertaken. In short, the inclusion of non-ideal effects remains an active area of research in plasma stability theory at the time of writing.

6.11.2 Enhanced Energy Release in Magnetic Reconnexion

An outstanding question in the physics of the Sun is the explosive release of energy in coronal solar flares, which observations have confirmed are localised phenomena. Magnetic reconnexion is the widely favoured mechanism for this phenomenon, but on its own the very low resistivity in the corona cannot account for the release rate [12]. However, it has been found that the reconnexion can be sufficiently fast for the observed rapid energy release if the Hall term is retained in the generalised Ohm's law (2.96), as discussed in the following Sect. 6.12. In this subsection, we consider the separate issue of the extremely large amount of energy released in solar flares, which can be accounted for by the dominant parallel ion viscosity since any reconnexion model typically exhibits strong plasma flows.

Let us adopt the simplest viscoresistive model, where homogeneous incompressible plasma (ρ constant and $\nabla\cdot\mathbf{v} = 0$) is described by

$$\rho\left(\frac{\partial \mathbf{v}}{\partial t} + \mathbf{v}\cdot\nabla\mathbf{v}\right) + \nabla p + \nabla\cdot\mathbf{t}_{\parallel} = \mathbf{j}\times\mathbf{B}, \tag{6.159}$$

$$\frac{\partial \mathbf{B}}{\partial t} = \nabla\times(\mathbf{v}\times\mathbf{B}) - \eta\nabla\times\nabla\times\mathbf{B}, \tag{6.160}$$

in addition to $\mathbf{j} = \mu_0^{-1}\nabla\times\mathbf{B}$ and $\nabla\cdot\mathbf{B} = 0$. The energy release is implicit in the energy transport equation

$$\frac{\partial}{\partial t}\left(\frac{1}{2}\rho v^2 + \frac{B^2}{2\mu_0}\right) + \nabla\cdot\left((p + \frac{1}{2}\rho v^2)\mathbf{v} + \mu_0^{-1}\mathbf{B}\times(\mathbf{v}\times\mathbf{B}) + \eta\,\mathbf{j}\times\mathbf{B} + \mathbf{t}_{\parallel}\cdot\mathbf{v}\right)$$
$$= -\mu_0\eta j^2 + \mathbf{t}_{\parallel} : \nabla\mathbf{v}, \tag{6.161}$$

derived from (6.159) in similar fashion to the derivation of (3.7) from (3.5) in Sect. 3.2. Moreover, for steady state solutions sustained by advective flows within a plasma volume V bounded by a closed surface S, the Divergence Theorem (1.60) identifies the kinetic, magnetic and dissipative (resistive plus viscous) fluxes through the boundary S with the additional resistive and viscous components represented by volume integrals obtained from the terms on the right-hand side of (6.161). On recalling the resistive Ohm's law (2.92), the terms on the right-hand side include the electromagnetic dissipation $\mathbf{j}\cdot(\mathbf{E}+\mathbf{v}\times\mathbf{B})$ in the energy Eq. (2.57) in addition to the power of the parallel viscous stress $\mathbf{t}_{\parallel} : \nabla\mathbf{v}$, and the term $\eta\,\mathbf{j}\times\mathbf{B}$ on the left-hand side can be interpreted as a Poynting flux involving the electric field in the moving plasma reference frame—cf. (2.92) and (5.27).

In several papers, Craig and Litvinenko have considered the energy dissipation due to anisotropic parallel ion viscosity during magnetic merging. For example, the stream function $\psi(x, y) = xf(y) + g(y)$ producing the velocity field $\mathbf{v} = (\partial\psi/\partial y)\,\hat{\mathbf{i}} - (\partial\psi/\partial x)\,\hat{\mathbf{j}} = (xf'(y) + g'(y))\,\hat{\mathbf{i}} - f(y)\,\hat{\mathbf{j}}$ defines two streams colliding at the plane $y = 0$, where a localised current layer is associated with piling up oppositely directed straight magnetic field lines $\mathbf{B} = B(y)\,\hat{\mathbf{i}}$—i.e. with $B(0-) = -B(0+)$ and $B(0) = 0$. The merging rate for this large-scale vortex flow may be characterised by the Alfvén Mach number $M_e = v_e/v_{Ae}$ where $v_{Ae} = B_e/\sqrt{\mu_0\rho}$ is the reference Alfvén speed, with v_e and B_e respectively denoting the inflow speed and magnetic field magnitude at some external boundary a distance $y = L$ away (the representative global length scale of the active solar region). From (6.157) we have $\mathbf{t}_{\parallel} = -\mu\, f'(y)\,[2\,\hat{\mathbf{i}}\,\hat{\mathbf{i}} - (\hat{\mathbf{j}}\,\hat{\mathbf{j}} + \hat{\mathbf{k}}\,\hat{\mathbf{k}})]$ such that $\nabla\cdot\mathbf{t}_{\parallel} = \mu\nabla f'(y)$ in this simple two-dimensional Cartesian steady state, hence for constant ρ and μ the curl of (6.159) implies

$$f\,f''' - f'\,f'' = 0 \quad\text{and}\quad f g''' - f'' g' = 0, \tag{6.162}$$

so that irrespective of the consequent form of $g(y)$ the familiar particular solution $f(y) = v_e \sin(\lambda y/L)/\sin\lambda$ $(0 \leq \lambda < L)$ obtained earlier for inviscid merging prevails in the presence of parallel ion viscosity. Thus neither the velocity profile nor the magnetic field profile (subsequently defined by the equation of magnetic induction) is altered, but the energy loss due to the viscous term

$$\mathbf{t}_{\|} : \nabla\mathbf{v} = -3\mu(f')^2 = -3\mu\left(\frac{v_e}{L}\frac{\lambda}{\sin\lambda}\right)^2 \cos^2\left(\frac{\lambda y}{L}\right) \tag{6.163}$$

in Eq. (6.161) is typically very much larger than that due to the resistive term $\mu_0\eta j^2$.

Enhanced dissipation rates due to parallel ion viscosity have also been demonstrated in planar reconnexion under compressible collapse at a magnetic field X-point. The predicted corrugated magnetic field profiles provide many locations where plasma can readily flow along the magnetic field and rapid wave damping may occur—again significantly faster than dissipation due to resistivity. Magnetic reconnexion models in three dimensions include socalled fan and spine structures corresponding to strongly localised current sheets and quasi-cylindrical tubes, respectively [12]. Spines may be less effective for fast energy release due to their small dissipative volume, and for current sheets Craig and Litvinenko found that both the associated large-scale advective flows and the coalescence of magnetic islands lead to similar substantially enhanced energy losses due to parallel ion viscosity. They have also recognised that parallel viscosity alone is probably inadequate in the near neighbourhood of a magnetic null, where recourse may be made to the exact form (2.69). Once again the magnetic field curvature may be significant in non-Cartesian geometry, such that additional terms in $\mathbf{t}$ (other than the collisional term) should also be retained as discussed in Sect. 2.10.

Exercises

(Q1) Derive the energy transport equation (6.161).

(Q2) Derive $\nabla\mathbf{v}$ for the two-dimensional flow defined by the stream function $\psi(x, y) = xf(y) + g(y)$, where $f(y)$ and $g(y)$ are arbitrary functions. Hence deduce that the flow is incompressible, and $\mathbf{t}_{\|} = -\mu f'(y)\,[2\hat{\mathbf{i}}\hat{\mathbf{i}} - (\hat{\mathbf{j}}\hat{\mathbf{j}} + \hat{\mathbf{k}}\hat{\mathbf{k}})]$ such that $ff''' - f'f'' = 0$ in the magnetic field $\mathbf{B} = B(y)\hat{\mathbf{i}}$. Show that this nonlinear differential equation in $f(y)$ can be solved in terms of two quite familiar linear second order differential equations, and identify the consequent solution types.

6.12 Hall Instability

We recall from Sects. 5.5 and 5.15 that the Hall term has often been retained in the MHD modelling of both conducting fluids and plasmas, and despite some null results in the literature it is now well known that the Hall effect can produce further instability. As previously discussed, Hall MHD recognises that there is inter-species diffusion in plasmas, where the electrons but not the ions are considered "frozen-in" to the prevailing magnetic field when Eq. (2.98) is reduced to $\mathbf{E} + \mathbf{v}_e \times \mathbf{B} = 0$. The equation of magnetic induction (5.123) is then adopted. However, we also noted that the Hall term may be retained in a resistive plasma model where the corresponding form of the magnetic induction equation is (5.126). We first consider the gravitational instability of an ion-electron plasma where (5.123) is invoked, but then the magneto-gravitational instability of a plasma where the ionisation is weak and the resistive form (5.126) is adopted.

6.12.1 Gravitational Interchange Instability in Ion-Electron Plasma

On recalling our observation in Sect. 5.15 that the Hall term introduces higher derivatives into the induction equation (5.123), we again anticipate that singular behaviour may render some new instability. Let us first write the fundamental MHD equations of continuity and motion (2.55) and (2.56) as

$$\frac{\partial \rho}{\partial t} + \nabla \cdot (\rho \mathbf{v}) = 0, \tag{6.164}$$

$$\rho \frac{d\mathbf{v}}{dt} + \nabla p = \rho \mathbf{g} + \mathbf{j} \times \mathbf{B}, \tag{6.165}$$

where $\mu_0 \mathbf{j} = \nabla \times \mathbf{B}$. From Sect. 3.2, for a barotropic fluid we have $\nabla p/\rho = \nabla h$ where $h = \int dp/\rho$ is the specific enthalpy, when the curl of (6.165) yields

$$\nabla \times \left(\frac{d\mathbf{v}}{dt} \right) = \nabla \times \left(\frac{1}{\rho} \mathbf{j} \times \mathbf{B} \right). \tag{6.166}$$

In a quasi-neutral ion-electron plasma in which $\rho \simeq n m_i$, where $n(\mathbf{r}, t)$ denotes the particle number density ($n_i = n_e = n$) and m_i the ion mass, the Hall term is therefore

$$\nabla \times \left(\frac{1}{ne} \mathbf{j} \times \mathbf{B} \right) = \frac{m_i}{e} \nabla \times \frac{d\mathbf{v}}{dt}. \tag{6.167}$$

Consequently, from (6.164)–(6.166) and (5.123) we have the following simple closed system of equations:

$$\frac{\partial n}{\partial t} + \nabla \cdot (n\mathbf{v}) = 0, \tag{6.168}$$

$$n m_i \frac{d\mathbf{v}}{dt} + T \nabla n = n m_i \mathbf{g} + \mu_0^{-1} (\nabla \times \mathbf{B}) \times \mathbf{B}, \tag{6.169}$$

as $\nabla p = T\nabla n$ with T constant and the Boltzmann constant suppressed for convenience, and the equation of magnetic induction rewritten as

$$\frac{\partial \mathbf{B}}{\partial t} = \nabla \times (\mathbf{v} \times \mathbf{B}) - \frac{m_i}{e} \nabla \times \frac{d\mathbf{v}}{dt}. \tag{6.170}$$

Note that variable density is not only accounted for in the equation of motion (6.169) but also in the equation of magnetic induction (6.170), corresponding to the retention of $n(\mathbf{r}, t)$ within the curl of the Hall term—cf. (6.167).

Let us once again consider a plasma vertically stratified under gravity in a magnetic field, in equilibrium where

$$\nabla n = \frac{dn}{dz}\hat{\mathbf{k}}, \quad \mathbf{g} = -g\,\hat{\mathbf{k}}, \quad \mathbf{v} = 0, \quad \mathbf{B} = B(z)\,\hat{\mathbf{j}} \tag{6.171}$$

and hence the initial pressure balance equation

$$\frac{d}{dz}\left(nT + \frac{B^2}{2\mu_0}\right) = -nm_i g \tag{6.172}$$

—i.e. an equilibrium in Cartesian geometry similar to that considered in Sect. 6.3, except that now $\mathbf{B}(z)$ is variable but uni-directional (there is no magnetic field shear).

Recalling the magnetic field stabilisation in Sects. 6.3 and 6.4, let us consider perturbations of Fourier form $f_1(\mathbf{r}, t) = f_1(z) \exp[i(kx - \omega t)]$ such that $\mathbf{k} \cdot \mathbf{B} = 0$, when the essential linearised perturbation equations from the Hall model (6.168)–(6.170) are

$$\begin{aligned}
i\omega n_1 &= n\frac{dv_{1z}}{dz} + iknv_{1x} + \frac{dn}{dz}v_{1z}, \\
i\omega n m_i v_{1x} &= ikTn_1 + i\mu_0^{-1}kB_{1y}, \\
i\omega n m_i v_{1z} &= T\frac{dn_1}{dz} + \mu_0^{-1}\left(B\frac{dB_{1y}}{dz} + B_{1y}\frac{dB}{dz}\right) + n_1 m_i g, \\
i\omega B_{1y} &= B\frac{dv_{1z}}{dz} - i\omega\frac{m_i}{e}\frac{dv_{1x}}{dz} + ikBv_{1x} + \left(\frac{dB}{dz} - \frac{m_i}{e}\omega k\right)v_{1z},
\end{aligned} \tag{6.173}$$

and the elimination of n_1 and B_{1y} between them produces two coupled first order differential equations in $\{v_{1x}, v_{1z}\}$ that of course can always be rendered as a single second order differential equation.

However, let us restrict our attention to modes that do not alter the magnetic energy—i.e. such that $\mathbf{B}_1 \cdot \mathbf{B} = 0$ so $B_{1y} = 0$. These modes are quasi-electrostatic (as $\mathbf{B}_1 \cdot \mathbf{B} = 0$ implies $\nabla \times \mathbf{E}_1 \cdot \mathbf{B} = 0$ and consequently $\mathbf{E}_1 = -\nabla_\perp \phi$, where ϕ is the relevant electric potential), and the magnetic field terms from the Lorentz force in the two perturbation equations obtained from the equation of motion prove to be unimportant. Moreover, we have $B \simeq$ constant if the equilibrium magnetic pressure $B^2/(2\mu_0)$ is presumed to be large compared to the plasma pressure and the hydrostatic pressure—indeed, the equilibrium pressure balance equation (6.172) rewritten as

$$\frac{dB}{dz} = -\beta\left(\frac{g}{c_s^2} + \frac{1}{n}\frac{dn}{dz}\right)$$

(where $c_s = \sqrt{T/m_i}$ is the sound speed) implies the initial magnetic field gradient is negligible for sufficiently small plasma beta ($\beta = \mu_0 nT/B^2 \ll 1$). Thus we set $B_{1y} = 0$ and omit the term involving dB/dz but retain the Hall contributions in the perturbation equation obtained from the equation of magnetic induction (6.170), such that the governing second order differential equation for the gravitational interchanges is (cf. Exercise below)

$$\frac{d}{dz}\left(n\frac{dV}{dz}\right) - k^2 nV = kD\,\frac{dn}{dz}\,V, \tag{6.174}$$

where $V = v_{1z} - i(\omega/\omega_{ci})v_{1x}$ and $D = (\omega + gk/\omega_{ci})^{-1}(\omega^2/\omega_{ci} + gk/\omega)$ involve the ion cyclotron frequency $\omega_{ci} = eB/m_i$ that characterises the Hall effect. For $V \to 0$ as $|z| \to \infty$, the generalised Rayleigh form obtained from (6.174) by integrating over $-\infty < z < \infty$ is therefore

$$\frac{\int n\left(|dV/dz|^2 + k^2|V|^2\right)dz}{\int k\,(dn/dz)\,|V|^2\,dz} = -D = -\frac{\omega^2/\omega_{ci} + gk/\omega}{\omega + gk/\omega_{ci}}, \tag{6.175}$$

so on denoting the integral quotient on the left-hand side by $Q(k; n, dn/dz)$ we have the dispersion relation (a cubic)

$$\left(\frac{\omega}{\omega_{ci}}\right)^3 + Q\left(\frac{\omega}{\omega_{ci}}\right)^2 + Q\,\frac{gk}{\omega_{ci}^2}\left(\frac{\omega}{\omega_{ci}}\right) + \frac{gk}{\omega_{ci}^2} = 0. \tag{6.176}$$

When ω_{ci} is so large that $gk/\omega_{ci}^2 \ll \omega/\omega_{ci} \ll 1$, the left-hand side of the dispersion relation (6.176) is approximated by the sum of the second and fourth terms that renders the ideal MHD dispersion relation $\omega^2 = -gk/Q$ in the limit of large wavelength (small k). However, when $gk/\omega_{ci}^2 \gg \omega/\omega_{ci} \gg 1$ the left-hand side is approximated by the sum of the first and third terms such that we obtain $\omega^2 = -gkQ$, the dispersion relation distinguishing the Hall instability in the limit of small wavelength (large k). Further, in a "local approximation" where $|dV/dz| \ll |V|$, the integral quotient reduces to $Q(k; n, dn/dz) \simeq kn/(dn/dz)$, so again writing $\sigma = -i\omega$ the first case renders the Rayleigh–Taylor growth rate $\sigma = \sqrt{g\kappa}$ and the second the growth rate $\sigma = k\sqrt{g/\kappa}$ of the Hall interchange mode (where we have introduced $\kappa = \rho^{-1}d\rho/dz = d(\ln\rho)/dz \simeq d(\ln n)/dz$ as in Sect. 6.4—cf. also (6.21) in the "local approximation" when $k_\parallel = 0$).[18]

[18]The Hall model (6.168)–(6.170) was first investigated by Huba and co-authors in a series of articles, and particular reference may be made to J.D. Huba, A.B. Hassan and P. Satyanarayana

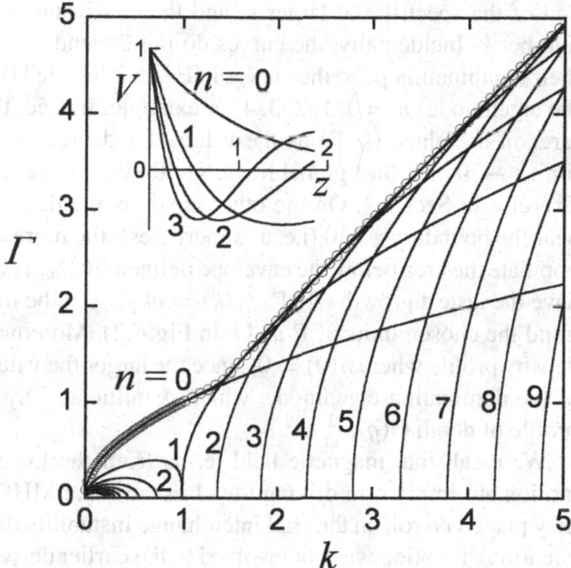

Fig. 6.2 Growth rates Γ (in units of ω_{ci}) versus wave number k (in units of ω_{ci}^2/g) for the first ten instability eigenmodes with $n = 0, 1, \ldots, 9$ for $s = 1$ (courtesy A.L. Velikovich)

Power law density profiles of form $n(z) = \bar{n} \exp(z/L)^s$ $(s > 0)$ in the half-space $z \geq 0$ have also been considered, where the solution to the low-beta equation (6.174) satisfies $V \to 0$ as $z \to \infty$ and is regular at its singular point $z = 0$. This solution is

$$V(z) = \exp(-kz) L_n^{s-1}(2kz), \tag{6.177}$$

where $L_n^\alpha(z)$ is the generalised Laguerre polynomial and $n = 0, 1, 2, \ldots$ denotes the number of the perturbation eigenmode, and the corresponding dispersion relation for each eigenmode is the cubic $D = -1 - 2n/s$. The spectra calculated for $s = 1$, $n = 0, \ldots, 9$ are shown in Fig. 6.2, where the growth rate $\Gamma = \Im(\omega)$ is in units of ω_{ci} and the wave number k is in units of ω_{ci}^2/g, together with the eigenfunctions $V(z)$ calculated for $s = 1$, $n = 0, 1, 2, 3$ in the insert. For each eigenmode, we have

$$\Gamma(k) = \begin{cases} \mu^{-1/2}\sqrt{gk}, & \text{if } |k| \ll \omega_{ci}^2/g, \\ \mu^{1/2}\sqrt{gk}, & \text{if } |k| \gg \omega_{ci}^2/g, \end{cases} \tag{6.178}$$

where $\mu = 1 + 2n/s$. The dispersion curves $\Gamma(k)$ are seen to be nested for the predominantly ideal smaller k and much more extensive for the predominantly Hall

(Footnote 18 continued)
(*Physics of Fluids* B **1**, 931, 1989). In passing, we note that the terminology "large Larmor radius" (LLR) was introduced for the regime where $\omega \gg \omega_{ci}$ and $\rho_i \gg L$ (with ρ_i the ion gyroradius and L the scale length of interest), of particular interest in space physics and elsewhere, when the electrons are "frozen-in" under the assumption $\mathbf{E} + \mathbf{v}_e \times \mathbf{B} = 0$ but the "un-magnetised" ions are not. These authors also refer to the reference length $L_n = (d(\ln n)/dz)^{-1}$, the reciprocal of κ that we again use here.

end of the spectrum at larger k, and there is instability for all values of the wave number k. Incidentally, the curves do not depend on the n and s separately but on their combination μ, so the curves in Fig. 6.2 labelled 0, 2, 4, 6, 8 also correspond to the eigenmodes $n = 0, 1, 2, 3, 4$ for example. Indeed, the successive curves become rarer in the plane (k, Γ) as the value of s decreases, until in the sharp boundary limit $s \to 0$ only the "global Rayleigh-Taylor" mode remains—cf. the answer to the Exercise in Sect. 6.4. On the other hand, as the density profile becomes smoother near the boundary $z = 0$ (i.e. as s increases), the more densely the dispersion curves populate the area below the envelope defined by $\Gamma_{max}(k) \simeq k$ at large k, such that we have the fastest growth rate $\Gamma_{max}(k) = gk/\omega_{ci}$ in the limit $gk/\omega_{ci}^2 \to \infty$ (bearing in mind the chosen units of Γ and k in Fig. 6.2). Moreover, this result follows for *any* density profile where $n(0) = 0$, since the larger the value of k the greater the number of the dominating eigenmode, which is influenced by the finite part of the density profile of depth $O(g/\omega_{ci}^2)$.[19]

We recall that magnetic field terms from the Lorentz force in the equation of motion are implicit in driving the classical ideal MHD waves and instabilities, but they played no role in the Hall interchange instability discussed above. Similarly, the equation of motion was not involved in the earlier derivation of the whistler and Hall drift wave dispersion relations—cf. Sect. 5.15, where in the Exercise we noted that the retention of the number density $n(\mathbf{r}, t)$ within the curl of the Hall term renders not only the contribution $-(ne)^{-1}\nabla\times(\mathbf{j}_1\times\mathbf{B})$ that produced whistlers but also the term $(n^2e)^{-1}(\nabla n)\times(\mathbf{j}_1\times\mathbf{B})$ responsible for Hall drift waves in an inhomogeneous plasma, which are related to the Hall instability discussed here. Finally, let us also recall that there is the component proportional to the magnetic field curvature in non-Cartesian geometry (in both homogeneous and inhomogeneous plasma), which has not been explored here.

6.12.2 Hall Effect in Kepler Disc Dynamics

In Sect. 6.8, it was shown that the stability of a differentially rotating plasma disc is significantly altered by a perpendicular magnetic field. Thus in ideal MHD, planar short wavelength self-gravitational modes can be stabilised, but in a Keplerian disc

---

[19] See A.L. Velikovich (*Physics of Fluids* B **3**, 492–494, 1991). Subsequent analysis for the Rayleigh–Taylor instability in a Hall plasma slab explicitly considered acceleration by the magnetic field, in the context of an imploding plasma liner—cf. A.V. Gordeev (*Plasma Physics Reports* **25**, 70–76 and 202–206, 1999; and *Plasma Physics Reports* **29**, 459–465, 2003). We also note that gk/ω_{ci} is the growth rate of the incompressible Hall instability when $\mathbf{k}\cdot\mathbf{B} \neq 0$ found much earlier for a density discontinuity under gravity, with the perturbations no longer constrained to satisfy $\nabla\times\mathbf{v}_1 = 0$ and $\mathbf{j}_1 = \mu_0^{-1}\nabla\times\mathbf{B}_1 = 0$ everywhere as the ideal MHD analysis in Sect. 6.3 requires—cf. R.J. Hosking (*Physical Review Letters* **15**, 344–345, 1965). Reference may also be made to the theory of Electron Magnetohydrodynamics (EMD), where the quasi-neutrality assumption is usually preserved and the ion component is "un-magnetised", subsequently developed by the Russian school at the Kurchatov Institute—cf. A.V. Gordeev, A.S. Kingsep and L.I. Rudakov (*Physics Reports* **243**, 215–315, 1994).

the magnetic field offsets favourable angular rotation such that axial modes can become unstable, with a maximum growth rate characterised by the angular velocity. However, the implicit "frozen-in" magnetic field condition is inappropriate (at least for the ions) when the density is high but the ionisation is low, as envisaged in protostellar discs for example. Moreover, the ideal magnetorotational instability is restricted to wavelengths defined by (6.121), typically rather longer than the disc thickness in protostellar models.

In a weakly ionised gas, although the more mobile electrons may be considered "frozen-in" to the magnetic field, the motion of the much more massive positive ions is expected to be largely coupled with that of the ubiquitous neutrals. Thus if $\mathbf{v}$ now denotes the velocity of the neutrals, and electrons and ions are again denoted by e and i subscripts, on assuming there is an equal number of electrons and singly charged ions in the gas ($n_e = n_i = n$) we have

$$\mathbf{v}_e = \mathbf{v} + (\mathbf{v}_e - \mathbf{v}_i) + (\mathbf{v}_i - \mathbf{v}) \simeq \mathbf{v} - \frac{\mathbf{j}}{ne}$$

provided the contribution from the socalled ambipolar diffusion term $\mathbf{v}_i - \mathbf{v}$ is negligible—e.g. at typical densities in protostellar discs, except perhaps in their outer reaches. Consequently, the resistive Hall induction equation (5.126) may be invoked at such low as well as the high ionisation levels previously envisaged. Moreover, there are now formerly quite familiar field equations *for the predominant neutrals* in weakly ionised gas—including the equation of motion (5.34) dominated by the neutrals, on equating the ion-neutral drag term with the Lorentz force $\mathbf{j} \times \mathbf{B}$ under the low inertia limit for the ion momentum. Let us therefore investigate the stability of the thin differentially rotating Kepler disc in the resistive Hall context, subject to an initial constant magnetic field $\mathbf{B} = B_\theta \hat{\mathbf{e}}_\theta + B_z \hat{\mathbf{e}}_z$. We consider local eikonal perturbations of the form $f_1(r) \exp[i(\mathbf{k} \cdot \mathbf{r} - \omega t)]$, where $\mathbf{k} = \alpha \hat{\mathbf{e}}_r + k_z \hat{\mathbf{e}}_z$ and the radial dependence of $f_1(r)$ is negligible. (The previous ideal MHD discussion in Sect. 6.8 assumed a perpendicular magnetic field $\mathbf{B} = B_z \hat{\mathbf{k}}$ and axial wave numbers such that $k \equiv k_z = k_\parallel$.) Thus in a Boussinesq-type approximation there are the following system of linearised perturbation equations in the neighbourhood of any radius r:

$$\alpha v_{1r} + k_z v_{1z} = 0$$

$$-i\omega v_{1r} - 2\Omega v_{1\theta} - i\frac{k_z B_z}{\rho} B_{1r} + i\frac{\alpha}{\mu_0 \rho}(B_\theta B_{1\theta} + B_z B_{1z}) + i\alpha \frac{p_1}{\rho} = 0$$

$$-i\omega v_{1\theta} + \frac{\kappa^2}{2\Omega} v_{1r} - i\frac{\mathbf{k} \cdot \mathbf{B}}{\mu_0 \rho} B_{1\theta} = 0$$

$$-i\omega v_{1z} - i\frac{\alpha B_z}{\mu_0 \rho} B_{1z} + i\frac{k_z}{\mu_0 \rho}(B_\theta B_{1\theta} + B_z B_{1z}) + ik_z \frac{p_1}{\rho} = 0$$

from continuity and the equation of motion, and (using $\nabla \cdot \mathbf{B}_1 = 0$)

$$-i\omega B_{1r} + \frac{\mathbf{k}\cdot\mathbf{B}}{\mu_0 ne} k_z B_{1\theta} - i\mathbf{k}\cdot\mathbf{B}\, v_{1r} + \eta k^2 B_{1r} = 0$$

$$-i\omega B_{1\theta} - \left(r\frac{d\Omega}{dr} + \frac{\mathbf{k}\cdot\mathbf{B}}{\mu_0 ne} k_z \right) B_{1r} + \frac{\mathbf{k}\cdot\mathbf{B}}{\mu_0 ne} \alpha B_{1z} - i\mathbf{k}\cdot\mathbf{B}\, v_{1\theta} + \eta k^2 B_{1\theta} = 0$$

$$-i\omega B_{1z} - \frac{\mathbf{k}\cdot\mathbf{B}}{\mu_0 ne} \alpha B_{1\theta} - i\mathbf{k}\cdot\mathbf{B}\, v_{1z} + \eta k^2 B_{1z} = 0$$

from the resistive Hall induction equation (5.126), where $k^2 = \alpha^2 + k_z^2$. The nontrivial solution of this system of equations yields the dispersion relation—viz. on again writing $\sigma = -i\omega$ (conveniently rendering all of its coefficients real):

$$\sigma^4 + 2\eta k^2 \sigma^3 + \mathcal{C}_2 \sigma^2 + 2\eta k^2 \left(\frac{k_z^2}{k^2}\kappa^2 + k_\parallel^2 c_\mathrm{A}^2 \right) \sigma + \mathcal{C}_0 = 0, \tag{6.179}$$

where

$$\mathcal{C}_2 = \frac{k_z^2}{k^2}\kappa^2 + 2k_\parallel^2 c_\mathrm{A}^2 + \eta^2 k^4 + \frac{k_z k_\parallel c_\mathrm{A}^2}{2\,\omega_{ci}\,\Omega} \left(r\frac{d\,\Omega^2}{dr} + 2\frac{k_z \Omega\, k_\parallel c_\mathrm{A}^2}{\omega_{ci}} \frac{k^2}{k_z^2} \right), \tag{6.180}$$

$$\mathcal{C}_0 = \eta^2 k_z^2 k^2 \kappa^2 + \left(k_\parallel^2 c_\mathrm{A}^2 + 2\frac{k_z \Omega\, k_\parallel c_\mathrm{A}^2}{\omega_{ci}} + \frac{k_z^2}{k^2} r \frac{d\,\Omega^2}{dr}\bigg|_R \right) \times \left(k_\parallel^2 c_\mathrm{A}^2 + \frac{\kappa^2}{2\Omega^2} \frac{k_z \Omega\, k_\parallel c_\mathrm{A}^2}{\omega_{ci}} \right), \tag{6.181}$$

κ is the epicyclic frequency (cf. Sect. 6.8), and each Hall term is again identified by the ion cyclotron frequency $\omega_{ci} = eB/m_i$ in its denominator.

The quartic (6.179) has at least one positive root when $\mathcal{C}_0 < 0$, such that when $d\,\Omega^2/dr < 0$ the ideal MHD instability condition (6.121) is replaced by

$$k_\parallel^2 c_\mathrm{A}^2 + 2\frac{k_z \Omega\, k_\parallel c_\mathrm{A}^2}{\omega_{ci}} < -\frac{k_z^2}{k^2} r \frac{d\,\Omega^2}{dr}, \tag{6.182}$$

so the Hall term

$$2\frac{k_z \Omega\, k_\parallel c_\mathrm{A}^2}{\omega_{ci}} = 2\frac{(\mathbf{k}\cdot\Omega)(\mathbf{k}\cdot\mathbf{B})}{\mu_0 ne}$$

is either destabilising or stabilising when the combination $(\mathbf{k}\cdot\Omega)(\mathbf{k}\cdot\mathbf{B})$ is negative or positive, respectively. The Hall effect is therefore destabilising or stabilising according as $\Omega \cdot \mathbf{B}$ is negative or positive, respectively.

6.12.3 Hall Reconnexion

We began this section with a discussion of the enhanced interchange instability in collisionless (non-resistive) Hall MHD, where the electrons are "frozen in" but the ions are not. The Hall model of the ion-electron plasma considered there is a simple case of a "two fluid" description, where the ion and electron number densities are assumed to be equal everywhere so the plasma remains quasi-neutral during the separate motion of the two species. We then discussed the application of resistive Hall MHD to partially ionised plasma, where the ions and neutrals but not the electrons tend to be strongly correlated. However, it has usually been in the context of fully ionised plasma that we contemplated the resistivity, where the "frozen-in" field concept of ideal MHD no longer applies but the species are again correlated on the Debye length scale and move as a single quasi-neutral fluid. Near the end of Sect. 5.10, we first mentioned the notion that magnetic field lines may break and reconnect, and in Sect. 6.10 we discussed the resistive tearing instability that renders typically large-scale topological changes to the magnetic field configuration involving magnetic field reconnexion. Associated self-organisation and relaxation to lower magnetic energy configurations were attributed to confinement devices such as the reverse-field-pinch mentioned in Sect. 6.11, where we discussed viscous stabilisation and demonstrated that the very large thermal energy release in solar flares can be explained by anisotropic viscous dissipation in the large plasma flows accompanying the magnetic reconnexion. In brief, it seems clear that not only global constraints but also local plasma processes in the reconnexion region are important.

Indeed, it was soon recognised that resistive reconnexion times are much too long in comparison to the very short timescales of the energy release in solar flares, unless the resistivity was somehow quite anomalous. It also became clear that the current layer envisaged in the traditional "Sweet-Parker" model (cf. Fig. 5.6 of Sect. 5.12), which Dungey had shown could form in the collapse of the magnetic field near a neutral point under ideal MHD, was incompatible with the collisionless length scale of reconnexion layers observed in the magnetosphere—i.e. the ion skin depth, or the ion gyroradius when the plasma pressure p is comparable to the magnetic pressure $B^2/(2\mu_0)$. It seems fitting to conclude this book with brief mention of the associated numerical simulation, which is a good case example of the importance of modern computational work in MHD—much as numerical simulation has proven to be indispensable in applications of fluid mechanics, such as the solution of the governing Navier–Stokes equations in various areas of engineering design. Thus extensive numerical simulations have consistently shown that the Hall effect is a key factor in producing significantly enhanced magnetic reconnexion rates, irrespective of other model assumptions. Rather than the double Y-point geometry associated with the "Sweet-Parker" model, the computer simulations demonstrated that there is an X-point geometry. The mechanism appears to be that the ions become "un-magnetised" as they enter a neutral sheet and sharply turn perpendicularly in the reconnexion plane, before they flow away from an X-point. On the other hand, the electrons are governed by the approximate generalised Ohm's law $\mathbf{E} + \mathbf{v}_e \times \mathbf{B} \simeq 0$ as they mainly flow inward along the separatrices toward the X-point where the magnetic field is weaker, so they are ejected with the correspondingly large velocity $\mathbf{v}_e = \mathbf{E} \times \mathbf{B}/|\mathbf{B}|^2$. The strong electron flow generates circular currents in the reconnexion plane, which produce an out-of-plane magnetic field with a quadrupole profile that is recognised as a "signature" of the Hall effect. The increased electric field caused by the Hall term produces a steady laminar cross-field electron current, and consequently fast moving magnetic flux lines in the reconnexion plane (i.e. the fast rate of magnetic reconnexion). There is also considerable supporting experimental evidence for the role of the Hall effect in fast magnetic reconnexion.[20]

Exercise

(Q1) For quasi-electrostatic modes and $B \simeq$ constant, show that the perturbation equations (6.173) to be considered reduce to

$$\frac{d}{dz}(nv_{1x}) - iknv_{1z} = i\,\frac{gk}{\omega^2}\left(\frac{d}{dz}(nv_{1z}) + iknv_{1x}\right),$$
$$B\frac{dv_{1z}}{dz} - i\omega\frac{m_i}{e}\frac{dv_{1x}}{dz} + ikBv_{1x} - \frac{m_i}{e}\omega kv_{1z} = 0.$$

Then introduce the ion gyrofrequency $\omega_{ci} = eB/m_i$, and derive the governing second order differential equation (6.174).
Hint: Combine $-i\omega^2/\omega_{ci}$ times the first equation with the derivative of ωn times the second equation.

Bibliography

1. G. Bateman, *MHD Instabilities* (MIT Press, Cambridge, 1978). (Graduate textbook on basic concepts in the tokamak context, but pre-dating ballooning theory)
2. D. Biskamp, *Magnetic Reconnection in Plasmas* (Cambridge University Press, Cambridge, 2000). (A theoretical introduction, with emphasis on solar and geophysical phenomena but some reference to fusion devices); and *Magnetohydrodynamic Turbulence* (Cambridge University Press, Cambridge, 2003). (On both macroscopic aspects and small-scale scaling properties, again with reference to astrophysical and laboratory applications)
3. S. Chandrasekhar, *Hydrodynamic and Hydromagnetic Stability* (Oxford University Press, Oxford, 1961). (The classic mentioned in the Preface and first referenced in Chapter 2)
4. P.G. Drazin, *Nonlinear Systems* (Cambridge University Press, Cambridge, 1992). (First referenced in Chapter 3)
5. J.W. Dungey, *Cosmic Electrodynamics* (Cambridge University Press, Cambridge, 1958). (On magnetohydrostatics, Alfvén waves, magnetic storms, solar and ionospheric phenomena)
6. J.P. Freidberg, *Ideal Magnetohydrodynamics* (Plenum, New York, 1987). (A highly recommended presentation on ideal MHD oriented to controlled thermonuclear fusion research, reprinted by Springer (2013)); also of interest *Plasma Physics and Fusion Energy* (Cambridge University Press, Cambridge, 2008). (Surveys world energy needs and the possible future contribution from fusion power)

[20] An excellent detailed survey of magnetic reconnexion is M. Yamada, R. Kulsrud and H. Ji (*Reviews of Modern Physics* **82**, 603–664, 2010).

7. I.M. Gelfand, S.V. Fomin, *Calculus of Variations* (Dover, New York, 2000). (On basic concepts and methods, followed by various applications and suggested further reading)
8. J.P. Goedbloed, S. Poedts, *Principles of Magnetohydrodynamics* (Cambridge University Press, Cambridge, 2004). (First referenced in Chapter 2)
9. R.D. Hazeltine, J.D. Meiss, *Plasma Confinement* (Dover, New York, 2003). (First referenced in Chapter 2)
10. R.M. Kulsrud, *Plasma Physics for Astrophysicists* (Princeton University Press, Princeton, 2004). (Mentioned in the Preface and referenced in Chapter 5)
11. P.D. Miller, *Applied Asymptotic Analysis* (American Mathematical Society, Providence, 2006). (First referenced in Chapter 3)
12. E.R. Priest, T.G. Forbes, *Magnetic Reconnection: MHD Theory and Applications* (Cambridge University Press, Cambridge, 2007). (An accessible discussion of relevant magnetospheric, solar and astrophysical phenomena)
13. H.-J. Stöckmann, *Quantum Chaos: An Introduction* (Cambridge University Press, Cambridge, 2007). (A graduate level presentation of the subject)
14. R.B. White, *Theory of Tokamak Plasmas* (Elsevier, New York, 1990). (Graduate textbook that discusses both ideal and resistive MHD instabilities in tokamaks)

Glossary

Normal Mode

A normal mode of a continuous system disturbed from an equilibrium (with or without flow) is a pattern of motion where all fields describing the system oscillate with the same frequency. This definition includes the case of nonlinear normal modes, where this frequency typically depends on amplitude. However, in this book we restrict attention to the *linear* case where the normal mode frequencies are eigenfrequencies of the operator describing linearised perturbations, and the normal modes are the corresponding eigenfunctions—so *eigenmode* could be used as a synonym for normal mode.

The nature of the frequency spectrum depends on the symmetries of the system. Historically, the term "normal mode" arose in acoustics (cf. J.W. Strutt, Baron Rayleigh, *The Theory of Sound*, Vol. 1, MacMillan, London, 1877), but "normal mode analysis" is now used in various engineering and physics applications—e.g. computing vibration modes in structures or analysing plasma instabilities as in this book. Quantum mechanics provides well-known examples and terminology, ranging from the complete set of "good quantum numbers" describing the hydrogen atom to the statistical characterisations of "quantum chaotic" spectra arising when all continuous symmetries are strongly broken—e.g. see H.J. Stöckmann, *Quantum Chaos: an Introduction* (Cambridge University Press, 1999).

In linear theory we can easily generalise to consider complex frequency, with the imaginary part determining the rate of exponential damping or growth of the mode.

Slab Model

A slab model is an idealised three-dimensional system bounded by two parallel planes (one of which may be at infinity), uniform in directions parallel to these planes—i.e. essentially one-dimensional in its undisturbed state.

R.J. Hosking and R.L. Dewar, *Fundamental Fluid Mechanics and Magnetohydrodynamics*, DOI 10.1007/978-981-287-600-3

The main virtue of a slab model is that a simple Cartesian coordinate system $\{x, y, z\}$ can be used. In this book, we use two common conventions depending upon the physical context, and the reader needs to be alert to the interchange of x and y in switching between these "gravitational" and "magnetic confinement" conventions as follows.

- Gravitational confinement convention. In geophysical and astrophysical systems, where there is a strong uniform gravitational field $\mathbf{g}$, we take the z-axis to be vertical (i.e. antiparallel to $\mathbf{g}$). Thus when the slab model is a local approximation to the neighbourhood of a point on the surface of a planet or star for example, the x-axis is in the latitudinal (east-west or *zonal*) direction and the y-axis in the meridional (north-south) direction.
- Magnetic confinement convention. In magnetospheric applications or fusion power research, where the plasma is confined by the magnetic field and gravity is weak or negligible, a slab model is often used as an approximation to a localised plasma subregion. Thus in considering a tokamak for example, the main longitudinal magnetic field is in the z-direction and the transverse magnetic field component is in the y-direction, and any equilibrium density or pressure variation is in the x-direction.

Analogies are sometimes made between gravitationally confined systems and magnetically confined systems, where the term "zonal" is often used to refer to the x-direction rather than to the y-direction!

Toroidal and Poloidal

Historically, the toroidal-poloidal terminology was introduced in the context of geophysical dynamo theory, with the adjective "toroidal" for magnetic fields directed east or west and "poloidal" to those directed north or south. (Clearly, "poloidal" refers to the directions of the Earth's poles—but as "toroidal" does not relate to the Earth's shape, the terminology presumably refers to the fact that axisymmetric toroidal fields lie in tori.) However, this terminology has been adopted to describe magnetically confined plasmas where the field lines lie in nested tori. In that context, the toroidal direction is taken to be the long way around a torus (cf. Fig. 5.3), with the corresponding coordinate denoted by ϕ or ζ in magnetic coordinates (cf. Sect. 5.11); and the poloidal direction is then the short way around the torus, with the corresponding coordinate denoted by θ in magnetic coordinates. On adopting the slab approximation, the toroidal coordinate is denoted by z and the poloidal coordinate by y in the laboratory plasma confinement literature—cf. item *Slab Model* above.

Appendix Tables

SI quantities and nomenclature		
Quantity	Name	Symbol
Length	metre	m
Mass	kilogramme	kg
Time	second	s
Temperature	Kelvin	K
Frequency	Hertz	Hz (s^{-1})
Force	Newton	N(kg m s^{-2})
Pressure	Pascal	Pa (N m^{-2})
Energy	Joule	J(N m)
Power	Watt	W (J s^{-1})
Electric charge	Coulomb	C
Electric potential	Volt	V
Electric current	Ampere	A (C s^{-1})
Magnetic field	Tesla (10^4 Gauss)	T(N m^{-1} A^{-1}) (10^4 G)

Physical constants		
Quantity	Symbol	Value
Gravitational constant	G	6.671×10^{-11} m^3 kg^{-1} s^{-1}
Boltzmann constant	k	1.381×10^{-23} JK^{-1}
Electron charge	e	1.602×10^{-19} C
Electron mass	m_e	9.109×10^{-31} kg
Proton mass	m_p	1.673×10^{-27} kg
Vacuum permittivity	ϵ_0	8.854×10^{-12} C $m^{-1}V^{-1}$
Vacuum permeability	μ_0	$4\pi \times 10^{-7}$ J $m^{-1}A^{-2}$
Speed of light	$c\ (= \sqrt{\epsilon_0/\mu_0})$	2.998×10^8 m s^{-1}

Fluid and plasma parameters	
Quantity	Symbol and definition
Thermal speed	$v_T = \sqrt{2kT/m}$
Sound speed	$c_s = \sqrt{\gamma p/\rho}$
Plasma frequency	$\omega_{pe} = \sqrt{ne^2/(\epsilon_0 m_e)}$
Debye length	$\lambda_D = \sqrt{\epsilon_0 k T_e/(ne^2)}$
Gyrofrequency (or cyclotron frequency)	$\omega_c = eB/m$
Gyroradius	$r_g = v_T/\omega_c$
Alfvén speed	$c_A = B/\sqrt{\mu_0 \rho}$
Magnetic energy density	$\mathcal{E} = B^2/(2\mu_0)$
Electron skin depth	$\delta_e = c/\omega_{pe}$
Electron-ion (proton) collision frequency, when $T_e = T_i$	$\nu_{ei} = 6\,(\ln \Lambda)\,\omega_{pe}/\Lambda$ where $\Lambda = 24\pi n \lambda_D^3$
Electric conductivity	$\sigma = \sqrt{\pi/6}\, ne^2/(m_e \nu_{ei})$

Note For plasma species the density ρ equals nm where n denotes the number density, and subscripts e and i are added to denote the respective electron and ion components.

Index

R.J. Hosking and R.L. Dewar, *Fundamental Fluid Mechanics and Magnetohydrodynamics*, DOI 10.1007/978-981-287-600-3

Printed in the United States
By Bookmasters